Geospatial Technology for Sustainable Oil Palm Industry

Oil palm cultivation is a significant contributor to the global agricultural industry, providing valuable resources for various products. While it has faced challenges, including concerns about deforestation and environmental degradation, it is crucial to understand the geographical distribution of oil palms to ensure responsible and sustainable management.

This book introduces the application of geospatial technology to the palm oil industry, demonstrating how these tools can address key issues while promoting responsible practices. The topics discussed in this book aim to tackle challenges such as illegal deforestation and the monitoring of environmental impacts associated with oil palm cultivation.

Key Features:

- The first book to explore the use of remote sensing and other geospatial technologies for mapping and monitoring oil palm plantations.
- Introduces cutting-edge techniques from the fourth industrial revolution, including the use of Internet of Things (IoT), Artificial Intelligence, big data and analytics, block chain, and 5G technology for effective oil palm tree monitoring.
- Addresses current environmental concerns affecting the industry, including issues such as floods, changes in precipitation, temperature, humidity, and El Niño events.
- Explores the scientific aspects of promoting sustainability in oil palm developments and the responsible conversion of rainforests into oil palm plantations.
- Provides a wealth of application data for early detection and continuous monitoring of environmental challenges.

This book serves as an insightful resource for those interested in geospatial technologies and their positive impact on monitoring and mitigating environmental changes in the context of the palm oil industry.

Geospatial Technology for Sustainable Oil Palm Industry

Kasturi Kanniah and Le Yu

CRC Press
Taylor & Francis Group
Boca Raton London New York

CRC Press is an imprint of the
Taylor & Francis Group, an **informa** business

Designed cover image: © I Net Spatial Sdn Bhd

First edition published 2024
by CRC Press
2385 NW Executive Center Drive, Suite 320, Boca Raton FL 33431

and by CRC Press
4 Park Square, Milton Park, Abingdon, Oxon, OX14 4RN

CRC Press is an imprint of Taylor & Francis Group, LLC

ISBN: 978-0-367-19009-5 (hbk)
ISBN: 978-1-032-72958-9 (pbk)
ISBN: 978-0-429-19981-3 (ebk)

DOI: 10.1201/9780429199813

Typeset in Times
by codeMantra

We dedicate this book to those who believe in the power of innovation and the potential of geospatial technology to create a more sustainable future for the palm oil industry.

Your dedication to responsible practices inspires us all.

Kasturi and Le Yu

Contents

About the Authors

Dr. Kasturi Devi Kanniah is a professor at the Department of Geoinformation, Faculty of Built Environment and Surveying, Universiti Teknologi Malaysia (UTM). Her academic journey is marked by an unwavering commitment to environmental sustainability. As a member of the Centre for Environmental Sustainability and Water Security at UTM and a valued member of the TropicalMap research group within the Sustainability Research Alliance, Dr. Kanniah's research primarily focuses on bridging field data with remote sensing technology. This synergy provides invaluable insights into large-scale changes in land use and land cover, as well as the intricate dynamics of vegetation processes. Her research extends across both rural and urban environments, shedding light on their pivotal role in the pursuit of low-carbon cities and the mitigation of atmospheric pollution. She is recognized as a fellow member of the Malaysia Academy of Sciences and a senior member of the Institute of Electrical and Electronic Engineers (IEEE), reflecting her exceptional standing in the academic and scientific circles. Beyond her research portfolio, Dr. Kanniah serves as the associate editor of the *International Journal of Remote Sensing* (*IJRS*) and holds an editorial board position at the *International Journal of Geoinformatics* (*IJG*). She has also curated special issues for *IJG*, *IJRS*, and *Remote Sensing*. Throughout her illustrious career, Dr. Kanniah has earned numerous national and international awards for her exceptional research and publications, further solidifying her position as a top-tier scientist in the world. Her awards include recognition as one of the world's top 2% scientists in 2021, 2022, and 2023. In recognition of her outstanding contributions, Dr. Kanniah was a Global Scholar Programme awardee at the Institute for Social Sciences Research, University of Dundee, UK (2022), a visiting researcher at Tsinghua University, China, in 2019 and an International Visiting Scholar at Massachusetts Institute of Technology (USA) under the MIT-UTM Malaysia Sustainable Cities Program for the academic year 2016-2017.

Dr. Le Yu is an associate professor at the Department of Earth System Science, Tsinghua University. His research has been on the use of geographical information techniques to monitor and model global land-use change, especially cropland, and to facilitate many applications, for example, food security, biodiversity conservation, and land system modelling. He particularly focuses on satellite-based methods to quantify the spatiotemporal change of land cover/use and understand their ecological, environmental, and socioeconomic impacts on sustainable development. He is Scientific Steering Committee (SSC) Member of the Global Land Programme (GLP), Fellow of the Royal Geographical Society (RGS-IBG),

Fellow of the Higher Education Academy, Senior Member of the Institute of Electrical and Electronics Engineers, and Prominent Visiting Researcher at the University of Technology Malaysia (UTM). He is currently Editor-in-Chief of *Geo: Geography and Environment*, Section Editor-in-Chief of *LAND*, Associate Editor of *International Journal of Remote Sensing*, and Editorial Board Member of *Global Sustainability and Geoscientific Model Development.*

Foreword

In the dynamic landscape of the global palm oil industry, sustainability has emerged as the core of our shared obligation. It is a commitment to balance economic growth with environmental stewardship and social well-being. This book, *Geospatial Technology for Sustainable Oil Palm Industry*, is a noteworthy contribution to this endeavour, addressing the urgent demand for innovative approaches to ensure long-term viability and environmental sustainability of the palm oil industry.

At FGV HOLDINGS, we have recognized the importance of geospatial technology in ensuring the sustainable management of oil palm plantations. This commitment is exemplified by the establishment of the geographical information system (GIS) unit within our research and development division in 2005. The science of remote sensing, GIS, and geomatics has evolved into an indispensable tool in our pursuit of sustainable practices. This book, with its comprehensive exploration of 17 chapters, explores these tools and techniques, providing invaluable insights into their application in the oil palm industry.

This book's coverage is extensive, addressing vital facets of sustainable oil palm cultivation, from tree detection and age classification to productivity and yield. Chapters on topography mapping and oil palm pest and disease detection emphasize the importance of precise data for informed decision-making. Moreover, the inclusion of climate suitability assessment, carbon cycle analysis, and discussions on environmental and social sustainability reaffirm the multidimensional nature of our commitment to responsible palm oil production.

This book is a stand out with its forward-looking approach. The concluding chapter on the future of geospatial technology for sustainable palm oil offers a roadmap towards a more equitable and responsible industry, fostering a greener and more sustainable world. It is the product of the researchers who share a deep passion for sustainability in the oil palm sector. Their commitment to advancing the integration of geospatial technology and sustainable practices is evident on every page.

As a professional within FGV FOLDING's sustainability division, I wish to express my appreciation to the authors for their remarkable work. I also commend the readers for their interest in exploring this important topic. It is my sincere hope that the knowledge shared within these pages will ignite a transformative shift, guiding us towards a more sustainable and responsible future for the palm oil industry.

Sustainability is our mutual pledge, and this book is a significant step forward in our collective journey towards a sustainable palm oil industry.

Sincerely,

K. Ilangovan
Senior General Manager
Head, Sustainability Technical Department
Group Sustainability Division
FGV HOLDINGS

Preface

In today's world, where technology and sustainability are top priorities globally, their combination gives hope to industries grappling with complex challenges. This book, *Geospatial Technology for Sustainable Oil Palm Industry*, has been crafted with the aim to connect modern geospatial technology with the changing needs of the sustainable palm oil industry. The palm oil industry plays a pivotal role in global food security and contributes significantly to the economies of producing nations. Nevertheless, it grapples with the imperative of achieving environmental and social sustainability. The road to sustainability is non-negotiable, and this book signifies a substantial stride on that transformative path.

This book is organized into 17 chapters, each addressing a specific aspect of geospatial technology in the context of sustainable oil palm cultivation. These chapters thoroughly explore the utilization of remote sensing, GIS, and geomatics technologies in the palm oil sector, demonstrating the transformational potential of these tools.

This book begins by establishing the groundwork for comprehending the essential principles and procedures of remote sensing, GIS, and geomatics within the context of oil palm. Moving forward, it broadens its scope to include topics such as land-cover mapping, tree detection, and age classification, presenting valuable insights on how to enhance resource management and productivity. Subsequent chapters on topography mapping, productivity, and yield explore the potential of geospatial technology to improve land utilization and overall profitability.

The critical issues of pest and disease detection, climate suitability, and the carbon cycle in oil palm plantations are addressed in subsequent chapters, highlighting the role of geospatial technology in mitigating challenges that jeopardize the industry's sustainability. This book also explores the environmental and social aspects of sustainable palm oil production, emphasizing the intricate interplay of these elements in our quest towards conscientious practices to achieve sustainable development.

The concluding chapter, "Future Outlook," presents a forward-looking view of geospatial and other cutting-edge technologies in palm oil industry to increase the yield and boost sustainability. It is a call for action, encouraging stakeholders in the palm oil industry to embrace innovation, implement optimal strategies, and cultivate a united vision for a more sustainable future.

It is our sincere hope that *Geospatial Technology for Sustainable Oil Palm Industry* serves as a valuable resource for everyone who wishes to contribute to the creation of a sustainable palm oil industry. Together, let us continue to explore the possibilities of geospatial technology to enhance the industry's profitability while aligning with the principles of environmental and social responsibility.

Sincerely,

Kasturi Devi Kanniah

Le Yu

Acknowledgements

We would like to express our heartfelt gratitude to the many individuals and organizations whose support and contributions have made the creation of this book, *Geospatial Technology for Sustainable Oil Palm Industry*, possible. First and foremost, in loving memory of Prof. Arthur P. Cracknell, whose invaluable contributions to this book will forever be etched in our hearts. Your passion, dedication, and wisdom live on in these pages, a lasting tribute to a remarkable collaborator and dear friend. We extend our deep appreciation to our families for their unwavering encouragement and understanding throughout this journey. Their support has been instrumental to our determination.

Kasturi wishes to acknowledge the Faculty of Built Environment and Surveying and Universiti Teknologi Malaysia (UTM) for granting her a six-month sabbatical leave to complete this book. The research presented in this book received financial support from UTM and the Ministry of Higher Education Malaysia through research grants R.J1300000.7827.4F725, and Q.J130000.7127.01J44. Kasturi also acknowledges funding from Earthworm Foundation, Switzerland (R.J130000.7352.4B406). Leyu would like to acknowledge the National Key R&D Program of China (grant numbers: 2019YFA0606601 and 2022YFE0209400), as well as the Tsinghua University Initiative Scientific Research Program (20223080017).

We are immensely thankful to the management of Genting Plantation, Sime Darby Plantation, Malaysia Palm Oil Board, and Kulim Plantation for granting permission to collect ground data and photos on their estates. Our extended gratitude also goes to the researchers, experts, and professionals who generously shared their knowledge and expertise to enrich the content of this book. We particularly thank Dr. Tan Kian Pang (Refinitiv, Singapore), Dr. Nazarin Ezzaty Mohd Najib, and Ms. Camalia Saini Hamsa for their contributions to research related to oil palm through their postgraduate studies. We also thank Ms. Nordiana Abd Aziz (Malaysia Palm Oil Board), Ms. Siti Hajar Md Nor Azam (Sime Darby), Ms. Haryati Abidin (Felda FGV), Dr. Farrah Melissa Muharam (Universiti Putra Malaysia), Christine Pohl (former Universiti Teknologi Malaysia professor), and Dr. Sanath Kumaran Kolandai from United Plantations Berhad for their knowledge sharing. I extend special thanks to Dr. Gobi Krishna from UTM for his invaluable assistance in collecting and interpreting social data utilized in the case study outlined in Chapter 15. Thanks are also extended to Ms. Nurul Asyikin (UTM), Dr. Nurul Amalin Fatihah (UTM), Mr. Qiang Zhao, Tsinghua University, and Dr. Juepeng Zheng, Sun Yat-sen University, China, for formatting this book and drafting some of the figures used in it.

Last but not least, we are grateful to our friends and pets (Nessa, Togo, Berry, Rin, Ikuya, and Haruka) who offered moral and emotional support throughout this endeavour. We deeply appreciate all contributions. It is our hope that *Geospatial Technology for Sustainable Oil Palm Industry* serves as a valuable resource in advancing sustainable practices and innovations in the palm oil industry.

We would also like to extend our gratitude to the readers of this book, whether you are a student, researcher, practitioner, or an advocate for sustainability in the palm oil industry. Your interest and engagement with this subject are essential for driving positive changes within the industry.

Thank you all for your role in making this book a reality.

1 Oil Palm as a Commercial Crop

1.1 INTRODUCTION

The palm oil industry has emerged as a significant global player in the agricultural sector, with its versatile product, palm oil, finding widespread applications in various industries. The high demand for palm oil from food, industrial applications, and bioenergy sector has increased palm oil production 35 times higher since 1970 (FAO, 2022). As a result, the land area dedicated to oil palm cultivation increased by nearly eightfold, reaching almost 29 million hectares in 2020, compared to the 1961 figure (Ritchie and Roser, 2021). The global palm oil production reached a staggering 77 million metric tons in 2021, accounting for nearly 40% of the world's vegetable oil production (FAO, 2022). The industry's economic impact is particularly significant in major palm oil–producing nations, where it contributes significantly to the gross domestic product (GDP), employment, and export revenues (Henson and Roscoe, 2020). The sector contributed to 10% and 5% of the agricultural export of Indonesia and Malaysia, respectively, in 2020 (Ritchie and Roser, 2021). Nevertheless, palm oil sector faces many challenges to attain an optimum yield production. Diseases and pest infestation, labour shortage, and the impact of climate change such as flooding and increased air temperature are a few to mention. Additionally, palm oil sector poses a threat to the environment and social sustainability through deforestation, fire, biodiversity loss, and human rights violation. Addressing these challenges and promoting sustainable practices within the industry are crucial for its long-term viability and minimizing its ecological footprint. These global challenges require an effective use of modern technology. Geospatial technology enables the collection of accurate and up-to-date spatial data, which are crucial for assessing the plant physiological conditions and understanding the dynamics of oil palm plantations and their impact on the environment. Through mapping, monitoring, environmental impact assessment, and supply chain transparency, geospatial tools provide valuable insights and support informed decision-making in increasing the productivity. Moreover, by harnessing the power of geospatial technology, stakeholders can work towards balancing economic development with environmental conservation, fostering a responsible and sustainable palm oil industry.

This chapter provides a concise exploration of the oil palm's historical background, its notable expansion, particularly within South East Asian nations, and the forces surrounding palm oil production and trade.

1.2 ORIGIN OF OIL PALM

The term 'oil palm' refers to the trees responsible for producing fruits from which oil (palm oil) is extracted. These oil palm trees are categorized into two distinct species,

DOI: 10.1201/9780429199813-1

each named after their respective origins. *Elaeis guineensis* is thought to have originated in the tropical rainforests of West Africa as far back as 3000 B.C. (Van Gelder, 2004; Obahiagbon, 2015). In early days, palm oil was used to burn street lamps, as building materials, as skin ointment, as an antidote to poisons, and as a drink (wine made from young inflorescence of the fruits) during ritual ceremonies in Nigeria.[1] People cultivated oil palm trees in small scales for many years until in the late 19th century, the plant was introduced to other regions particularly to Latin America and Southeast Asia by European traders and explorers (Corley and Tinker, 2016). Dura, Pisifera, and Tenera are distinct varieties of the *E. guineensis* oil palm species. Dura palms produce thick-shelled fruits, and they are resistant to certain diseases. In contrast, Pisifera palms yield shell-less fruits, and they are primarily used as male pollinators for Dura and Tenera palms. Tenera palms are created through the cross-breeding of Dura and Pisifera palms, resulting in fruits with thinner shells, which in turn yield a higher oil output. Thus, commercial plantations prefer Tenera over Dura palms (Corley and Tinker, 2016).

Elaeis oleifera is another species of oil palm that originated in South and Central America but is not widely recognized for commercial oil production. It is primarily used for producing oil consumed locally. However, these two species are mixed to produce oil palm trees that are disease-resistant and can yield high-quality oil with a high percentage of unsaturated fatty acids (Avila-Diazgranados et al., 2016). Oil is produced by squeezing and crushing the fruits' flesh (crude palm oil; CPO) which accounts for 90% of the oil, while the remaining 10% is derived from the kernels or seeds (palm kernel oil; PKO) (Figure 1.1). To obtain refined, bleached, and deodorized palm oil, often referred to as RBDPO, both of these oil types undergo additional refinement processes. RBDPO is used in various food products such as cooking oil, bakery products, margarine, and industrial applications such as soap, candles, pharmaceuticals, biofuels within the energy industry, and many more.

FIGURE 1.1 Oil palm fruit where the flesh is denoted by orange colour, while kernel is the soft, inner part of the seed. (Source: CC BY-NC-ND 2.0.)

The international significance of palm oil grew as it found diverse applications in both the food industry and the industrial sectors. This transformation was propelled by factors such as the British industrial revolution and the expansion of global trade. The Europeans invested in palm oil production in the African countries before expanding their business to Southeast Asia. In Malaysia, the first commercial scale plantation was founded in 1917 in the state of Selangor (Teoh, 2002). Oil palm has expanded rapidly into various regions due to its ability to yield more oil with less land compared to other crops like soybeans, sunflowers, olives, and rapeseed. Palm trees contribute to 36% of global oil production while utilizing less than 9% of the land, in contrast to soybeans, which require 39% of the land to produce only 25% of the oil (Ritchie and Roser, 2021). It also requires relatively less fertilizer and pesticides and is resistant to oxidation. The cheap price of palm oil compared to other oils, coupled with its long shelf-life, and versatility increased its demand in the global market. Oil palm trees can be cultivated on limited land areas throughout most of the year. This makes it an appealing choice for both large-scale growers and smallholders, ensuring a consistent source of income from palm oil production, earning it the title of a cash crop. The geographical distribution and factors that drove the expansion into other countries are explained next.

1.3 GEOGRAPHICAL DISTRIBUTION OF OIL PALM

Oil palm is cultivated in many countries around the world particularly in the Southeast Asia, Africa, and South America (Figure 1.2). Using Sentinel 1 and 2 satellite imageries at 10 m spatial resolution, Descals et al. (2021) mapped a total of 21 million ha of oil palm planted in 49 countries. Majority of the areas (73%) are

FIGURE 1.2 Land areas planted with oil palm in 2021. (Source: Ritchie and Roser, 2021.)

planted with industrial oil palm while, the rest are covered by smallholder plantations. Predominantly, Southeast Asia with Indonesia and Malaysia accounts for the largest portion (89% or 18.7 million ha) of the total planted area, whereas the rest of the countries accounts for the balance of 2.3 million ha (Descals et al., 2021). Other countries that also cultivate substantial areas of oil palm include Nigeria, Colombia, Brazil, and Papua New Guinea. Additionally, some smaller producing countries include Vanuatu, Gabon, and the Solomon Islands. Regarding palm oil production, Indonesia and Malaysia are the leading producers (Figure 1.3), accounting for approximately 84% (66 million metric tonnes) of the global supply in 2023.[2] They are followed by Thailand, Colombia, and Nigeria, each producing more than a million metric tonnes (Table 1.1).

Oil palm cultivation is categorized based on the total area covered, distinguishing between industrial plantations and smallholder plantations. Industrial plantations typically cover thousands of hectares and are usually owned and managed by large companies such as Wilmar Plantations, GAR, Sime Darby, IOI Plantations, among others. These companies also operate their own palm oil mills to process fresh fruits and extract oil. Smallholder plantations, on the other hand, refer to areas where oil palm is cultivated on a smaller scale, typically <50 ha according to the Roundtable on Sustainable Palm Oil (RSPO 2020) definition. In some countries, the land size for smallholders is even smaller, for instance, only 2–4 ha in Papua New Guinea, <4 ha in peninsular Malaysia, and 5–10 ha in Peru.[3] These plantations are often managed by individuals or family members either independently or in collaboration with large companies.

The role of smallholders has become increasingly significant in the palm oil industry due to their contributions both in terms of the land area planted with oil palm and their influence on issues related to deforestation and sustainability.

FIGURE 1.3 Palm oil production in 2021. (Source: Ritchie and Roser, 2021.)

TABLE 1.1
Major Palm Oil–Producing Countries as of 2023

Country	Production (000 metric tonne)	Country	Production (000 metric tonne)
Indonesia	47,000	Congo	300
Malaysia	19,000	Peru	275
Thailand	3,450	Costa Rica	275
Colombia	1,800	Peru	278
Nigeria	1,400	Mexico	235
Guatemala	920	Philippines	100
Papua New Guinea	800	Sierra Leone	75
Côte d'Ivoire	600	Benin	70
Honduras	595	Angola	55
Brazil	585	Dominican Republic	53
Ecuador	480	Guinea	50
Cameroon	465	Liberia	45
India	305	Senegal	14
Ghana	300	Togo	9
		Venezuela	8

Source: https://www.indexmundi.com/.

Thailand, for example, has a notably high proportion, with three-quarters (71.5%) of oil palm plantations categorized as smallholder-owned, in contrast to Malaysia and Indonesia, where 40% of oil palm cultivation is attributed to smallholders (Descals et al., 2021).

The areas dedicated to oil palm cultivation have exhibited a consistent upward trend over the years. Between 1997 and 2018, there was a twofold increase, expanding from 10 to 21 million ha. Concurrently, CPO production also tripled, surging from 100 to 300 million tons during the same period (FAOSTAT, 2019). However, this extensive growth presents significant challenges to the local environment. Oil palm expansion into forests jeopardizes tropical ecosystems, puts wildlife habitats and protected areas at risk, and contributes to biodiversity loss and carbon emissions from land-use changes (Meijaard et al., 2020).

1.4 CONDITIONS FAVOURABLE FOR OIL PALM GROWTH

Oil palm trees thrive in tropical climates typically found within a 10° range on either side of the equator. However, there are exceptions, such as their cultivation extending as far as 13°S in Zampalm, Zambia, and even as far as 18°S in Madagascar. In the Americas, they can be grown as far north as 16°N in southern Mexico and Guatemala (Baskett, 2016). Oil palm trees are adaptable to mineral and peat soils that are loamy, alluvial, rich in organic matter, and well-drained, and they can thrive in soils with acidity levels ranging between pH 4.0 and 8.0 (Corley and Tinker, 2016). An essential factor for successful growth is a sufficient soil depth of at least 1 m, which is crucial for root development (Corley and Tinker, 2016). Soils that are too alkaline or waterlogged can impede root development and hinder tree growth.[4] This favourable climate predominantly exists within the equatorial regions of Africa, South America, and Southeast Asia. In recent times, the industry has expanded to encompass other areas, including Africa and Latin America, thereby contributing to the increasing global expanse dedicated to oil palm cultivation (Lesage et al., 2021).

1.5 LIFE CYCLE OF OIL PALM

The life cycle of the oil palm spans several years, from its beginnings as a seed to the establishment of mature palm plantations (Figure 1.4). This life cycle plays a crucial role in the global production of palm oil, carbon stock, climate change, and various environmental concerns.

Seeds and Seedlings: Oil palm trees are grown from seeds, with a typical productive lifespan of 25–30 years. High-quality seeds like Tenera are preferred for their high oil yield and disease resistance. Seeds are nurtured in nurseries for about a year before planting in plantations typically with 9 m × 9 m × 9 m spacing in Malaysia or 7.5–10 m apart in Indonesia.

Growth: Oil palm growth can be categorized into four stages: young (up to 3 years), intermediate (4–10 years), productive (up to 20 years), and mature (beyond 20 years). The palm tree is monoecious, with both male and female parts on the same tree. Growth stages influence frond development and productivity, therefore pruning is essential for efficient harvesting. Palm trees continue to grow taller but without developing branches as they age.

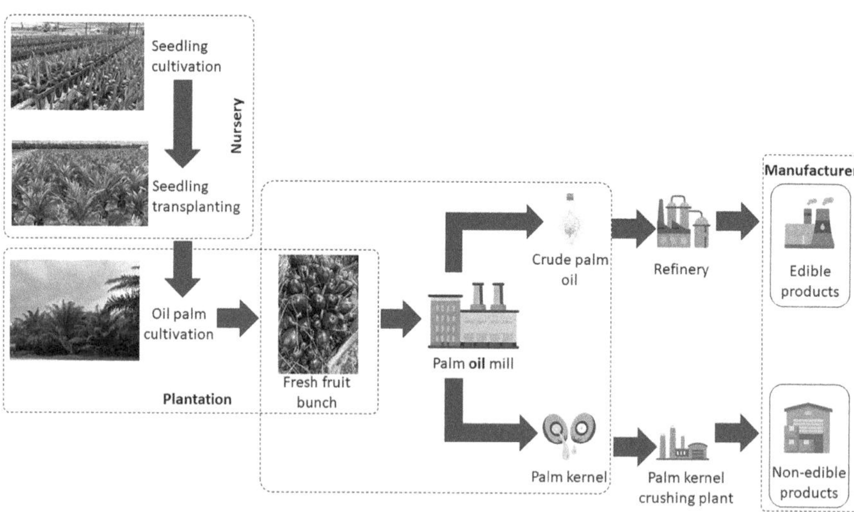

FIGURE 1.4 Life cycle of oil palm.

Condition for Growth and Productivity: Ideal conditions for oil palm growth include soil suitability, terrain, and proper drainage. Growth is maintained through nutrient supply, weed removal, and pest management.

Fruiting: Fruiting begins around 3 years after planting and mature in 5–6 months, with peak production around 10 years.

Yield Production: Oil palm yields around 3 tonnes of palm oil per hectare (current average of global yield), making it one of the most productive oil crops. Yield varies among countries and production systems, depending on soil, terrain, drainage, and management factors.

Harvesting: Harvesting is labour-intensive, usually done manually or with machines in some plantations. At harvesting, each fruit can contain oil and the rest parts are consisting of kernel. The fruit contains up to 25%–30% of oil and the rest is kernel, surrounded by mesocarp.

Milling: Harvested fruit bunches are processed in mills within 50 km of plantations. Oil is extracted from the flesh and kernels. Empty fruit bunches and waste are recycled as fertilizers. Milling can produce effluents that require proper treatment.

Refineries: CPO is transported to refineries for purification before distribution.

Products from Palm Oil: A significant portion of palm oil production is used in food products while PKO is used in industrial applications and consumer products such as soaps and detergents. A small percentage is used for bioenergy.

1.6 CONTRIBUTION OF PALM OIL INDUSTRY TO ECONOMY AND SOCIETY

The palm oil industry significantly contributes to the economies and social well-being of the main producing countries, including Indonesia, Malaysia, Thailand, as well as various nations in Africa and South America. Indonesia and Malaysia stand as the world's leading palm oil exporters, with their national economies heavily reliant on this commodity. Indonesia annually exports palm oil worth USD 35 billion, having shipped 28.45 million metric tons of palm oil in 2022/2023 (USDA, 2023). Palm oil ranks as Indonesia's second-largest export, constituting 9%–17% of its GDP.[5] In 2021, palm oil contributed 2.5% to Malaysia's GDP (Statista, 2023). The palm oil industry is a cornerstone of Malaysia's economy, with 16 million metric tons of palm oil and palm-based products exported in 2022/2023 (USDA, 2023). Meanwhile, Thailand, the third-largest palm oil exporter, shipped 900,000 metric tons in 2022/2023. Other palm oil exporting countries include Guatemala (860,000), Papua New Guinea (780,000), and Colombia (700,000), with most other nations exporting less than 20,000 metric tons during the same period (USDA, 2023). The primary importing countries encompass India, China, the European Union (EU), Pakistan, the United States, Bangladesh, Egypt, the Philippines, and Kenya (USDA, 2023).

Furthermore, the palm oil industry has generated job and business opportunities, boosting the income and purchasing power of people living in impoverished conditions. It plays a pivotal role in alleviating poverty in Indonesia, Malaysia, Thailand, Nigeria, and Colombia, where 17% of the world's impoverished population resides (https://www.worldbank.org/en/topic/poverty). In Indonesia, the industry offers employment to 8 million people, accounting for approximately 3.5% of Indonesia's workforce, and has lifted 2.6 million Indonesians out of poverty (Edwards, 2019). The palm oil–producing countries generally benefit millions of smallholders, elevating their living conditions. Nevertheless, the industry encounters various challenges that affect production and exports, particularly in relation to the EU, as outlined below.

1.7 CHALLENGES FACED BY PALM OIL INDUSTRY

While the palm oil industry significantly contributes to uplifting the economies and social well-being of major producing countries worldwide, it faces various issues that affect both its production and trade, especially with European countries. The industry has been criticized for causing detrimental environmental and social impacts, including deforestation, biodiversity loss, greenhouse gas emissions, air and water pollution, biowaste production, and social injustice (Gaveau and Sheil, 2016). These issues are discussed in detail in Chapters 16 and 17. As a result of these concerns, market demands and consumer preferences have evolved, leading the EU and the United States to impose restrictions on palm oil imports, particularly from Indonesia and Malaysia. The EU has enacted stringent regulations requiring evidence of sustainable palm oil production that does not harm the environment. Major producing countries are grappling with the challenge of collecting evidence across the entire industry's supply chain to demonstrate their commitment to deforestation-free production. Meeting these demands necessitates certification and compliance with sustainability standards, such as those established by the Roundtable on Sustainable Palm Oil (RSPO). However, achieving widespread adoption of sustainable practices throughout the industry remains a challenge (refer to Chapter 16 for a detailed review).

Another critical challenge is increasing yield and productivity while controlling the spread of diseases and implementing improved crop varieties (refer to Chapters 11 and 12). Disease and pest infestations pose significant threats to oil palm plantations. *Ganoderma boninense*, a deadly disease, can reduce yields by 50%–80% (Corley and Tinker, 2016). This disease is widespread in Southeast Asian countries, particularly in Malaysia (Peninsular Malaysia and Sarawak) and Indonesia (Sumatra, and Java). The disease's expansion to new locations is a concern. Detecting the disease at an early stage to prevent its spread remains a challenge for the industry. Major pests in oil palm plantations include defoliators like bagworms (*Metisa plana* and *Pteroma pendula*), nettle and slug caterpillars (*Darna* spp. and *Setora nitens*), moths, rhinoceros beetles, and others (Kalidas, 2013). These threats are exacerbated by climate change, which can impact crop performance in unpredictable ways.

A declining oil extraction rate from fresh fruit bunches (FFBs) due to low-quality seeds is another important problem faced by the industry. The use of high-quality seeds, such as Dura x Pisifera (DxP) and Topaz seeds (Asian Agri, 2023), is crucial not only for maximizing palm oil production but also for optimizing land use. High-quality seeds can produce the same amount of oil with less land, promoting sustainability. However, accessing quality plant materials remains a challenge for many smallholders.

Social issues, including land conflicts with indigenous communities and labour exploitation in some areas, also pose challenges to the industry. Additionally, labour shortages in plantations, exacerbated during the Covid-19 pandemic when many labourers returned to their home countries, resulted in significant productivity losses for Malaysia and Indonesia. Addressing these challenges requires a multi-stakeholder approach and the utilization of modern technology to promote sustainable palm oil production practices that balance economic growth with environmental and social concerns.

1.7.1 Geospatial Technology

Geospatial technology plays a crucial role in advancing sustainability within the palm oil industry. Geospatial technology encompasses a range of tools and techniques for capturing, managing, analysing, and presenting geospatial data. It includes Geographic Information Systems (GIS), remote sensing, Global Positioning Systems (GPS), and spatial modelling. These technologies are fundamental for mapping and monitoring oil palm plantations and collecting accurate, up-to-date spatial data, such as land cover, land use, and environmental parameters, which are essential for comprehending the dynamics of oil palm plantations and their environmental impact. Integrating geospatial technology into the palm oil industry offers numerous advantages for sustainable management. Through remote sensing, including satellite and airborne imagery, data captured by unmanned aerial vehicles (UAV), and light detection and ranging (LiDAR), we can map the extent of oil palm plantations; delineate individual palm trees; assess their age; detect their health conditions, infestation, and disease attack; estimate carbon stock; identify biodiversity hotspots; and aid in risk assessment for ecosystems. Remote sensing enables accurate and efficient land cover mapping, including oil palm plantations and other land uses. This aids in identifying suitable expansion areas and monitoring land use changes over time. By analysing historical and current data, stakeholders can evaluate deforestation risks, monitor illegal encroachment, and measure environmental impact (Gaveau et al., 2016). GIS enables spatial analysis, facilitating the identification of expansion areas, optimizing plantation layouts, and assessing environmental impacts. Geospatial technology enhances supply chain transparency in the palm oil industry. By integrating geospatial data with other relevant sources, stakeholders can trace the origin and movement of palm oil products throughout the supply chain. This allows for the verification of sustainable sourcing practices, certification compliance, and identification of potential risks or breaches in responsible production. By harnessing the power of geospatial technology, stakeholders can strive to balance economic development with environmental conservation, fostering a responsible and sustainable palm oil industry.

1.7.2 Aim and Objectives

This book is designed to serve as a valuable resource for a diverse audience, including palm oil industry stakeholders, environmentalists, researchers, and policymakers, who share an interest in harnessing geospatial technology to promote sustainable palm oil production. Our primary goal is to introduce the practical application of geospatial technology within the palm oil industry and to demonstrate its potential to enhance sustainability practices.

In the initial chapters (Chapters 2–6), we lay the foundation by introducing geospatial technology and its essential components, elucidating their relevance to the sustainable management of oil palm plantations. Throughout the book, we provide a series of compelling case studies and examples (Chapters 7–14) that showcase the successful utilization of remote sensing techniques to bolster the industry's sustainability efforts. We emphasize the integration of geospatial data with field-based surveys and various data sources from extensive databases to facilitate comprehensive analyses, ultimately empowering informed decision-making processes.

Additionally, we dedicate sections of the book (Chapters 14–16) to review existing sustainability standards and evaluate their effectiveness in advancing environmental, social, and economic sustainability within the industry. Acknowledging that geospatial technology, while powerful, may have limitations specific to the palm oil sector, we address these constraints and provide insightful recommendations for maximizing the technology's efficacy (Chapter 17).

By achieving these objectives, our aim is to raise awareness and equip all stakeholders with actionable insights, enabling them to leverage geospatial technology effectively for the pursuit of sustainable palm oil production.

NOTES

1 Zi, Lin. 2023. "China Dialogue | China Environment and Climate News." China Dialogue. November 2, 2023. https://chinadialogue.net/en.
2 "IndexMundi - Country Facts." https://www.indexmundi.com/.
3 Roundtable on Sustainable Palm Oil (RSPO). 2023. "Get Involved as a Smallholder - Roundtable on Sustainable Palm Oil (RSPO)." October 21, 2023. https://rspo.org/as-a-smallholder/.
4 Sergieieva, Kateryna. 2023. "Oil Palm Plantation: Cultivation and Management." *EOS Data Analytics* (blog). July 7, 2023. https://eos.com/blog/oil-palm-plantation/.
5 Indonesia Palm Oil Facts. 2022. "G20 - Indonesia Palm Oil Facts." November 11, 2022. https://www.indonesiapalmoilfacts.com/g20/.

2 Remote Sensing

2.1 REMOTE SENSING TECHNOLOGY

Remote sensing is a detection technology that involves identifying objects or phenomena without direct physical contact. The method of remote sensing involves detecting electromagnetic energy emitted or reflected from the Earth's surface at a distance using observation devices. Remote sensors can be classified as passive or active, depending on the source of electromagnetic energy. Passive remote sensing relies on detecting naturally occurring electromagnetic energy, such as sunlight. Active remote sensing, on the other hand, relies on artificial light sources, such as radio detection and ranging (RADAR) and light detection and ranging (LiDAR). Remote sensing platforms are typically categorized into three main types: (i) Ground-based platforms, which encompass vehicle-borne systems, hand-held devices, fixed installations, and mobile-elevated platforms; (ii) airborne platforms, including both unmanned aerial vehicles (UAVs or drones) and manned aerial vehicles, as well as balloons; and (iii) spaceborne remote sensing platforms, which consist of artificial Earth satellites, space shuttles, space stations, and rockets.

These platforms are utilized to acquire a wide range of accurate, up-to-date, remotely sensed data and derived geospatial information products data that allows for the observation and monitoring of vast areas of oil palm plantations and its environment quickly and efficiently. The sensors mounted on the platforms can capture data across various electromagnetic spectra, enabling the analysis of diverse features, from palm tree health to detecting individual trees. Remote sensing can be a cost-effective and non-invasive solution for applications in the palm oil sector, minimizing environmental impact and reducing the need for extensive fieldwork. Satellite-based remote sensing systems particularly provide the ability to collect data over time, facilitating the study of land use changes caused by palm oil industry or to study the impact of climate change on palm oil yield. Remote sensing technology provides essential support for avoiding deforestation, carbon monitoring, and biodiversity conservation, thus contributing to the sustainable development of the palm oil industry.

2.2 REMOTE SENSING PLATFORM

The remote sensing platform refers to the vehicle of the remote sensor, which is usually composed of remote sensor, data recorder, attitude controller, communication system, power supply system, thermal control system, etc. The remote sensing platform provides a working environment and technical support for the remote sensor to monitor the object or environment. According to the purpose, object, and technology of remote sensing, remote sensing platforms can be divided into three types, that is, ground-based, airborne, and spaceborne remote sensing platform. Some popular remote sensing platforms, such as tripod stands, towers, cranes, tall buildings or scaffolding, balloons, aircraft, rockets, and satellites, have been widely used in scientific

DOI: 10.1201/9780429199813-2

TABLE 2.1
Typical Sensor and Platform Configurations with Main Operational Parameters (Toth and Jóźków, 2016)

Applicability and Operation Aspects	Data Acquisition Platforms				
	Satellite (Spaceborne)	Airborne	UAV	Mobile/Static (Ground)	Crowd-Sensed
Manoeuvrability	No/limited	Moderate	High	Limited	High
Observation space	Worldwide	Regional	Local	Local	Local
Sensor diversity	MS/HSI/SAR	MS/HSI/ LiDAR/SAR	MS (LiDAR/ HSI)	MS/LiDAR (HSI)	MS/video
Environment	Outdoors	Outdoors	Outdoors/ indoors	Outdoors/ indoors	Outdoors/ indoors
Scale (inverse sensor range)	Small	Small/ medium	Medium/large	Medium/large	Medium/large
Ground coverage	Large (10 km)	Medium (1 km)	Small (100 m)	Small (50 m)	Small (10 m)
FOV	Narrow	Wide	Wide/super wide	Wide/super wide	Wide/super wide
Repeat rate	Day	Hours	Minutes	Minutes	Minutes/ seconds
Spatial resolution (GSD)	0.30–300 m	5–25 cm	1–5 cm	1–5 cm	1–5 cm
Spatial accuracy	1–3 m	5–10 cm	1–25 cm	3–50 cm	1–2 m
Deployability	Difficult	Complex	Easy	Moderate	Simple
Observability	Vertical/ oblique	Vertical/ oblique	Vertical/ oblique/360°	Oblique/360°	Oblique
Operational risk	Moderate	High	Low	Moderate	No
Cost	$$$$$	$$$	$	$$	No

HSI, hyperspectral image, LiDAR, light detection and ranging; MS, multi-spectral; SAR, synthetic aperture radar.

research and production. The platform can better help the observation of the area of interest, and the type of platform depends on the requirements of the measurements to be performed. Ground-based and airborne remote sensing platforms are sufficient for small-scale infrequent monitoring or calibration purposes, while large-scale observation requires the support of spaceborne remote sensing platforms (Table 2.1).

2.2.1 Satellite Platforms

The idea of using orbiting spacecraft to make observations of the ground was first explored by Herman Potočnik. Earth observation (EO) provides an effective way of exploring the physical, chemical, and biological information related to the Earth and of supporting the sustainable development of human civilization and the global environment by monitoring and assessing natural and artificial environmental conditions and changes (Gomes et al., 2020; Song and Wu, 2021). EO satellites monitor the Earth's land, ocean, atmosphere, cryosphere, and carbon cycle from space in real time and constantly transmit that information to the ground. Several important or commonly used space-based remote sensing platforms are listed in Table 2.2. Vanguard 2, designed by the United States Naval Research Laboratory, was the first satellite specifically designed for EO. Since then, many important EO satellites and sensors, including Landsat, AVHRR (Advanced Very High Resolution Radiometer), MODIS (Moderate Resolution Imaging Spectroradiometer), and Sentinel, have been launched (Vali et al., 2020). According to the database of the Committee on Earth Observation Satellites (CEOS), in October 2020, 327 EO satellite missions and 845 instruments were either currently operating or were planned for launch in the next 15 years.

After more than 60 years of development, EO satellites have flourished. The significant advances in the spatial, spectral, and temporal resolutions of remote sensing data brought about by improvements in remote sensing technology have dramatically changed how we observe the Earth (Vali et al., 2020). Multi-spectral and hyper-spectral technologies, satellite radar and SAR (Synthetic Aperture Radar), as well as lasers on EO satellites, now satisfy the needs of many types of research (Li et al., 2012; Lodhi et al., 2018). Also, as the cost of instruments and launching satellites has fallen, EO data have become more widely available. Thus, in many ways, EO is positively affecting our lives. There are two main classes of orbits, namely geosynchronous orbit and the polar orbit.

The polar or near-polar orbit is an orbit with an inclination angle of 90° or near 90° (Table 2.2). Satellites in this orbit must pass over or near the two poles of the Earth in each circle. At present, most EO satellites operate in near-polar orbits at a height range from 600 to 2,000 km. These satellites with an operating period of about 100 minutes always pass over the same position in multiple cycles. If such an orbit passes over the poles of the Earth, then the satellite in this orbit would always pass through the same points on the Earth's surface. If the orbits are slightly inclined away from the poles, these orbits are called near-polar orbits. Due to the angle between the satellite orbit and the Earth's rotational axis, the orbit would precess with respect to the Earth. Such an orbit needs several circles to pass over the same point on the Earth, and the time required to complete this process is the satellite revisit time. The sun-synchronous orbit is a special near-polar orbit whose precession is the same as that of the Earth's sun. In such an orbit, satellites take about the same amount of time to image in the same position on the Earth, so the angle of the sun and the illumination and shadow are approximately the same. Meteorological satellites and Earth resource satellites generally operate this orbit, which is particularly suitable for monitoring highly dynamic features, such as vegetation.

The geosynchronous orbit is a prograde orbit with the same period as the rotation period of the Earth. In fact, the orbital height of a geosynchronous satellite is about 35,786 km, and the orbital period of a satellite is approximately 23 hours 56 minutes and 4 seconds (Table 2.2). The geostationary orbit is a special geosynchronous orbit; the inclination of this orbit is zero (over the equator). As a result, the satellites in this orbit are stationary to the Earth and appear to remain in position over a point on the equator. It is more advantageous to use the geostationary orbit for communication,

TABLE 2.2
Important Satellite-Based Remote Sensing Platforms and Sensors

Satellite	Sensor	Bands	Spatial Resolution	Revisit Time (Days)
ALOS	PALSAR	L-band SAR	7–100 m	46
	PRISM	MS 3, Pan 1	2.5 m	
	AVNIR-2	MS 4	10 m	
ALOS-2	PALSAR-2	L-band SAR	3–100 m	14
NOAA/AVHRR	AVHRR	MS and TIR 4–5	1.1 km	1 (Vis), 2 (IR)
IRS-P5 (CartoSat-1)	PAN	PAN 1	2.5 m	5
IRS-P7 (CartoSat-2)	PAN	PAN 1	1 m	5
CartoSat-2A	PAN	PAN 1	1 m	5
CartoSat-2B	PAN	PAN 1	1 m	5
Cartosat-2C	PAN	PAN 1	0.65 m	5
	HRMX	MS 4	2 m	
Cartosat-2D	PAN	PAN 1	0.65 m	4
	HRMX	MS 4	2 m	
Cartosat-2E	PAN	PAN 1	0.65 m	4
	HRMX	MS 4	2 m	
Cartosat-3	PAN	PAN 1	0.25 m	4
	MX	MS 4	1 m	
COSMO-SkyMed (4 satellites)	SAR 2000	X-band SAR	1–100 m	16
Fengyun (FY)-1A	MVISR	5	1.1 km	55
FY-1B	MVISR	5	1.1 km	59
FY-2A	VISSR-1	3	1.44–5.76 km	
FY-1C	MVISR	10	1.1 km	10.61
FY-2B	VISSR-1	3	1.44–5.76 km	
FY-1D	MVISR	10	1.1 km	10.61
FY-2C	VISSR-2	5	1.25–5 km	
FY-2D	VISSR-2	5	1.25–5 km	
FY-3A	IRAS	26	17 km	5.5
	MERSI-1	20	0.25–1 km	
	MWHS-1	5	15 km	
	MWRI	10	15–85 km	
	MWTS-1	4	50–75 km	
	TOU	6	50 km	
	SBUS	12	200 km	
	VIRR	10	1.1 km	

(*Continued*)

TABLE 2.2 (*Continued*)
Important Satellite-Based Remote Sensing Platforms and Sensors

Satellite	Sensor	Bands	Spatial Resolution	Revisit Time (Days)
FY-2E	VISSR-2	5	1.25–5 km	
FY-3B	IRAS	26	17 km	5.5
	MERSI-1	20	0.25–1 km	
	MWHS-1	5	15 km	
	MWRI	10	15–85 km	
	MWTS-1	4	50–75 km	
	TOU	6	50 km	
	SBUS	12	200 km	
	VIRR	10	1.1 km	
FY-2F	VISSR-2	5	1.25–5 km	
FY-3C	IRAS	26	17 km	5.5
	MERSI-1	20	0.25–1 km	
	MWHS-2	15	16–32 km	
	MWRI	10	15–85 km	
	MWTS-2	13	32 km	
	TOU	6	50 km	
	SBUS	12	200 km	
	VIRR	10	1.1 km	
FY-2G	VISSR-2	5	1.25–5 km	
FY-4A	AGRI	14	0.5–4 km	
	GIIRS	1650	16 km	
	LMI	1	7.8 km	
FY-3D	GAS	4	10 km	5.5
	HIRAS	1370	16 km	
	MWRI	10	15–85 km	
	MWTS-2	13	32 km	
	MERSI-2	25	0.25–1 km	
	MWHS-2	15	16–32 km	
FY-2H	VISSR-2	5	1.25–5 km	
FY-4B	AGRI	14	0.5–4 km	
	GIIRS	1650	16 km	
	LMI	1	7.8 km	
FY-3E	MERSI-2	25	0.25–1 km	5.5
	MWHS-2	15	16–32 km	
	MWTS-2	13	32 km	
	HIRAS	1370	16 km	
Gaofen 1 (4 satellites)	PMC	PAN 1 and MS 4	PAN 2 m and MS 8 m	41
	WFI	4	16 m	

(*Continued*)

TABLE 2.2 (*Continued*)
Important Satellite-Based Remote Sensing Platforms and Sensors

Satellite	Sensor	Bands	Spatial Resolution	Revisit Time (Days)
Gaofen 2	PMC-2	PAN 1 and MS 4	PAN 1 m and MS 4 m	69
Gaofen 3 (2 satellites)	C-SAR	C-band SAR	1–500 m	3
Gaofen 4	MWIR	1	400	
	VNIR	5	50	
Gaofen 5	EMI	4	48 km × 13 km	
	GMI	4		
	DPC	8	3.5 km	
	AIUS			
	AHSI			
	VIMS	12	20–40 m	
Gaofen 6	PMS	PAN 1 and MS 4	PAN 2 m and MS 8 m	2
	WFV	8	16 m	
Gaofen 7	MUX			<60
	BWD			
	FWD			
Gaofen 8				
Gaofen 9 (5 satellites)				
Gaofen 10				
Gaofen 11 (2 satellites)				
Gaofen 12 (2 satellites)				
Gaofen 13				
Gaofen 14				
IRS-1A	LISS-I	4	72.5 m	26
	LISS-II A/B	4	36.25 m	
IRS-1B	LISS-I	4	72.5 m	26
	LISS-II A/B	4	36.25 m	
IRS-P1 (IRS-1E)				
IRS-P2	LISS-II M	4	32 × 37 m	22
IRS-1C	LISS-III	4	23.5–70 m	24
	PAN		5.8 m	
	WiFS	2	188 m	

(*Continued*)

TABLE 2.2 (*Continued*)
Important Satellite-Based Remote Sensing Platforms and Sensors

Satellite	Sensor	Bands	Spatial Resolution	Revisit Time (Days)
IRS-P3	WiFS	3	188 m	24
	MOS-A B C	18	520–1,500 m	
IRS-1D	LISS-III	4	23.5–70 m	24
	PAN		5.8 m	
	WiFS	2	188 m	
IRS-P4	OCM	8	360 × 236 m	2
(OceanSat-1)	MSMR	4	105 × 68, 66 × 43, 40 × 26, 34 × 22 km	
IRS-P6	LISS-IV	3	5.8 m	26
(ResourceSat-1)	LISS-III	4	23.5 m	
	AWiFS	3	70 m	
IRS-P5 (CartoSat-1)	PAN	1	2.5 m	5
Landsat (LS)-1	MSS	4	80 m	18
(ERTS-1)	RBV	3	80 m	
LS-2 (ERTS-2)	MSS	4	80 m	18
	RBV	3	80 m	
LS-3	MSS	5	80 m	18
	RBV	3	80 m	
LS-4	MSS	4	80 m	16
	TM	7	30–120 m	
LS-5	MSS	4	80 m	16
	TM	7	30–120 m	
LS-6	ETM	PAN 1, MS 6, and TIR 1	15–120 m	16
LS-7	ETM+	PAN 1, MS 6, and TIR 1	15–60 m	16
LS-8	OLI	PAN 1 and MS 8	15–30 m	16
	TIRS	2	100 m	
LS-9	OLI-2	PAN 1 and MS 8	15–30 m	16
	TIRS-2	2	100 m	
Terra (EOS AM-1)	MODIS	MS and TIR 36	250–1,000 m	16
Aqua (EOS PM-1)	MODIS	MS and TIR 36	250–1,000 m	16
Radarsat-1	SAR	C-band SAR	8–100 m	24
Radarsat-2	SAR	C-band SAR	0.8–24 m	24
RCM-1	SAR	C-band SAR	1–100 m	12
RCM-2	SAR	C-band SAR	1–100 m	12
RCM-3	SAR	C-band SAR	1–100 m	12

(*Continued*)

TABLE 2.2 (*Continued*)
Important Satellite-Based Remote Sensing Platforms and Sensors

Satellite	Sensor	Bands	Spatial Resolution	Revisit Time (Days)
Sentinel-1A	C-SAR	C-band SAR	9–50 m	12
Sentinel-1B	C-SAR	C-band SAR	9–50 m	12
Sentinel-2A	MSI	13	10–60 m	10
Sentinel-2B	MSI	13	10–60 m	10
Sentinel-3A	OLCI	21	300 m	27
	SLSTR	9	500–1,000 m	
	SRAL	C-band SAR	300 m	
Sentinel-3B	OLCI	21	300 m	27
	SLSTR	9	500–1,000 m	
	SRAL	C-band SAR	300 m	
Sentinel-5P	TROPOMI	8	7 km	16
SPOT-1	HRV	PAN 1 and MS 3	PAN 10 m and MS 20 m	26
SPOT-2	HRV	PAN 1 and MS 3	PAN 10 m and MS 20 m	26
SPOT-3	HRV	PAN 1 and MS 3	PAN 10 m and MS 20 m	26
SPOT-4	HRVIR	PAN 1 and MS 4	PAN 10 m and MS 20 m	26
	VEGETATION	4	1.15 km	
SPOT-5	HRG	PAN 1 and MS 4	PAN 2.5 m, MS 10 m, and SWIR 20 m	26
	HRS	PAN 1	PAN 10 m	
	VEGETATION	4	1.15 km	
SPOT-6	NAOMI	PAN 1 and MS 4	PAN 1.5 m and MS 6 m	26
SPOT-7	NAOMI	PAN 1 and MS 4	PAN 1.5 m and MS 6 m	26
TerraSAR-X	X-Band SAR	X-band SAR	1–16 m	11
TanDEM-X	X-Band SAR	X-band SAR	1–16 m	11
WorldView (WV) -1	WV60	PAN 1	PAN 0.5 m	1.7
WV-2	WV110	PAN 1 and MS 8	PAN 0.46 m and MS 1.85 m	1.1
WV-3	WV-3	PAN 1 and MS 16	PAN 0.31 m, MS 1.24 m, and SWIR 3.7 m	1
	CAVIS	12	30 m	
WV-4 (GeoEye-2)	SpaceView™ 110	PAN 1 and MS 4	PAN 0.31 m and MS 1.24 m	<1

(*Continued*)

TABLE 2.2 (*Continued*)
Important Satellite-Based Remote Sensing Platforms and Sensors

Satellite	Sensor	Bands	Spatial Resolution	Revisit Time (Days)
Ziyuan (ZY)-1 01 (CBERS 1)	HRCC	PAN 1 and MS 4	20 m	26
	IRMSS	PAN 1, MS 2, and TIR 1	PAN, SWIR 80 m, and T 160 m	
	WFI	2	260 m	
ZY-1 02 (CBERS 2)	HRCC	PAN 1 and MS 4	20 m	26
	IRMSS	PAN 1, MS 2, and TIR 1	PAN, SWIR 80 m, and T 160 m	
	WFI	2	260 m	
ZY-1 02B (CBERS 2B)	HRCC	PAN 1 and MS 4	20 m	26
	HRPC	PAN 1	2.7 m	
	WFI	2	260 m	
ZY-1 02C	PMS	PAN 1 and MS 3	PAN 5 m and MS 10 m	26
	HRC	PAN 1	2.36 m	
ZY-1 02D	PMS	PAN 1 and MS 3	PAN 5 m and MS 10 m	26
	HRC	PAN 1	2.36 m	
ZY-1 03 (CBERS 3)				
ZY-1 04 (CBERS 4)	MUXCam	4	20 m	26
	PanMUX	PAN 1 and MS 3	PAN 5 m and MS 10 m	
	IRS	4 (MWIR 1, SWIR 2, and TIR 1)	40 and T 80 m	
	WFI	4	64 m	
ZY-1 04A (CBERS 4A)	MUX	4	16 m	31
	WFI	4	55 m	
	WPM	PAN 1 and MS 3	PAN 2 m and MS 8 m	
ZY-2 01 (Jianbing-3 1)				
ZY-2 02 (Jianbing-3 2)				

(*Continued*)

TABLE 2.2 (*Continued*)
Important Satellite-Based Remote Sensing Platforms and Sensors

Satellite	Sensor	Bands	Spatial Resolution	Revisit Time (Days)
ZY-2 03 (Jianbing-3 3)				
ZY-3 01	TAC	PAN 1	2.1 m	59
	MSC	4	5.8 m	
ZY-3 02	TAC	PAN 1	2.1 m	59
	MSC	4	5.8 m	
ZY-3 03	TAC	PAN 1	2.1 m	59
	MSC	4	5.8 m	

MIR, middle wavelength infrared; MS, multi-spectral; PAN, panchromatic; SAR, Synthetic Aperture Radar; SWIR, short wavelength infrared; TIR, thermal infrared.

broadcasting, and meteorological satellites. There are several meteorological satellites, such as Meteosat, GOES (Geostationary Operational Environmental Satellite), GOMS (Geostationary Operational Meteorological Satellite), and INSAT (Indian National Satellite System) are all distributed around the equator. Compared with other satellite orbits, there is only one geostationary orbit, which is very precious.

2.2.2 Airborne Platforms

Airborne remote sensing refers to a remote sensing technology that uses various aircraft, airplanes, and balloons as sensor vehicles in the air. In recent years, UAV remote sensing has been developed rapidly, and it is one of the important airborne remote sensing platforms. The height of the vehicle is always below 80 km. The flying height and the flight lines can be altered to adjust the scale and coverage area of the photo or image or to avoid cloud cover. The airborne platform has a low flying height, is flexible in the sensing process, is not restricted by ground conditions, has a short period of sense, and is convenient for transmission, so a great deal of remote sensing research has been conducted using airborne platforms. However, the cost of flying is high, the images sensed are prone to distortion, and the process is not reproducible.

Because airborne remote sensing can quickly obtain high-resolution images, it is widely used in land cover mapping and land cover change monitoring (Adão et al., 2017). Especially in tropical areas with high cloud coverage, airborne remote sensing has significant advantages compared to the EO satellites and can be applied to oil palm monitoring effectively. At present, many commercial oil palm plantation companies have developed drone teams to regularly collect images of oil palm plantations for the management of oil palm plantations (Hoffmann et al., 2016). However, due to the high cost and small or medium scale of the airborne platform, the sensed data provided by airborne platforms is only used in the local oil palm studies.

2.2.3 Ground-Based Platforms

Ground-based remote sensing refers to the remote sensing technology system with high towers, vehicles, and ships as platforms. Spectrometers or sensors can be installed on these ground-based platforms to perform spectral measurements. Ground-based remote sensing platforms can be divided into mobile mapping systems (MMS) and static installation of remote sensing systems. MMS refers to those sensors installed on a moving vehicle, such as cars, trucks, to acquire geospatial data in a highly effective way. Static installation of remote sensing systems refers to those sensors installed on a static vehicle. A representative example is a webcam, which has provided a great deal of time-lapse and real-time visual data from many interesting locations. With the advance of imaging technology and the reduction of costs, static ground-based remote sensing platforms provide unprecedented temporal resolution and observation capabilities. In addition, ground remote sensing experiment is a significant support for sensor calibration, process modelling, and information extraction.

In the research of oil palm, ground-based remote sensing is a significant support for geometric correction of airborne and spaceborne remote sensing images, especially airborne images, to improve the accuracy of research results. Ground-based remote sensing has been successfully used in the detection and classification of oil palm basal stem rot disease and is an important monitoring method for oil palm diseases (Husin et al., 2020).

2.3 RESOLUTIONS OF REMOTE SENSING DATA

Spatial, spectral, radiometric, temporal, and angular resolutions are key factors used to assess the quality of remote sensing data. Resolution in remote sensing refers to the capability of remote sensing devices to differentiate and distinguish different objects/features on the Earth. The resolution of remotely sensed data can significantly impact mapping and various applications related to oil palm research and analysis. Table 2.3 lists various remote sensing sensors and their resolution characteristics.

2.3.1 Spatial Resolution

Spatial resolution is a measure of sharpness or fineness of spatial details, usually specified as the size of a pixel that is recorded in a raster image. It describes the ability of a sensor to distinguish the smallest size of objects/features on the Earth. In other words, spatial resolution is the smallest distance between distinguishable patterns or objects that can be separated. Therefore, the smaller the distance, the higher the spatial resolution of the image, and the richer the spatial content contained in the image (Lachowski et al., 1995). The spatial resolution of optical sensors can be roughly divided into high (<1 m), medium (1–100 m), and low (>100 m). High spatial resolution images can offer more detailed spatial information about the observed objects and effectively distinguish between objects (Figure 2.1). Figure 2.1 illustrates the impact of high-resolution Google Earth imagery in effectively distinguishing oil palm trees when compared to images from Sentinel, Landsat Operational Land

TABLE 2.3
Spectral, Spatial and Temporal Resolutions of Different Instruments

Satellite	Sensor	Bands	Spectral Resolution	Spatial Resolution	Revisit Time (Days)
NOAA	AVHRR	MS and T 4–5	~0.40 µm to ~0.1 cm	1.1 km	1 (Vis), 2 (IR)
Landsat-1 (ERTS-1)	MSS	4	~0.40 to ~1.3 µm	80 m	18
Landsat-2 (ERTS-2)	MSS	4	~0.40 to ~1.3 µm	80 m	18
Landsat-3	MSS	5	~0.40 to ~1.3 µm	80 m	18
Landsat-4	MSS	4	~0.40 to ~1.3 µm	80 m	16
	TM	7	~0.40 to ~15.0 µm	30–120 m	
Landsat-5	MSS	4	~0.40 to ~1.3 µm	80 m	16
	TM	7	~0.40 to ~15.0 µm	30–120 m	
Landsat-6	ETM	PAN 1, MS 6, and TIR 1	~0.40 to ~15.0 µm	15–120 m	16
Landsat-7	ETM+	PAN 1, MS 6, and TIR 1	~0.40 to ~15.0 µm	15–60 m	16
Landsat-8	OLI	PAN 1 and MS 8	~0.40 to ~3.0 µm	15–30 m	16
	TIRS	2	~6.0 to ~15.0 µm	100 m	
Landsat-9	OLI-2	PAN 1 and MS 8	~0.40 to ~3.0 µm	15–30 m	16
	TIRS-2	2	~6.0 to ~15.0 µm	100 m	
Terra (EOS AM-1)	MODIS	MS and T 36	~0.40 to ~15.0 µm	250–1,000 m	16
Aqua (EOS PM-1)	MODIS	MS and T 36	~0.40 to ~15.0 µm	250–1,000 m	16
SPOT-1	HRV	PAN 1 and MS 3	~0.40 to ~1.3 µm	PAN 10 m and MS 20 m	26
SPOT-2	HRV	PAN 1 and MS 3	~0.40 to ~1.3 µm	PAN 10 m and MS 20 m	26
SPOT-3	HRV	PAN 1 and MS 3	~0.40 to ~1.3 µm	PAN 10 m and MS 20 m	26
SPOT-4	HRVIR	PAN 1 and MS 4	~0.40 to ~3.0 µm	PAN 10 m and MS 20 m	26
	VEGETATION	4	~0.40 to ~3.0 µm	1.15 km	

(Continued)

TABLE 2.3 (*Continued*)
Spectral, Spatial and Temporal Resolutions of Different Instruments

Satellite	Sensor	Bands	Spectral Resolution	Spatial Resolution	Revisit Time (Days)
SPOT-5	HRG	PAN 1 and MS 4	~0.40–~3.0 μm	PAN 2.5 m, MS 10 m, and SWIR 20 m	26
	HRS	PAN 1	~0.40 to ~0.75 μm	PAN 10 m	
	VEGETATION	4	~0.40 to ~3.0 μm	1.15 km	
SPOT-6	NAOMI	PAN 1 and MS 4	~0.45 to ~0.89 μm	PAN 1.5 m and MS 6 m	26
SPOT-7	NAOMI	PAN 1 and MS 4	~0.45 to ~0.89 μm	PAN 1.5 m and MS 6 m	26
WV-1	WV60	PAN 1	~0.45 to ~0.90 μm	PAN 0.5 m	1.7
WV-2	WV110	PAN 1 and MS 8	~0.40 to ~1.04 μm	PAN 0.46 m and MS 1.85 m	1.1
WV-3	WV-3	PAN 1 and MS 16	~0.40 to ~2.37 μm	PAN 0.31 m, MS 1.24 m, and SWIR 3.7 m	1
	CAVIS	12	~0.40 to ~2.25 μm	30 m	
WV-4 (GeoEye-2)	SpaceView™ 110	PAN 1 and MS 4	~0.45 to ~0.92 μm	PAN 0.31 m and MS 1.24 m	<1

Imager (OLI), and the Moderate Resolution Imaging Spectrometer (MODIS) sensors onboard Terra and Aqua satellites, which have coarser spatial resolutions.

2.3.2 Spectral Resolution

Spectral resolution refers to a specific wavelength range that the sensor can record (mostly for satellite and airborne sensors). In general, a wider wavelength range results in lower spectral resolution, whereas a narrower wavelength range enables the sensor to detect specific objects more effectively since different objects typically reflect distinct wavelengths. Additionally, a greater number of bands enhances the spectral information provided by the sensor about the object. Hyperspectral remote sensing holds significant importance in the analysis of an object's chemical composition, as it provides abundant spectral information, and the appropriate spectral resolution allows for the differentiation of specific materials. The wavelength range covered is typically determined by the number of frequency bands recorded by the platform. For instance, the current Landsat OLI collection includes nine bands, some

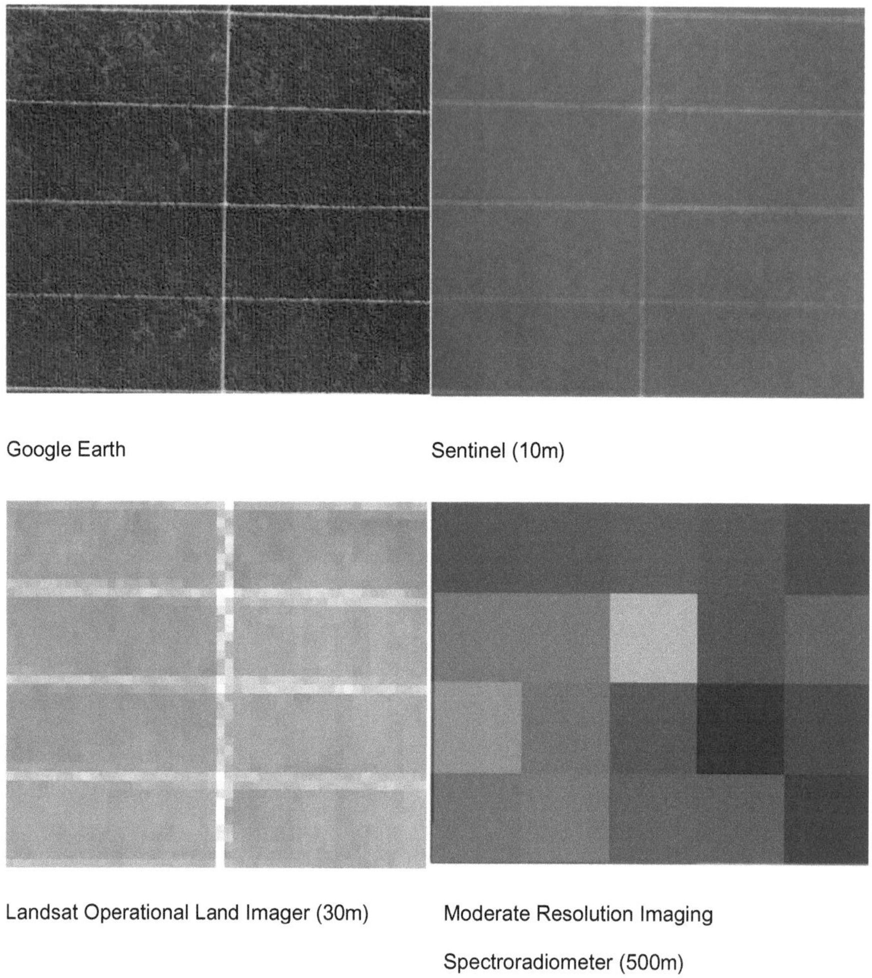

FIGURE 2.1 The oil palm on different spatial resolutions at location 0°06′33.00″ N 111°38′06.96″ E.

of which are within the infrared spectrum, spanning from a spectral resolution of 0.4 to 3 μm. In contrast, the MODIS sensor on Terra and Aqua offers 36 bands covering a wide range from 0.4 to 15 μm (Table 2.3).

2.3.3 Radiometric Resolution

Radiometric resolution often refers to contrast and indicates the sensor's capacity to distinguish between different radiation intensities. It measures the sensor's ability to detect changes in electromagnetic signals from objects, typically ranging from 8 to 14 bits. This corresponds to 256 levels of grayscale or up to 16,384 shades of colour in each band and is influenced by instrument noise. Radiometric resolution determines a sensor's ability to differentiate objects with similar reflectance, essentially its capacity to detect and identify objects within their surroundings. Sensors with higher radiometric resolution can detect and identify many objects that have distinct radiation intensities. In remote sensing images, radiometric resolution describes the quantization level of radiation for each pixel. The maximum brightness of an image depends on the number of bits in the recorded electromagnetic signal. For example, sensors recording data in 8 bits yield a maximum brightness of $2^8 = 256$, with values ranging from 0 to 255. Landsat missions, such as the previous generation sensor ETM+ (Enhanced Thematic Mapper Plus), supports 8-bit data products (i.e., 256 data values). In contrast, sensors like OLI, OLI-2, TIR (Thermal Infrared Sensor), and TIR-2 offer 12-bit radiometric resolution, allowing for discrimination among over 4,096 data values. Consequently, imagery provided by OLI, OLI-2, TIR, and TIR-2 can capture finer distinctions between objects and prevent saturation, even with very bright targets such as snow.

2.3.4 Temporal Resolution

Temporal resolution refers to the minimum time interval between two consecutive sensor observations in the same area, also known as the revisit period. It refers to a series of images captured by the same sensor over a specific time span, with a higher temporal resolution corresponding to a greater number of captured images. For example, the MODIS sensor, with a revisit period of twice daily, offers high temporal resolution compared to Landsat, which revisits the same area every 16 days (Table 2.3). Temporal resolution is crucial in change detection studies, such as monitoring changes in forest cover or tracking the expansion of oil palm plantations and infestations within them.

2.3.5 Angular Resolution

Angular resolution refers to a sensor's capability to observe the same area from various angles, allowing for the examination of object details from different viewing perspectives. In an ideal scenario, we assume that objects exhibit Lambertian reflection, meaning the electromagnetic signal remains unaffected by the observation angle. However, in practice, this is not always the case, especially on surfaces with strong bidirectional reflectivity effects, where the detected electromagnetic signal varies with the observation angle. Multi-angular observation is a valuable approach to simulate these effects, enabling a more accurate electromagnetic characterization of objects. Numerous satellites are now equipped with multi-angular sensors, such as ERS-2 (European Remote Sensing Satellite) with ATSR-2 (Along Tract Scanning Radiometer) and Terra with MISR (Multi angle Imaging Spectroradiometer). MISR is a highly sophisticated multi-angular instrument that can provide electromagnetic signals from nine observation angles of the same area.

The potential of incorporating remotely sensed angular information to enhance the accuracy of land cover classification has been demonstrated. A spatial and angular fusion (SAF) model was developed for fusing MISR and Landsat ETM + data. The results indicated the effectiveness of leveraging angular information to improve classification accuracy (Huang et al., 2012).

2.4 REMOTE SENSING TOOLS

Remote sensing tools refer to software applications and technologies used to process, analyse, visualize, and interpret remotely sensed data collected from various platforms. These tools play a crucial role in transforming raw remote sensing data into actionable information for a wide range of applications. They assist in tasks such as image calibration, geometric correction, and radiometric enhancement. Many remote sensing tools integrate with geographic information systems (GIS) and offer geospatial analysis capabilities. This allows users to perform tasks like image classification, land cover mapping, change detection, and terrain analysis. Many modern remote sensing tools incorporate machine learning and artificial intelligence (AI) algorithms for automated feature extraction, classification, and pattern recognition in large datasets. Some popular remote sensing tools include ArcGIS, QGIS (Quantum GIS), ENVI (environment for visualizing images), ERDAS IMAGINE, eCognition, Google Earth Engine, SNAP (Sentinel Application Platform), and various specialized software for LiDAR data processing. These tools are capable of reading popular remote sensing file formats, including GeoTIFF, NITF, JPEG 2000, ECW, MrSID, HDF, and NetCDF, which contain remote sensing image data, georeferencing information, and sensor metadata. With the advancement of remote sensing technology, the volume of remote sensing data has significantly increased. Cloud computing has emerged as a popular tool for processing remote sensing big data. Consequently, many remote sensing platforms are currently being developed based on cloud computing, such as Google Earth Engine, Microsoft Azure, and Amazon Web Services.

These tools have considerably reduced the time and manpower required for traditional palm tree monitoring and palm plantation management. For instance, ArcGIS provides client software, server software, and online GIS services. It is used for creating and using maps, as well as compiling, analysing, sharing, and managing geographic data. In contrast to traditional remote sensing tools, eCognition innovatively employs an object-oriented classification method, which greatly enhances the automatic classification accuracy of high spatial resolution data. eCognition also offers a unique oil palm application tool called Trimble eCognition Oil Palm Application. This software features an easy-to-use and intuitive graphical user interface, providing a powerful and straightforward deep learning-driven palm tree detection approach for production-oriented users. With deep-learning-based object detection capabilities, ERDAS IMAGINE streamlines the process of identifying and mapping oil palm growth locations from aerial and satellite imagery.

ERDAS IMAGINE is a comprehensive image processing software package that enables users to process both geospatial and other types of imagery, as well as vector data. It has the capability to handle hyperspectral imagery and LiDAR data from various sensors. ERDAS also includes a 3D viewing module called VirtualGIS and a vector module for modelling. Its native programming language is EML (ERDAS Macro Language). ERDAS is seamlessly integrated with other GIS and remote sensing applications, and its imagery storage format is compatible with many other applications through *.img files. Leica Geosystems also acquired ER Mapper to enhance their mapping software. ERDAS IMAGINE is known for its tight integration within the GIS ecosystem, which sets it apart from other image processing software packages and offers distinct advantages.

Introduced by Google in 2005, Google Earth has emerged as the foremost and highly successful virtual globe tool (Yu and Gong, 2012). This program maps our planet by overlaying satellite imagery, aerial photography, and GIS data onto a 3D globe, enabling users to explore cities and landscapes from diverse perspectives. Navigating this virtual globe is user-friendly, with options to input addresses, coordinates, or simply using a keyboard or mouse. Google Earth offers a variety of overlaid images on Earth's surface and functions as a Web Map Service client. In a recent announcement, Google disclosed that Google Earth now covers over 98 per cent of the world and has amassed 10 million miles of Street View imagery. While Google Earth has risen to prominence as the most favoured choice, offering advantages in visualization, ease of access to diverse geospatial data, and a unified coordinate system, it still faces limitations in terms of extensibility (Yu and Gong, 2012).

In recent years, Google Earth has become an indispensable tool for the remote sensing community, playing a pivotal role in various research endeavours related to global environmental change. Its user-friendly visualizations and high-spatial resolution imagery have positioned it as an exceptional platform for observing and validating the growth of oil palm, making it an invaluable validation tool for oil palm research (Figure 2.2).

Google Earth Engine, launched by Google in 2010 (Amani et al., 2020), is a robust cloud computing platform designed for processing satellite imagery and various geospatial and observation data. It offers substantial computational capabilities and has proven to be an invaluable tool for the remote sensing community. Google Earth Engine provides users with access to publicly downloadable EO and geospatial data on a petabyte scale, advanced algorithms for analysing large geographic datasets, and an interactive programming environment (Python and JavaScript application programming interfaces (APIs)). All of these services are provided at no cost, which is particularly significant in less-developed countries. Additionally, it hosts extensive time-series records of EOs, playing a vital role in environmental monitoring and analysis (Amani et al., 2020). This platform utilizes APIs for server requests.

Currently, Google Earth Engine stands as the most widely adopted cloud computing platform in Earth system science. It has been extensively utilized for data processing across numerous fields related to environmental change, including agriculture. It supports global or large-scale oil palm research (Chapter 7) and the sharing of research outcomes. It efficiently processes large-scale images and geospatial data, allowing researchers to quickly upload their findings to Google Earth Engine for easy dissemination (Figures 2.3 and 2.4).

2.5 REMOTE SENSING FOR OIL PALM

Remote sensing has been used in oil palm monitoring since the 1990s. Progress has been made in oil palm mapping and change detection, including (i) data sources from optical satellite earth observations (Lee et al., 2016; Srestasathiern and Rakwatin, 2014) to microwave datasets such as phased array type L-band Synthetic Aperture Radar (PALSAR) (Cheng et al., 2018; Dong et al., 2015), (ii) spatiotemporal

FIGURE 2.2 Google Earth used for monitoring of oil palm.

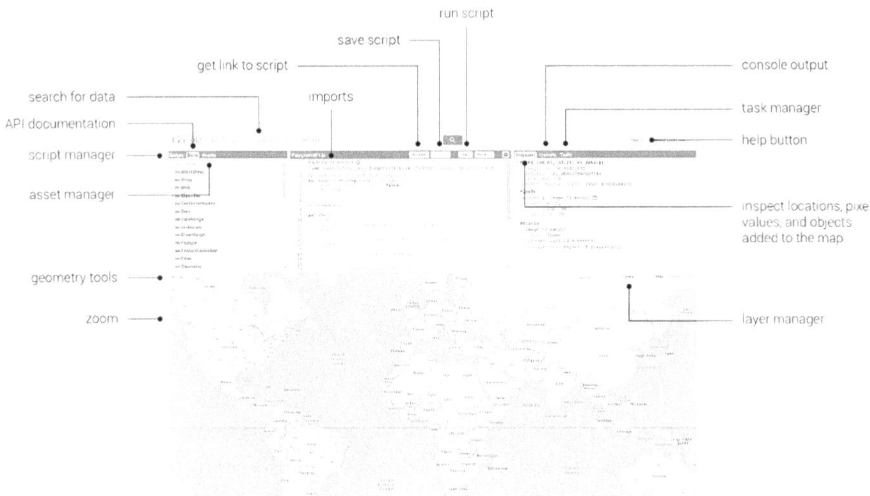

FIGURE 2.3 Google Earth Engine.

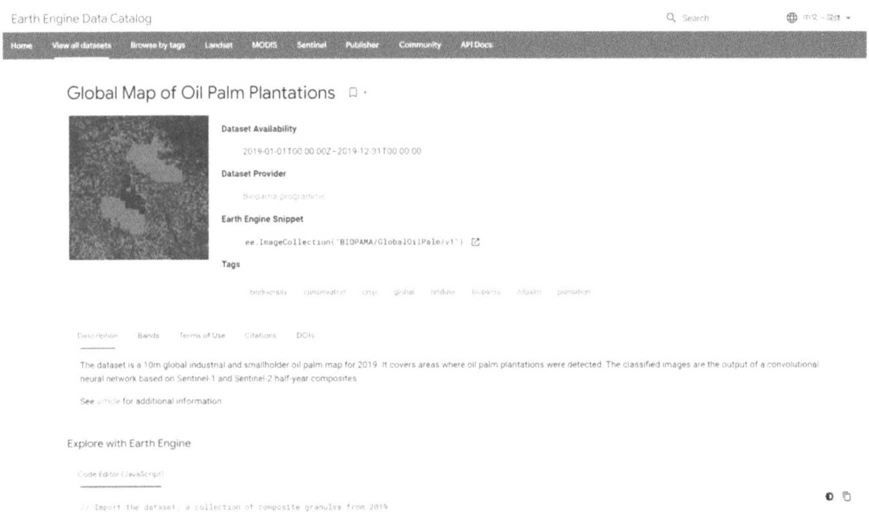

FIGURE 2.4 Google Earth Engine used for oil palm research (Descals et al., 2021).

resolutions from regional to national scales (Miettinen et al., 2019) and from single to multi-decadal mapping (Gaveau et al., 2016; Miettinen et al., 2016), (iii) interpretation methods from manual to semi-automatic identification (Cheng et al., 2017; Ordway et al., 2019), and (iv) products going from oil palm land cover maps to more detailed datasets on plantation structure, for example, tree counting (Li et al., 2019), age and yield estimation (Balasundram et al., 2013; Tan et al., 2013). A few studies also focused on the continuous oil palm change detection (Carlson et al., 2013; Gaveau et al., 2016; Vijay et al., 2018). These studies adopted visual or semi-automatic interpretation for oil palm plantation, which is labour-intensive and not appropriate for long-term annual oil palm plantation monitoring. Automatic identification can overcome this difficulty by using classification algorithms based on Landsat and PALSAR/PALSAR-2 data, which were successfully applied to produce the 2015 land cover map of insular Southeast Asia with discrimination of oil palm plantation (Miettinen et al., 2017). So far, however, the annual dynamics of oil palm plantations (expansion and shrinkage) remains unquantified for Malaysia and Indonesia.

During the last two decades, satellite remote sensing technique has been adopted in many kinds of applications for oil palm plantation monitoring and management (Chong et al., 2017), such as oil palm plantation mapping (Cheng et al., 2016; Dong et al., 2019; Mohd Najib et al., 2020), oil palm tree detection (Li et al., 2017; Freudenberg et al., 2019), age, carbon and yield estimation (Kanniah et al., 2012; Balasundram et al., 2013; Tan et al., 2013; Cracknell et al., 2015), and change detection (Pittman et al., 2013; Cheng et al., 2019). Nowadays, as rapid development of remote sensing techniques, it is potential to automatically and accurately detect oil palm trees from high-resolution remotely sensed images instead of time-consuming, labour-exhausting and costly manual work.

Various types of remote sensing data (such as optical, SAR and high-resolution remote sensing data) have been used in the monitoring of oil palm (ElRaey et al., 1996; Nooni et al., 2014). The specific spectral and texture characteristics of oil palm plantations help to be recognized in medium and high-resolution optical images (Jansen and Di Gregorio, 2002; Srestasathiern and Rakwatin, 2014) and help to derive the spatial distribution of oil palm plantations automatically (Shafri et al., 2011). However, oil palm is usually planted in cloudy and rainy tropical areas, resulting in very limited available optical remote sensing data. In this case, it is difficult to monitor the distribution and changes of oil palm on a global scale using optical remote sensing. But now, with the increase in the frequency of observation of EO systems (such as Sentinel-2 compared to Landsat missions), the impact of weather is constantly decreasing (Xu et al., 2018; Zhao et al., 2019).

SAR can provide all-weather observation data, and oil palm plantation can be well recognized from evergreen broad-leaved forests at specific polarization bands (Koh et al., 2011; Morel et al., 2011). At present, SAR satellite systems in orbit include ALOS-2 PALSAR-2 (L-band), Radarsat-2 (C-band), Sentinel-1 (C-band), COSMO-SkyMed (X-band), TerraSAR-X (X-band), TanDEM-X (X-band), GaoFen-3 (C-band), and so on. Compared with optical remote sensing data, SAR data have obvious advantages in monitoring oil palm (Dong et al., 2015; Teng et al., 2015). Thus, it has advantages over optical remote sensing instruments in monitoring oil palm, and more and more oil palm studies are carried out using PALSAR data. For example, Cheng et al. (2018) mapped the distribution of oil palm plantations in Malaysia in 2015 using ALOS PALSAR and ALOS2-PALSAR2 data. However, SAR data with fewer bands and lower signal to noise ratio compared with optical remote sensing data, which brings difficulties to detecting oil palm (Cheng et al., 2016).

Fusion of optical remote sensing data and SAR data may be an important way to improve the accuracy of mapping oil palm. Some scientists have detected oil palm using the fusion data of optical and SAR, and the results show higher accuracy

TABLE 2.4
Large-Scale Oil Palm Plantation Mapping Products

Study Area	Remote Sensing Data	Spatial Resolution (m)	Year	References
Indonesia (Riau)	Sentinel-1, Sentinel-2	10, 20	2018	Descals et al. (2019)
Dawna Tenasserim Landscape	Landsat 8, Sentinel-1, Sentinel-2	30	2018	Poortinga et al. (2019)
Latin America	MODIS	250	2014	Furumo and Aide (2017)
Malaysia	PALSAR	100	2016	Cheng et al. (2018)
Global (15 countries)	PALSAR, Google Earth images	100	2015	Cheng et al. (2018)
Malaysia, Indonesia	MODIS, PALSAR	250	2010	Koh et al. (2011)

(Cheng et al., 2016; Koh et al., 2011). For example, Koh et al. (2011) used the fusion data of PALSAR-1 data and MODIS data to map oil palm plantations in Malaysia and Indonesia in 2010 with a 250-m spatial resolution, however, small oil palm plantations were missed due to the coarse spatial resolution.

At present, the remote sensing data used in large-scale oil palm plantation mapping are recorded by Landsat, MODIS, Sentinel, and ALOS-PALSAR (Table 2.4). Transitioning from the utilization of remote sensing data and techniques in mapping oil palm plantations, the next chapter introduces the integration of Geographic Information Systems (GIS) for further analysis and management within the palm oil industry.

3 Geographical Information System

3.1 INTRODUCTION TO GEOGRAPHICAL INFORMATION SYSTEM

A geographical information system (GIS) is a computer system used to collect, store, analyze, and display geospatial data. Geospatial data encompasses information related to Earth's surface, atmosphere, and oceans, all with location references. It both describes the geometric characteristics of spatial objects (attribute data) and provides complete information about them. What distinguishes GIS from other information systems is its ability to process and analyze georeferenced data. Geospatial data are typically stored using specific file formats, often created with GIS software like ArcGIS or by government agencies such as the United States Geological Survey (USGS). Common GIS file formats include vector data (e.g., KML and KMZ for Google Earth), shapefile data (e.g., SHP developed by ESRI – Environmental Systems Research Institute), raster data (e.g., TIF), and tabular data (e.g., DBF). For more details on GIS data formats, please visit https://sites.udel.edu/gis/file-formats-for-gis/.

Similar to remote sensing technology, GIS plays a crucial role in promoting the sustainable operation of the palm oil industry (Baharim et al., 2022). GIS technology allows for the analysis and management of oil palm spatial information, supporting activities such as oil palm mapping, land suitability assessment, plantation management, disease detection, and pest control. For instance, Rakib et al. (2017) utilized GIS to analyze the spatial correlation of copper, zinc, and stem rot in peat plantations, while Phua et al. (2018) combined GIS with remote sensing images to investigate the occurrence of rats in oil palm plantations. One notable application of GIS is the exploration of suitable land for potential oil palm expansion while ensuring sustainability. Pirker et al. (2016) conducted an assessment of globally suitable land for oil palm development and identified 243 million hectares of land meeting eight criteria for sustainable oil palm expansion.

3.2 SPATIAL ANALYSIS

Spatial analysis refers to the examination of topological, geometric, or geographic characteristics of geospatial entities or phenomena. The ability to perform spatial analysis is a key differentiator between GIS systems and traditional information systems, serving as a critical indicator of a GIS system's success. Spatial analysis stands out as the most significant capability of GIS and serves as a powerful tool for extracting profound insights from geospatial data. Its primary objective is to reveal the spatial attributes such as location, distribution, morphology, distance, and relationships among geospatial entities. Commonly used spatial analysis techniques include spatial queries and

DOI: 10.1201/9780429199813-3

measurements, spatial transformations, reclassification, buffer analysis, overlay analysis, network analysis, spatial interpolation, and spatial statistical classification.

Spatial queries and measurements provide the quantitative foundation for conducting advanced spatial analysis. Their primary tasks involve querying the necessary spatial data (based on location or attribute information of the spatial entity) and performing calculations related to centroids, geometry, and shape. Consequently, spatial analysis typically commences with spatial queries and measurements.

In general, different layers in the same area are often related in spatial locations or attributes. They can be transformed through a series of logical or algebraic operations to meet the needs of different spatial analysis (Figure 3.1). Such a process is called spatial transformation.

Reclassification involves the reorganization and transformation of original data, serving as a vital tool for extracting in-depth information from geospatial data.

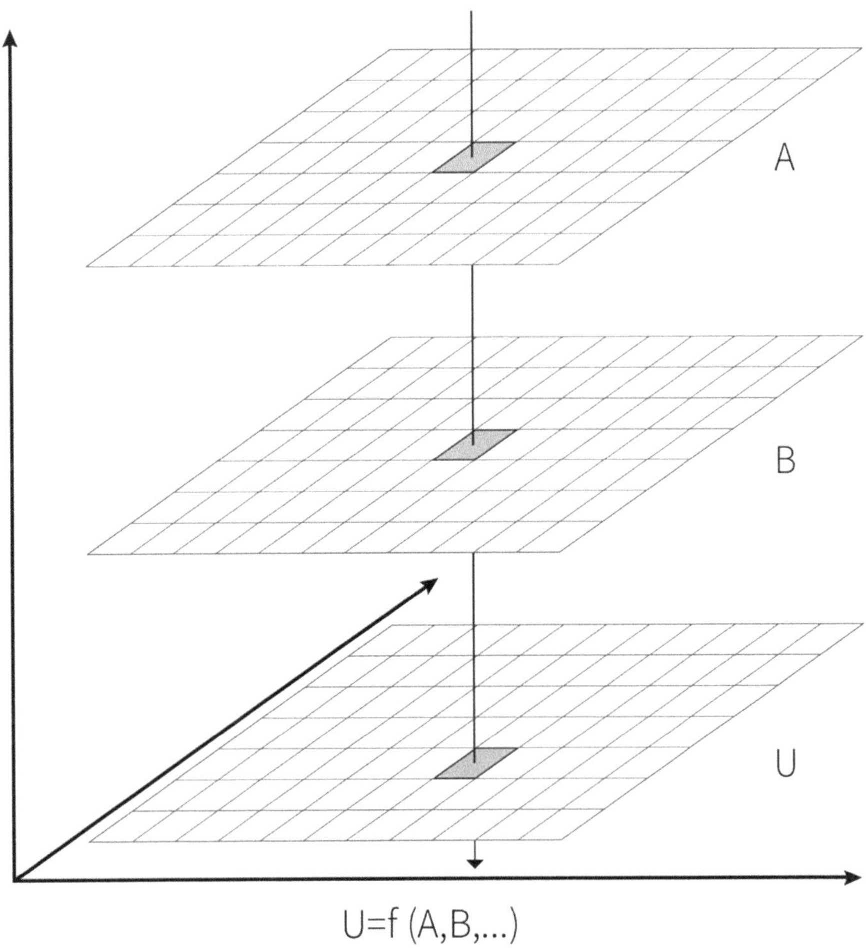

FIGURE 3.1 The spatial transformation of different layers.

It allows for the reassignment of individual values or groups of values based on specific properties of geospatial entities, such as area and slope. Reclassification enables the replacement of values with new information, grouping of values, or the adjustment of values within a set of rasters to bring them to a common scale (Figure 3.2).

In geography, numerous geospatial entities exhibit spatial relationships, indicating geographic similarities or differences based on spatial proximity. Proximity serves as a critical measure in spatial analysis, addressing various issues such as

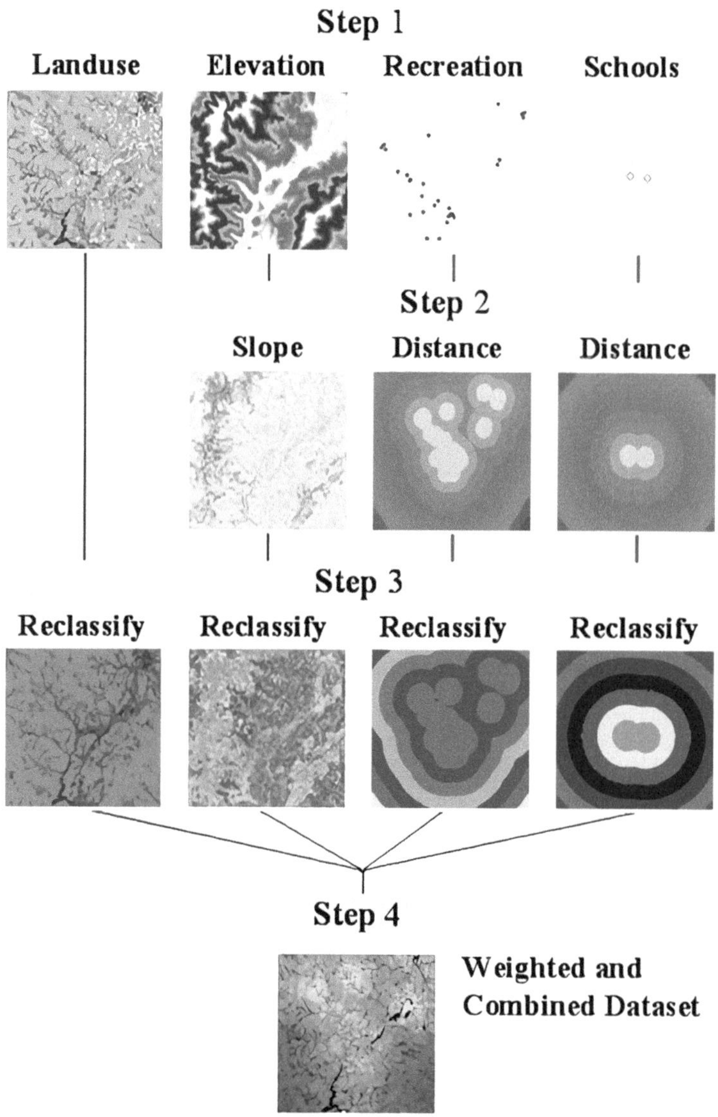

FIGURE 3.2 Example of using reclassification in a Weighted Overlay workflow (ESRI, 2023).

those related to the degree of transportation convenience offered by traffic routes, the service radius of amenities such as supermarkets, hospitals, and banks, and more.

Buffer analysis is an essential technique employed to tackle these proximity-related problems (Figure 3.3). It involves the creation of a buffer zone around existing geospatial entities, which can be point, line, or surface entities. This buffer zone serves the purpose of identifying geographical features and characteristics of the surrounding spatial entities. Additionally, buffer analysis is valuable in exploring spatial similarities and disparities.

Overlay analysis is a technique used to superimpose multiple layers of datasets in order to analyze or identify relationships between different geospatial data. The primary objective of overlay analysis is to generate new geometries or geospatial entities by combining data from two or more input datasets. It finds applications in various types of analysis, such as determining suitable land for expanding oil palm plantations by overlaying climate, soil and palm plantation data.

There are five common types of overlay analysis:

Network analysis is a crucial component of scientific geography and has been widely integrated into GIS platforms as analytical tools (Huang et al., 2017). Its primary objectives involve geographical analysis and modelling of various types of networks, including transportation networks and urban infrastructure networks like network cables. Network analysis aims to optimize performance or minimize costs in network projects. In essence, it focuses on determining the optimal spatial locations to achieve the best outcomes, which holds significant implications for geographic research. Spatial interpolation is the process of estimating values at unknown points based on known values at other points. There are two primary types of spatial interpolation: interpolation within an existing area and extrapolation to areas beyond. In interpolation, known values at certain points are used to estimate values at other

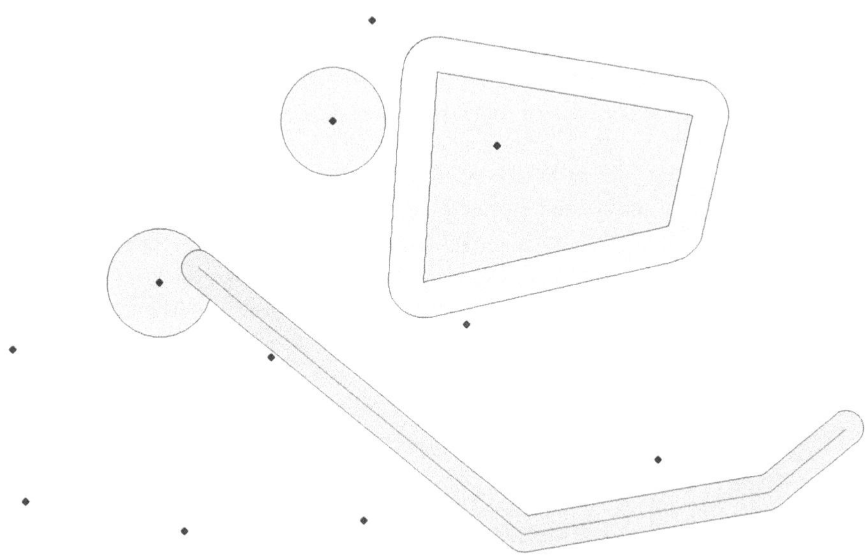

FIGURE 3.3 Creation of buffers around points, lines and polygons.

points within the same area, while extrapolation involves estimating values for points in different areas. In GIS applications, in essence, spatial interpolation entails generating polygons or raster layers from points with known values or converting raster layers into new raster layers.

3.3 GIS FOR OIL PALM STUDY

At present, GIS technology is mainly used for oil palm change mapping, suitability analysis, plantation management, and monitoring of pests and diseases. Remote sensing technology was widely used to detect the oil palm land changes, but it is difficult to predict the future expansion trend of oil palm plantation. GIS can simulate future land cover changes using spatiotemporal simulation models, supporting for optimizing the spatial pattern of oil palm and other land-cover types (Nourqolipour et al., 2015).

Nourqolipour et al. (2015) applied overlay analysis in GIS to assess the suitability of land for oil palm, and the results are an important basis for future oil palm expansion and sustainable development. Research has showed that climate is the main factor limiting the growth of oil palm, and topography, soil and socio-economic all limiting the expansion of oil palm (Ogunkunle, 1993; Olaniyi et al., 2015; Pirker et al., 2016; Xin et al., 2021). The oil palm area in Malaysia and Indonesia is about to exceed or has exceeded its sustainable oil palm plantation area (Pirker et al., 2016), and it is urgent to explore suitable land for oil palm cultivation in other regions.

Based on GIS techniques, such as overlay analysis and network analysis, the oil palm suitability can be assessed. Pirker et al. (2016) assessed the global suitable land for cultivation of oil palm based on overlay analysis and found that climatic conditions are the main limiting factor, and the suitable land is highly overlapped with high carbon storage forests and protected areas, resulting in very little remaining land for oil palm cultivation. Soil is the basis of all plant growth. Ogunkunle (1993) evaluated the four suitable soils for cultivation of oil palm in Nigeria and found that soil fertility and particle size were the main constraints. Socio-economic condition is also an important factor for the assessment of suitable land for oil palm cultivation, and it will be more reasonable of assessment result taking into account socio-economic conditions (Xin et al., 2021). Olaniyi et al. (2015) assessed the suitable land for oil palm planting in Malaysia based on overlay analysis and network analysis under the consideration of socio-economic, climatic, and environmental conditions, and found that the most suitable areas for oil palm planting are Sabak Bernam district and Kuala Langat in the state of Selangor. In the following section, we present a case study related to oil palm that utilized remote sensing and GIS (spatial analysis) to identify areas of high conservation value that had been encroached upon by oil palm cultivation (Yu et al., 2022).

3.4 OIL PALM EXPANSION IN CONSERVATION PRIORITY ZONES (CPZs)

We used the data on global conservation priority zones (CPZs) which is derived from a spatial meta-analysis of seven global biodiversity prioritization templates including Crisis Ecoregions, Biodiversity Hotspots, Endemic Bird Areas, Key

Biodiversity Areas, Centers of Plant Diversity, Global 200 Ecoregions, and Intact Forest Landscapes. According to the classification of CPZs, areas covered by three or more templates are defined as Level 1 CPZs, those covered by two templates are defined as Level 2 CPZs, and areas covered only by one template are defined as Level 3 CPZs (Yang et al., 2020). We then employed GIS to overlay the data on oil palm plantation in 2018 with global CPZs to assess to what extent oil palm plantation may conflict with biodiversity conservation. We also overlaid oil palm plantation with different levels of CPZs. In addition, we overlay the oil palm planation with each of the seven global biodiversity templates to understand the specific type of potential ecological impacts of oil palm plantation.

We overlay the oil palm planation (in year 2018) with CPZs. There is as high as 99% of oil palm plantation located in CPZs, which indicates that almost all oil palm plantation is located in regions important for biodiversity conservation, and has great possibilities to cause conflict with nature conservation. Of all oil palm plantations, there are 47.6%, 36.9%, 14.4% located in the Level 1, Level 2, Level 3 CPZs, respectively, indicating that nearly half of the oil palm plantation is located in regions that are covered by at least three global biodiversity templates.

We overlay the oil palm plantation with seven global biodiversity templates. We find that there are 91.7%, 66.4%, 65.2%, 13.1%, 3.4% of oil palm plantation located in Biodiversity Hotspots, Global 200 Ecoregions, Crisis Ecoregions, Endemic Bird Areas, and Key Biodiversity Areas, respectively, while there are very few (<1%) oil palm plantation located in Centers of Plant Diversity and Intact Forest Landscapes. This shows that oil palm plantation and expansion may have very high possibilities to cause conflicts in Biodiversity Hotspots. According to the results, the specific type of potential ecological impacts of each oil palm planation region could be identified, which has important policy implications for biodiversity conservation and sustainable development.

3.4.1 Protected Areas Invaded by Oil Palm Plantation

We chose three types of protected areas from the World Database on Protected Areas (WDPA), including Strict Nature Reserve (category Ia), Wilderness Area (category Ib), and National Park (category II). We combined Ia and Ib and collectively referred to as a category-I protected area. The reason for choosing these three types of protected areas is that they have the strictest protection, allowing only very limited human activities and usually prohibiting agricultural production (Jiang et al., 2019). We overlaid the global oil palm plantation map for year 2018 with protected areas data to identify protected areas which have oil palm plantations within its boundaries. Then, the high-resolution images of Google Earth were used to check the protected areas one by one, and those misjudged due to mapping errors were excluded. Finally, we calculated the oil palm plantation's occupation in protected areas.

In the 15 countries we analysed, there are 164 protected areas which have been invaded by oil palm plantations to varying degrees. Malaysia has a maximum of 64 protected areas that are invaded, followed by Thailand with 49 and Indonesia with 29. Papua New Guinea, Guinea, the Democratic Republic of the Congo, and Brazil's protected areas have not been invaded by oil palm plantations. In terms of area, a total of 165,123 ha of oil palm plantations are distributed in protected areas, accounting for 0.59% of the total area of global oil palm plantations. Among them, Indonesia's protected areas have been invaded the most, with 93,064 ha of protected land being occupied, accounting for 56.36% of the global total, followed by Malaysia with 19,619 ha (11.88%).

From the perspective of protected area categories, 84 category-I protected areas are invaded by oil palm plantations, involving a total area of 54,365 ha of protected lands, of which Malaysia has the largest number which is 52 (10,720 ha), and Indonesia has the largest invaded area of 36,525 ha (18 in protected areas). Eighty category-II protected areas are invaded by oil palm plantations, involving a total area of 110,758 ha of protected lands, of which Thailand has the largest number which is 36 (5,265 ha), and Indonesia has the largest invaded area of 56,539 ha. Although Honduras is not the largest country in terms of the number and area of invaded category-II protected areas, the area of oil palm plantations in its protected areas accounts for 7.96% of the total area of oil palm plantations in the country. As shown in Figure 3.4, Kalimantan Island is currently the most severe region where protected areas are invaded by oil palm plantations.

Figure 3.5 shows the situation of oil palm plantations expansion in six protected areas around the world. These six protected areas were designated between 1984(b) and 1994(e). According to the year of expansion of the oil palm plantations inside the

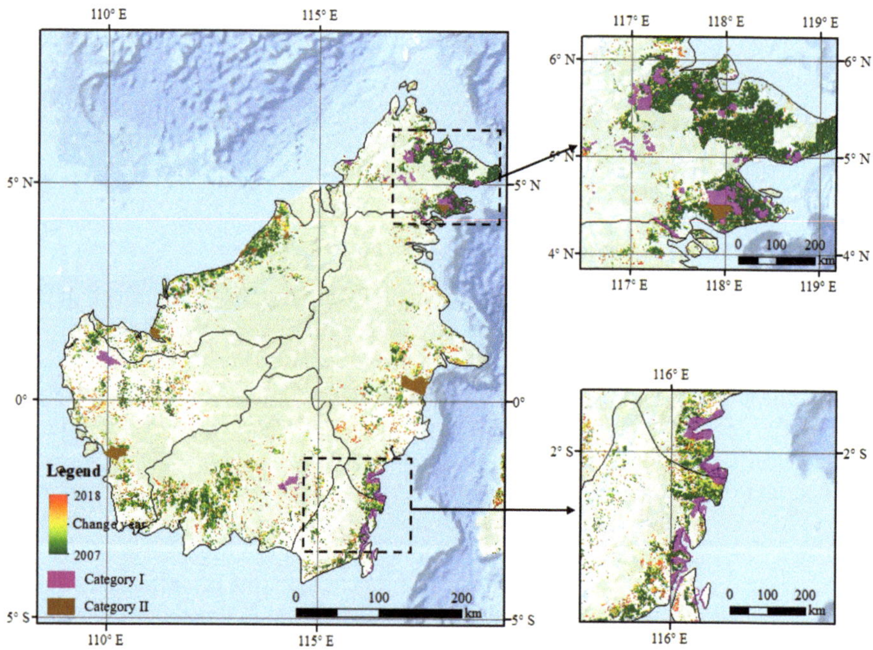

FIGURE 3.4 Protected areas invaded by oil palm plantation in Kalimantan Island. In Kalimantan, there are 59 of the category-I protected areas invaded by oil palm plantations, accounting for 70.23% of the global total. Of the 23 category-II protected areas invaded in Malaysia and Indonesia, 11 are located on Kalimantan. These invaded protected areas are mainly distributed in the northeast (a) and southeast (b) of the island.

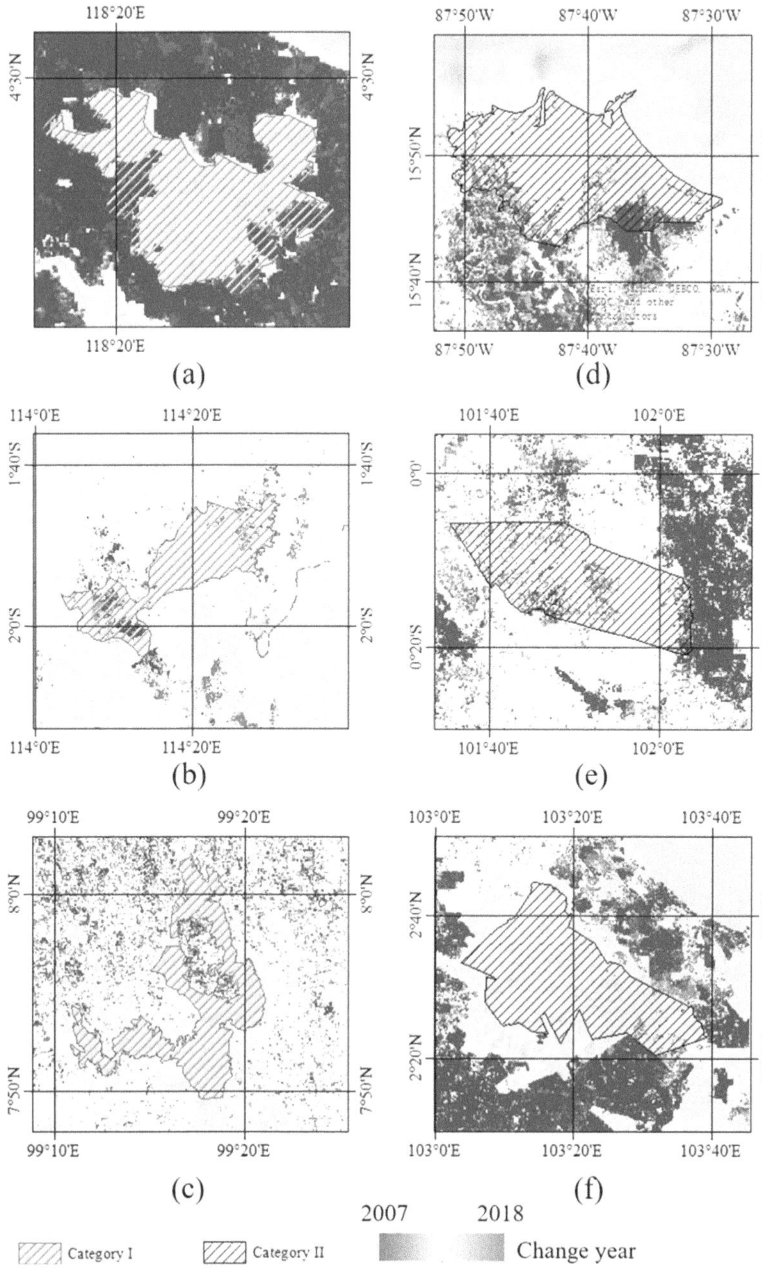

FIGURE 3.5 Invasion of oil palm plantations into category-I and category-II protected areas in different regions. (a) Mount Pock Protection Forest Reserve (designated in 1984); (b) protected area Sungai Barito (designated time unknown); (c) Khao Pra Bangkhram Wildlife Sanctuary (designated in 1993); (d) Blanca Jeannette Kawas (Punta Sal) National Park (designated in 1994); (e) Tesso Nilo National Park (designated in 2004); (f) Endau Rompin (Johor) National Park (designated in 1993).

protected area, situation can be divided into two types. The first type, the management of the protected area, is relatively effective, such as (a) most oil palm in the protected area existed before 2007, and there has been almost no new oil palm plantation expansion since then. The second type is that the management effectiveness of the protected area is low. There is obvious new expansion of oil palm plantations in the protected area since 2007 (such as (b)–(f)), especially (e) Tesso Nilo National Park Reserve in Honduras, where oil palm plantations gradually expanded and occupied from the edge of to the interior part of the protected area. The next chapter introduces Geomatics and its application in the palm oil sector.

4 Geomatics

4.1 INTRODUCTION TO GEOMATICS

Geomatics is a field that focuses on gathering, storing, analysing, processing, and presenting geographic data, with a primary emphasis on understanding the Earth and its surroundings. It encompasses various subdisciplines such as geodesy, photogrammetry, engineering surveying, marine surveying, and cartography, evolving over time with the use of instruments like total stations and image analysers.

In recent years, new technologies like Global Navigation Satellite Systems (GNSS), remote sensing (RS), Geographic Information Systems (GIS), digital mapping, and virtual reality have transformed geomatics. GNSS, in particular, has revolutionized geodesy, navigation, and positioning by providing precise and convenient location information through artificial satellites worldwide. GNSS has become an essential tool in agricultural sector where it aids in automatic oil palm plantation management, disease detection, pest control, and spatial data collection. While it enhances oil palm monitoring accuracy, serves as a valuable tool for precision farming. GNSS supports tasks such as oil palm tree counting, vacancy detection, and the creation of detailed digital elevation models, enabling more efficient oil palm plantation management.

In this chapter, we provide an introduction to several noteworthy GNSS and their associated augmentation systems. Building upon the insights from the preceding chapters (Chapters 2 and 3), we explain the concept of 3S-based precision farming, which leverages GNSS technology.

4.2 GLOBAL NAVIGATION SATELLITE SYSTEMS (GNSS)

GNSS represents an advanced technology for precise geospatial positioning. It enables the determination of three-dimensional location (longitude, latitude, and altitude/elevation) with high precision, utilizing small receivers or even smartphones. As of 2023, several fully operational GNSS systems exist (Figure 4.1), including the United States' Global Positioning System (GPS), Russia's Global Navigation Satellite System (GLONASS), China's BeiDou Navigation Satellite System (BDS), and the European Union's Galileo. Each of these systems comprises a constellation of 24–35 Medium Earth Orbit (MEO) satellites distributed across various orbits.

GPS stands as the pioneering GNSS, having been initially launched in 1978. It continues to provide stable and extensive services to countries worldwide. BeiDou boasts the most extensive satellite constellation, with a fully operational system comprising 35 satellite constellations in orbit. Consequently, BeiDou offers the most precise location information and the broadest coverage.

Furthermore, other nations are actively developing their GNSS systems, such as France's Doppler Orbitography and Radio-positioning Integrated by Satellite (DORIS), India's Indian Regional Navigation Satellite System (IRNSS), and Japan's Quasi-Zenith Satellite System (QZSS). Nevertheless, these GNSS systems,

DOI: 10.1201/9780429199813-4

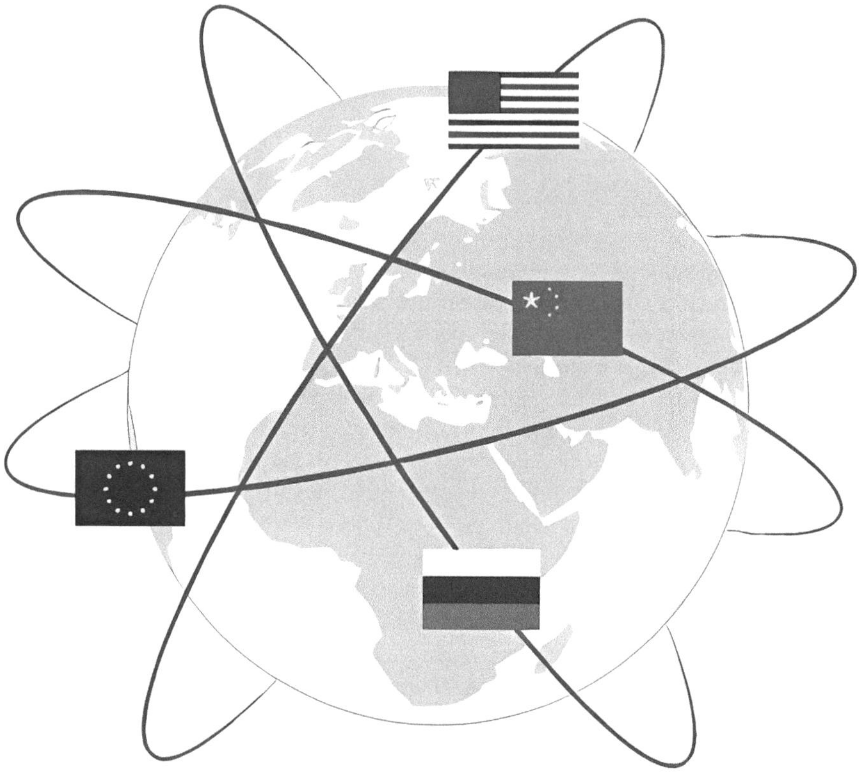

FIGURE 4.1 Galileo, GPS, GLONASS, and Beidou GNSS Systems (Galileo GNSS, 2014)

while valuable, do not currently offer global positioning and navigation coverage (Johnston et al., 2017). Achieving global coverage and accuracy will necessitate substantial advancements and additional satellite constellations.

The GNSS is an inclusive term for various satellite navigation systems and their augmentation systems, which provide precision positioning, navigation and timing services for receivers at any time and site. GNSS can cooperate when more than one constellation is used at the same time, providing unique services of each GNSS and more precise location information. Therefore, a brief description of four GNSS constellations, that is, GPS, GLONASS, BeiDou, and Galileo, is provided below.

4.2.1 GPS

GPS was first launched in 1978 by the United States and was the pioneer of GNSS. It was made available for global use in 1994. GPS was originally invented for an independent military navigation system. It consists of satellite constellation, ground segment and user receiver. At present, GPS is the most utilized satellite navigation system and currently consists of 32 satellites across 6 orbital planes, these satellites are constantly retired or replaced (Johnston et al., 2017). The position accuracy of GPS is 0.3–5 m, and in some case can even be within 2 cm (GPS, 2022). It provides significant support for car navigation, logistics industry, earthquake prediction, as well as farming and mapping (Wang et al., 2022).

4.2.2 GLONASS

GLONASS, Russia's counterpart to GPS, received support from Russia and had its inaugural launch in 1982. Its development began in 1976 under the Soviet Union's auspices. GLONASS achieved full global coverage in 1995 with its constellation of 24 satellites. Nevertheless, when compared to GPS, GLONASS lacks significant advantages (Kiliszek and Kroszczyński, 2020). Its coverage is not as extensive as GPS, and due to its smaller operational satellite count of 24, it offers positioning accuracy only within the range of 2–4 m. Certainly, GLONASS serves as a valuable backup to GPS.

4.2.3 BeiDou

BeiDou, the Chinese satellite navigation system, was initially launched in 2000. It comprises three distinct satellite constellations: BeiDou-1, which includes 4 BeiDou navigation experiment satellites; BeiDou-2, which consists of 20 BeiDou navigation satellites; and BeiDou-3, encompassing 35 BeiDou navigation satellites. BeiDou has launched a total of 55 navigation satellites, with the fully operational system now featuring 35 satellites. Global positioning and navigation services by BeiDou became available in December 2018. The expanded constellation of fully operational satellites, BeiDou, offers a positioning accuracy of 0.474 m (Yang et al., 2020).

4.2.4 Galileo

The Galileo positioning system, developed by the European Union in collaboration with the European Space Agency (ESA), had its initial launch in 2011. It commenced global positioning and navigation services in December 2016 and reached full operational capability in 2020, boasting a constellation of 30 satellites distributed across six orbital planes.

Galileo comprises three key segments: the space segment (comprising satellite constellations), the ground segment (responsible for controlling the entire constellation), and the user segment (consisting of receivers). Galileo is fully compatible with GPS and GLONASS and can provide even more precise and stable positioning information when used in conjunction with these systems (Kiliszek and Kroszczyński, 2020). Presently, Galileo serves a crucial role in offering traffic safety and emergency response services within Europe, enhancing overall navigational capabilities and safety within the region (Fernandez-Hernandez et al., 2022).

4.3 GNSS AUGMENTATION SYSTEMS

Positioning accuracy often falls short of meeting the requirements for various industrial applications and military navigation needs when relying solely on navigation satellite signals (Meng et al., 2018). To address this limitation, GNSS augmentation systems play a vital role in enhancing the accuracy, reliability, and availability of navigation systems (Langley et al., 2017). These systems optimize calculations by incorporating external information about potential error sources, including orbit errors (±2.5 m), tropospheric delay (±0.5 m), receiver noise (±0.3 m), satellite clocks (±2 m), and ionospheric delay (±5 m) (HEXNGON, 2023).

Augmentation systems are classified into two main categories: satellite-based augmentation systems and ground-based augmentation systems, depending on how they broadcast error information (Li et al., 2020). Satellite-based systems transmit correction information via geostationary orbit (GEO) satellites, while ground-based systems disseminate corrections through fixed stations with known coordinates.

4.3.1 Ground-Based Augmentation System

Ground-based augmentation systems utilize monitoring stations with known coordinates to enhance GNSS positioning accuracy. These stations receive signals from the core satellite constellation and calculate corrections to provide more precise location data to receivers. Theoretically, a higher number of reference stations on the ground results in more accurate positioning (Pervan, 2021). Ground-based systems offer local coverage and can achieve high accuracy, with position errors typically below 1 m. However, their coverage is limited, and receivers must be within the station's coverage area. Moreover, certain areas such as seas, deserts, mountains, and high-altitude locations pose challenges for establishing reference stations (Marini-Pereira et al., 2021). As a result, ground-based augmentation systems are not well-suited for large-scale utilization. They find applications in mapping, monitoring, driver training and testing, precision farming, and are notably successful at airports.

4.3.2 Satellite-Based Augmentation System

Satellite-based augmentation systems are designed to provide high-precision positioning across larger geographic regions. These systems calculate corrections at reference stations and transmit them to one or more GEO satellites. The augmentation information, including corrections and integrity data for orbit errors, satellite clock errors, ionospheric delays, and tropospheric delays, is then broadcast globally to end-users (Cheng et al., 2021). Satellite-based augmentation systems offer solutions for a broader range of precise positioning and navigation needs, making them suitable for various applications.

4.4 GEOMATICS FOR OIL PALM

Precision farming represents the forefront of innovation within the agricultural realm and has materialized due to substantial technological progress (Shaheen et al., 2020). These advancements encompass GNSS, GIS, RS, management information systems, and computer technology. Geospatial technology, encompassing GNSS, GIS, and RS, plays a pivotal role in precision farming by enabling the creation of spatial variability maps (McBartney and Pringle, 1999) leading to intelligent oil palm cultivation.

Geomatics plays a crucial role in oil palm–related research, furnishing precise spatial information that assists farmers in enhancing oil palm cultivation, increasing production efficiency, minimizing environmental impact, and fostering sustainable development. For instance, it aids in delineating the boundaries of oil palm planting plots to facilitate plot management and monitoring. It also enables accurate soil testing and sample collection and is employed in tandem with drone technology to monitor pests and diseases affecting oil palm plants, among other applications (refer Chapter 9). Furthermore, when combined with other RS and GIS technologies, geomatics enables more comprehensive oil palm production management.

5 Mapping Oil Palm Land Use

5.1 INTRODUCTION

Mapping the spatial distribution of oil palm plantations is crucial for various purposes. Categorizing items based on their land-cover characteristics aids in distinguishing oil palm plantations from the surrounding landscape elements, including but not limited to forests, infrastructures, open terrain, water bodies, and other agricultural fields. This process facilitates a clear delineation of boundaries and the precise calculation of the extent of oil palm cultivation. When employed in a temporal analysis, this approach becomes a valuable tool for identifying the growth of oil palm cultivation and associated land-related activities.

Existing oil palm plantation datasets covering a large area, such as that produced by Cheng et al. (2016), Mohd Najib et al. (2020), and Dong et al. (2020), developed oil palm plantations maps for Malaysia. Commonly used Earth observation data for mapping oil palm plantations are from Landsat, SPOT (Systeme Probatoire d'Observation de la Terre), QuickBird, Sentinel, ALOS-PALSAR (Advanced Land Observing Satellite Phased Array L-band Synthetic Aperture Radar), and Terra/Aqua. In terms of classification methods, the traditional classification method such as Maximum Likelihood (ML) is widely used, but the accuracy is insufficient to support further research (Dong et al., 2015; Santos and Messina, 2008). Machine learning techniques such as Random Forest (RF), Support Vector Machine (SVM), and Artificial Neural Network (ANN) emerged as classification methods with convincing accuracy (Cheng et al., 2019; Shaharum et al., 2019). Cheng et al. (2019) used ML, RF, and SVM to map oil palm in Malaysia, and the results showed that RF (91.07%) and SVM (90.41%) had better mapping accuracy than ML (89.72%).

Monitoring the dynamics of oil palm plantations is crucial for assessing the impact of their expansion on changes in land use and ecological environments. However, there have been limited studies focused on continuous detection of oil palm changes over large areas. Many of these studies have relied on visual interpretation techniques, which are labor-intensive and not suitable for long-term annual mapping of oil palm plantations (Carlson et al., 2013; Gaveau et al., 2016; Vijay et al., 2018). Moreover, annual mapping of oil palm plantations in tropical regions requires access to a larger volume of Earth observation data and more advanced data processing methods to address issues related to cloud cover. De Petris et al. (2019) estimated the oil palm plantation expansion in Central Kalimantan from 2000 to 2018 based on MODIS enhanced vegetation index datasets; the result showed an overall accuracy of 94%. The automatic identification using classification algorithms based on Landsat and PALSAR data has been used in producing

DOI: 10.1201/9780429199813-5

the 2015 land-cover map of insular Southeast Asia with classification of different oil palm types (Miettinen et al., 2019).

In this chapter, we present the results of the Annual Oil Palm Datasets (AOPDs), which have a spatial resolution of 100 m and cover Indonesia and Malaysia from 2001 to 2016. The study encompasses both mature and young oil palm trees and includes both industrial plantations and smallholders with land holdings larger than 1 ha.

5.2 METHODOLOGY TO PRODUCE ANNUAL OIL PALM DATASETS

The AOPDs were developed using a two-stage method. First, we mapped oil palm plantations in Indonesia (Kalimantan and Sumatera) and Malaysia for the years 2007, 2008, and 2009 using ALOS PALSAR/PALSAR-2 data and the RF classifier. Second, building upon the oil palm plantation map developed in the initial stage, we extended the mapping to the missing years (2001–2006 and 2011–2014) using MODIS NDVI (Normalized Difference Vegetation Index) data, employing the Breaks for Additive Season and Trend (BFAST) breakpoint detection method (Verbesselt et al., 2010). After thorough post-processing, we obtained the AOPDs for the target years at 100 m (Figure 5.1). While Landsat images offer a superior spatial resolution of 30 m, they were not utilised in this study due to insufficient coverage of available cloud-free images (as shown in Figure 5.2). Only 47.12% (49) of the total Landsat footprints had at least one scene available per year with a cloud cover below 30%.

The global 25-m resolution PALSAR/PALSAR-2 images utilized in this study were graciously provided by JAXA, the Japan Aerospace Exploration Agency (http://www.eorc.jaxa.jp/ALOS/en/palsar_fnf/data/index.htm). For the years (2007, 2008, 2009, 2010, 2015, 2016) in which PALSAR data were available, we used a total of 15 patches of $1° \times 1°$ PALSAR grids to cover the entire study area. We calculated normalized backscattering coefficients, expressed in decibels (dB), using the following formula (Rosenqvist et al., 2007):

$$\sigma^0(\text{dB}) = 10 \times \log_{10} \text{DN}^2 + \text{CF} \tag{5.1}$$

where DN is the digital number of PALSAR mosaic, and CF is the calibration factor with a default value of −83.0 dB (Shimada et al., 2009).

The HH and HV polarization images of the ALOS PALSAR data were processed to map oil palm by calculating the Difference and Ratio layers as follows:

$$\text{Difference} = \text{HH} - \text{HV} \tag{5.2}$$

$$\text{Ratio} = \text{HH/HV} \tag{5.3}$$

While PALSAR and PALSAR-2 are distinct sensors, the backscatter signals they record exhibit a degree of stability, as indicated in studies by Cheng et al. (2019) and Qin et al. (2017). Cheng et al. (2018) recommended a 100-m resolution for optimal identification of oil palm plantations, primarily due to the reduction in noise, known as the "salt and pepper" effect. Consequently, we resampled the PALSAR images to a spatial resolution of 100 m to enhance accuracy.

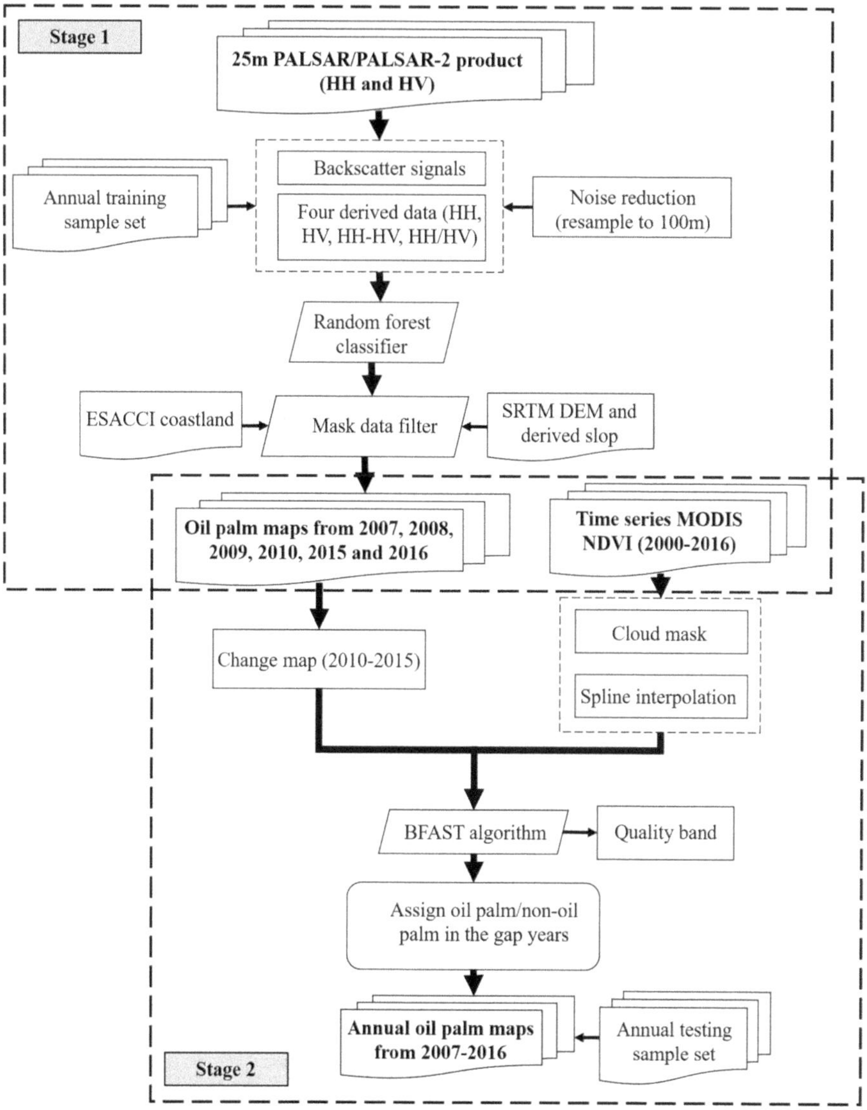

FIGURE 5.1 Workflow of the annual oil palm mapping procedure. Stage 1 (2007–2010 and 2015–2016) and Stage 2 (2011–2014).

MODIS features seven bands specifically designed for terrestrial applications, offering substantial support for global LULC mapping. The MODIS NDVI dataset (available for download at https://lpdaac.usgs.gov/) is part of the MOD13Q1 collection 6, spanning from 2000 to 2015. It provides a spatial resolution of 250 m and a temporal resolution of 16 days. We utilized this dataset to create the AOPDs during the data gaps in the years 2000–2006 and 2011–2014, employing change detection algorithms.

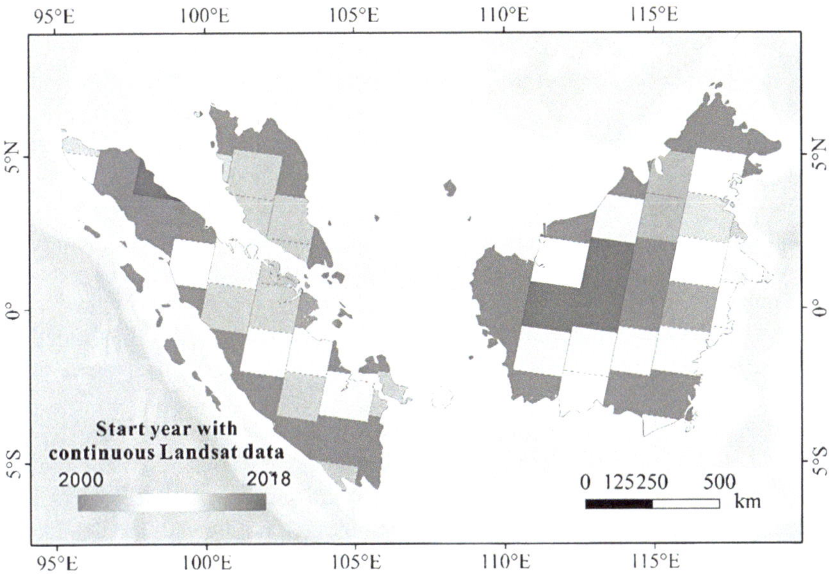

FIGURE 5.2 The Landsat data availability from 2000 to 2018.

After unifying the projection and resampling, we utilized the pixel quality and reliability layers from MOD13Q1 to exclude poor-quality pixels. Figure 5.3 illustrates the availability of the MODIS NDVI dataset from 2010 to 2015. Lastly, we completed the 16-day NDVI profiles for MODIS through the spline interpolation method.

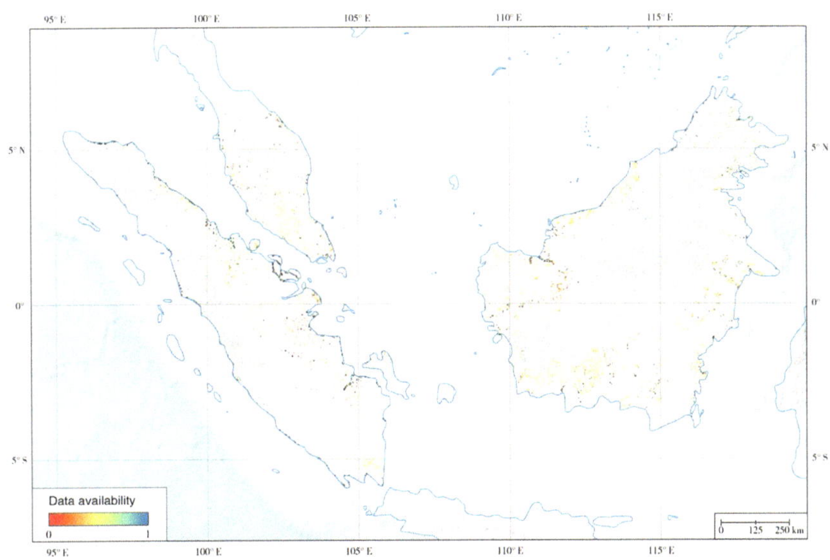

FIGURE 5.3 The availability of MODIS NDVI data from 2010 to 2015 is represented in the data availability map, which considers factors such as cloud cover and data quality.

The oil palm maps derived from PALSAR images for the years 2010 and 2015, were used to identify change pixels of the 'from-to' type, and subsequently, we extracted change maps for the gap years (2011–2014). For the oil palm maps from 2001 to 2006, we used the extent of oil palm in 2007 as the potential expansion area. These maps were generated under the assumption of either unidirectional or bi-directional expansion of oil palm. With this, we had datasets ready for mapping oil palm plantations from 2001 to 2016.

The training datasets for Indonesia was interpreted using a training sample set from Malaysia (Cheng et al., 2018) and applied the same interpretation method, which involves an equal-area hexagonal sampling approach implemented in the Discrete Global Grid Generation (DGGRID) software (Sahr et al., 2003) (as depicted in Figure 5.4). We cross-verified the training sample datasets by comparing them with Google Earth and PALSAR images. These datasets encompassed four LULC types: oil palm (both mature and young), water, other vegetation (including forests, shrubland, and other plantations such as rubber), and other categories (including impervious surfaces, cropland, and bare land). Further details are provided in Table 5.1.

The oil palm plantations were first mapped by creating the original AOPDs for the PALSAR years (2007–2010 and 2015–2016) using pre-processed PALSAR data. Then, we employed the RF classifier, a robust machine learning algorithm, for mapping oil palm plantations in Indonesia and Malaysia. The input data for the RF classifier included HH, HV, HH-HV, and HH/HV images. Due to the data gap in the PALSAR records, we were unable to produce AOPDs for the missing years (2001–2006 and 2011–2014). To address this gap, and with the aid of change maps

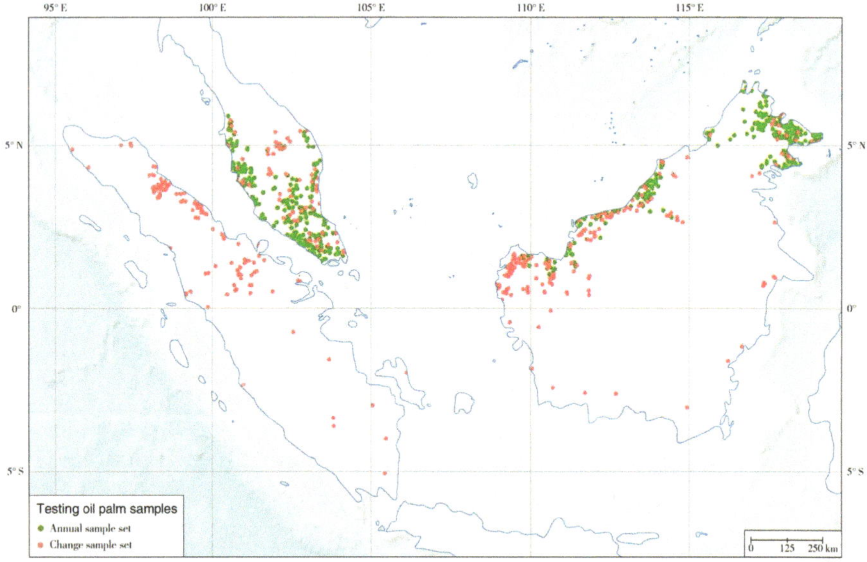

FIGURE 5.4 Map illustrating oil palm samples interpreted for the years 2007, 2008, 2009, 2010, 2015, and 2016. This includes 2,986 samples in the annual sample set and 370 samples in the change sample set.

TABLE 5.1
The Distribution of Training and Testing Data (Unit: pixel) in Malaysia

	Oil Palm		Other Vegetation		Water		Other		Total	
	Train	Test	Train	Test	Train	Test	Train	Test	Train	Test
2007	1,228	436	2,970	2,338	570	68	185	76	4,953	2,918
2008	1,279	463	2,994	2,337	570	70	185	77	5,028	2,947
2009	1,387	483	3,179	2,338	570	70	185	77	5,321	2,968
2010	1,405	483	3,228	2,338	570	70	185	77	5,388	2,968
2015	1,475	507	3,430	2,340	570	77	185	77	5,660	3,001
2016	1,475	507	3,430	2,340	570	77	185	77	5,660	3,001

that illustrated potential oil palm conversions during the missing period, our next objective was to pinpoint the exact year when these changes occurred. To achieve this, we leveraged MODIS NDVI data and applied the BFAST change-detection algorithm. BFAST is adept at identifying significant breakpoints in the NDVI series and recording the timing of these transitions (DeVries et al., 2015; Xu et al., 2020). Furthermore, considering the assumption of either unidirectional or bi-directional expansion of oil palm, we derived two versions of maps. Consequently, we developed oil palm plantation maps for the missing years, thus achieving a seamless and continuous AOPD dataset spanning from 2001 to 2016.

To improve the accuracy of the AOPDs derived from PALSAR images, we applied mode filtering as a post-processing step to handle very small patches, mainly single pixels. Additionally, we excluded oil palm plantations mapped in areas with elevations exceeding 1,000 m using the Shuttle Radar Topography Mission (SRTM) 30-m digital elevation model (DEM), intact forest landscapes (Potapov et al., 2008), and mangrove areas identified in the Global Mangrove Atlas (Spalding, 2010).

Furthermore, for the missing years, we identified oil palm changes based on the detected change times, enabling us to create continuous AOPDs spanning from 2001 to 2016. We employed three distinct structural change detection methods namely OLS-MOSUM, Lagrange multiplier statistics (SupLM), and Bayesian information criterion test (BIC) within the BFAST framework to ensure the accuracy of change time detection (Zeileis, 2005). The reliability of the detected change year increases with the consensus of multiple methods, as illustrated in Figure 5.5.

5.3 SPATIAL AND TEMPORAL CHARACTERISTICS OF OIL PALM EXPANSION

Throughout the study period, oil palm plantations exhibited rapid expansion within the study area. As depicted in Figure 5.6, these plantations are primarily concentrated in the coastal regions of peninsular Malaysia, Sumatra, and Borneo. There are relatively fewer oil palm plantations in the northwest of the southwestern coastal regions of Sumatra. On a regional scale, Sumatra witnessed substantial oil palm plantation expansion during the earlier years, while Kalimantan experienced even more rapid

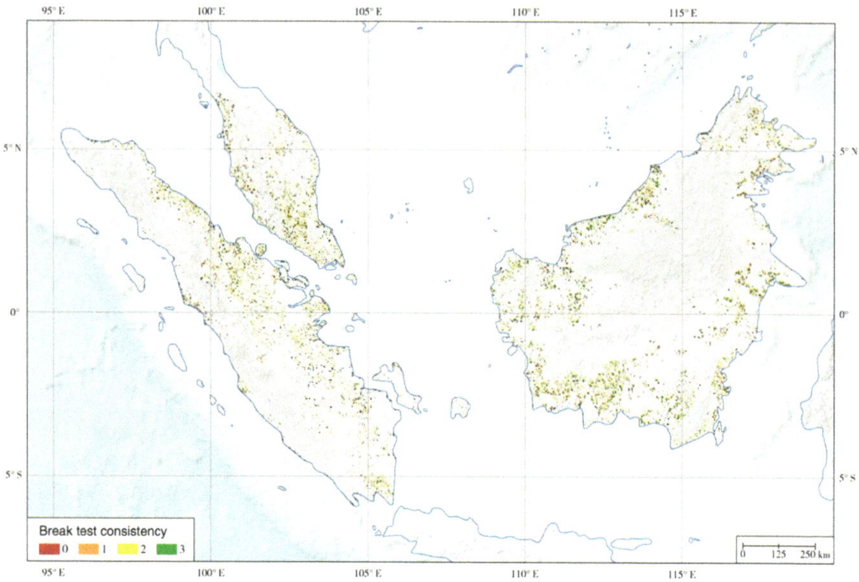

FIGURE 5.5 Quality maps showing change time detection from 2011 to 2014. Break test consistency indicates the count of methods detecting the same change year over the entire period.

expansion in recent years. Over 90% of the oil palm plantations were situated in areas with elevations greater than 250 m and slopes less than 2.5°. Conversely, only a very small number of oil palm plantations were distributed in areas with elevations exceeding 500 m and slopes greater than 5°. Additionally, the elevation and slope restrictions on land covered by oil palm plantations were slightly relaxed in 2016 compared to 2007 (elevation: 228.98 m in 2007 and 230.10 m in 2016; slope: 1.97° in 2007 and 1.99° in 2016).

While we have developed shrink maps of oil palm plantations, we cannot ensure their accuracy from 2001 to 2006. Consequently, we have excluded the shrink maps for the period 2001–2006. In contrast, the shrink map for the period 2011–2014 is reliable, as it is supported by 2 years of PALSAR mapping results (2010 and 2015) that provide constraints on its accuracy.

The area covered by oil palm plantations in Indonesia and Malaysia were more than tripled between 2002 and 2016, as shown in Figure 5.7. However, due to uncertainties in change detection, stemming from factors like the low spatial resolution of NDVI images, mapping errors from base maps, and cloud contamination, there is a 9.45% margin of uncertainty associated with the total changes between 2010 and 2014. The upper and lower boundary lines on the oil palm area curve represent when the change occurred in 2010 and 2014, respectively. The uncertainty is even greater between 2001 and 2006 due to the absence of a PALSAR-derived map of oil palm plantations in 2000.

The performance of the AOPDs was assessed using the test oil palm sample set for the years 2007–2010 and 2015–2016. As presented in Table 5.2, the *F*-score consistently exceeds 85, demonstrating the high accuracy of the AOPDs for these years.

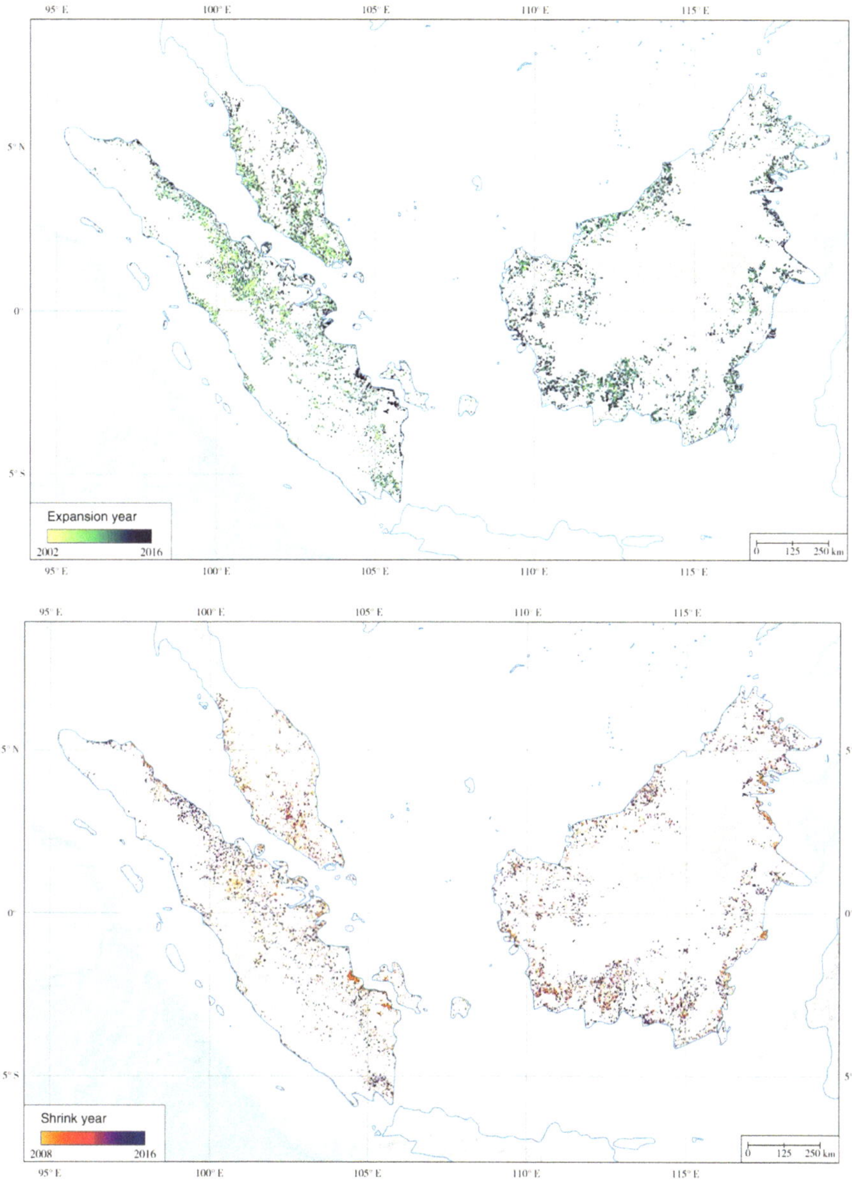

Figure 5.6 Oil palm changes at 100-m resolution from 2002 to 2016 (a) Expansion year, 2002–2016 and (b) Shrink year, 2008–2016.

Moreover, both user's accuracy (UA) and producer's accuracy (PA) of our results exhibit strong performance, with UA averaging around 90%. Our results demonstrate significantly better reliability compared to the previous maps developed by Cheng et al. (2019).

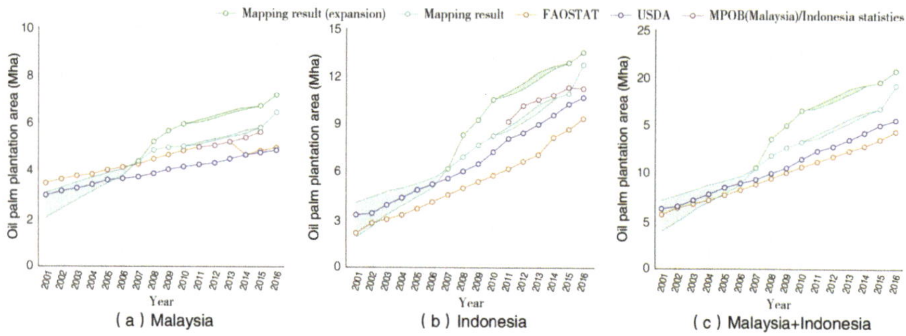

FIGURE 5.7 Comparison of annual oil palm areas between this study and estimates provided by FAOSTAT, USDA, and MPOB for (a) Malaysia, (b) Indonesia, and (c) Malaysia and Indonesia from 2001 to 2016.

TABLE 5.2
The Comparison of the Oil Palm Accuracy between Our Mapping Results and Cheng et al. (2019) for the Six Mapping Years

Year	*F*-Score	UA (%)	PA (%)
2007	86.21	93.40	80.05
2008	87.77	93.22	82.91
2009	86.26	92.12	81.10
2010	85.25	93.89	78.06
2015	85.96	92.08	80.59
2016	85.89	87.47	84.36

PA, producer's accuracy; UA, user's accuracy.

Mapping oil palm plantations is crucial for comprehending deforestation and its environmental consequences. Therefore, the AOPDs we have created for Indonesia and Malaysia, spanning from 2001 to 2016 at a 100-m resolution, hold substantial significance. This dataset attains a high level of accuracy and offers valuable potential for supporting further research on the expansion of oil palm plantations, including investigations into carbon storage changes, biodiversity impacts, and encroachments into protected areas. However, the spatial resolution of the products developed in this study could potentially be enhanced in the future as more cloud-free data become accessible from sources like ALOS, Landsat, and Sentinel images. These datasets can also be fused to generate moderate-resolution, cloudless datasets. Enhanced spatial information play a pivotal role in promoting the sustainable development of the palm oil industry.

6 Oil Palm Tree Detection

6.1 OIL PALM TREE CROWN DETECTION

Oil palm tree detection and counting not only benefits oil palm industries and smallholders in yield prediction, irrigation and fertilization planning, and smart management but also contributes to the protection of native species and ecological environment due to more reasonable oil palm plantation planning. However, only detecting the tree crown is not enough for smart management of oil palm plantation. Monitoring the growing status and disturbances (e.g., pest, disease, and mismanagement) of oil palm trees is crucial for both governments and industries. For example, farmers can replant dead palms, weed the fields to provide better growing environment for mismanaged palms, and duly observe the health conditions of oil palms, preventing pests and diseases from spreading to the whole plantation.

Existing tree crown detection algorithm can be grouped into three types, including classical image processing methods (Wulder et al., 2000; Pouliot et al., 2002; Ardila et al., 2012; Chemura et al., 2015; Gomes et al., 2018), traditional machine learning methods (Hung et al., 2012; Pu and Landry, 2012; Dalponte et al., 2014; Wang et al., 2019), and deep learning methods (Li et al., 2017; Dong et al., 2019; Freudenberg et al., 2019; Osco et al., 2020; Puttemans et al., 2018; Xia et al., 2019). Following the success of deep learning, many tree crown detection algorithms utilized convolutional neural network (CNN) to attain promising tree crown detection results in complex scenes and large study area. Generally, existing deep learning-based tree crown detection approaches can be categorized into CNN classification, semantic segmentation and object detection methods.

6.1.1 CNN-Based Tree Crown Detection

In recent times, the majority of tree crown detection methods based on deep learning have employed a CNN classification approach in combination with the sliding window technique. Li et al. (2017) employed a deep learning method to classify each image patch (17 × 17 pixels) as either background or a palm tree, subsequently predicting oil palms using the sliding window technique. This approach achieved an overall accuracy (OA) exceeding 96% in three distinct study regions. Following this, they introduced a two-stage (TS) CNN method (TSCNN) to enhance oil palm detection accuracy across larger areas and reduce confusion with buildings or other vegetation (Li et al., 2019). Furthermore, Mubin et al. (2019) utilized a geographic information system (GIS) tool to partition images into 26 × 26 and 31 × 31-pixel segments for mature and young oil palms, training a classification model with LeNet. Some researchers also employed deeper CNN architectures such as ResNet-classifiers or GoogLeNet-classifiers to achieve improved detection outcomes (Guirado et al., 2017). Notably, the images utilized in the aforementioned studies were sourced from satellites like WorldView-3, QuickBird, and Google Earth.

DOI: 10.1201/9780429199813-6

In a separate study, Csillik et al. (2018) harnessed unmanned aerial vehicle (UAV) images for the detection of citrus trees, employing a simple CNN algorithm followed by a classification refinement using the sliding window technique. This method exhibited strong performance in a relatively complex environment situated in Tulare County, California, achieving an impressive OA of 96.24%.

Although CNN classification-based tree crown method generally performs much better than classical image processing methods and traditional machine learning methods in regions with overlapping and crowded trees, they almost adopted sliding window approach to complete final detection, which is a time-consuming technique because of the quite large amount of detected candidates of various sizes. Additionally, these methods require not only the annotation of tree crown samples but also extra annotation efforts for other types such as other vegetation and impervious surfaces.

6.1.2 Semantic Segmentation–Based Tree Crown Detection

Semantic segmentation-based tree crown detection is an end-to-end method without requiring the time-consuming sliding window technique. Morales et al. (2018) utilized semantic segmentation to extract *Mauritia flexuosa* palm trees by the Deeplab v3+ architecture. They achieved an accuracy of 98.14% over the test set located in the north of Peru. However, they just detected the location but did not count the number of tree crowns, as the detected tree crowns are connected with each other. Freudenberg et al. (2019) adopted semantic segmentation for treetop detection using WorldView-2 satellite images. They used the U-Net to detect oil palms and coconut palms in Indonesia and India, respectively. Using a semantic segmentation architecture similar to fully convolutional network (FCN), Xiao et al. (2020) automatically produced pseudo labels from unsupervised tree crown detectors to train a semantic segmentation architecture similar to FCN, which saved the manual annotations and achieved a comparable performance to supervised training process using manually labelled samples. Recently, Osco et al. (2020) presented a new CNN method for counting trees using UAV images. They transferred the tree counting problem to estimating the confidence map, which indicates the probability that a tree crown occurs in each pixel. Their proposed method achieved an F1-score of 95%.

Semantic segmentation method is more efficient than the CNN method because it allows the detection of several trees at once. It is suitable for tree species mapping that do not require tree counting task. On the contrary, the performance of semantic segmentation methods gets worse for areas containing trees that appear to overlap with each other, accounting for detecting several overlapping or touching trees as only one tree. Apart from that, the output of semantic segmentation methods is a "probability map" or a "confidence map", indicating the probability that a pixel belongs to the tree crown type. These methods require a post-processing step to generate the final tree counting and detection results, like a local maximum detection (Freudenberg et al., 2019; Osco et al., 2020). Furthermore, the semantic segmentation methods focus on tree crown detection and counting, while it is difficult to measure the exact tree crown size.

6.1.3 Object Detection-Based Method for Tree Crown Identification

In the past few decades, substantial object detection methods have been explored to detect various geospatial objects, including tree crown using high-resolution remotely sensed images (Li et al., 2020). In 2018, Puttemans et al. firstly adopted object detection method for detecting trees. They selected "You Only Look Once version 2" (YOLOv2) deep learning technique for coconut tree detection in aerial images and achieved an OA of 92.95%, outperforming other two traditional object detection methods (Random Forest and Support Vector Machine). Santos et al. (2019) compared three state-of-the-art object detection algorithms including Faster R-CNN, YOLOv3 and RetinaNet for individual tree (Fabaceae) detection using UAV images. Experiments showed that RetinaNet had the highest average precision, while YOLOv3 achieved the highest detection efficiency. Similarly, Xia et al. (2019) implemented and evaluated five state-of-the-art object detection approaches (i.e. Faster R-CNN, VGG-SSD, YOLOv3, RetinaNet, and Mobilenet-SSD) for detecting oil palms using UAV images. Experimental results showed that VGG-SSD had the highest detection accuracy of 90.91% and Mobilenet-SSD had the fastest detection speed. For the tree crown detection using satellite images, our previous work applied object detection-based method for counting trees using 0.6 m-resolution QuickBird images, which performed better than CNN classification-based method (Zheng et al., 2019). In general, object detection-based method is faster and more robust than other types of tree crown detection methods, efficiently alleviating the performance drop caused by complex topography, confusions with other vegetation, etc.

Lately, deep learning-based approaches have emerged as the predominant trend in tree crown detection, consistently yielding promising and satisfactory results. Nevertheless, existing tree crown detection studies exhibit certain limitations. First, many of them concentrate solely on single-class tree detection without considering subcategories. While some researchers have addressed mature and young oil palm detection separately, their methods necessitate predefined regions for mature and young oil palms using distinct sliding window sizes (Mubin et al., 2019). Furthermore, existing pest or disease detection studies typically focus on mapping diseased trees rather than precisely pinpointing individual diseased trees. Although some studies investigate disease recognition for individual trees captured by multi-spectral sensors (Mahlein et al., 2013; Giannakis et al., 2019), their primary focus is on disease research rather than comprehensive, large-scale monitoring of tree or crop growth status. To the best of our knowledge, the fine-grained monitoring of individual tree growth status has yet to be explored.

In this study, our primary focus lies in the precise detection of oil palm trees and the differentiation of their fine-grained growth stages through UAV images. Our approach introduces an individual oil palm tree detection method that not only attains high accuracy in identifying oil palm trees but also excels in the fine-grained monitoring of the growth status of each oil palm, even in cases with severely imbalanced training samples.

6.2 INDIVIDUAL OIL PALM TREE COUNTING IN PAPUA, INDONESIA

Papua is located on the southwest of Indonesia (Figure 6.1), which hosts 2% of all oil palm plantations nationwide in 2015 and is regarded as a new leading oil palm development area in the future (Austin et al., 2017). An UAV image acquired with three bands (RGB) at 8-cm spatial resolution was used in this study. The size of the UAV image is 64, 273 × 27, 839 pixels. There are various land cover types in the study area, such as oil palm plantation, rivers, buildings, and other vegetation.

In order to automatically delineate individual palm trees from the UAV image, first a total of 130,019 oil palms were manually marked by experts. Each oil palm is marked/recorded as one of five classes: dead palm, healthy palm, mismanaged palm, smallish palm, and yellowish palm (Figures 6.1 and 6.2). Healthy palms denote the palm that is normally cultivated. Dead palms no longer have vitality and usually have grey tree crown. Mismanaged palms are generally surrounded by other vegetation (e.g. weeds or other trees) and have difficulty in obtaining sufficient soil nutrition and sunlight. Smallish palms mean the diameter of the palm crown is relatively smaller than healthy and mature oil palms. Some of them are under seeding stage and others are under abnormal growing stage. Yellowish palms often suffer from pest or disease and have yellow spots on their leaves in UAV images. We list some examples of these five classes of oil palms in Figure 6.2.

We selected three regions from the whole UAV image as our training regions (original training dataset), represented by three red squares in the right of Figure 6.1. The size of each region is 5,000 × 5,000 pixels and the percentage of training regions in the whole image is smaller than 5%. We randomly split each training region to 1,000 images with 1,024 × 1,024 pixels as our training dataset

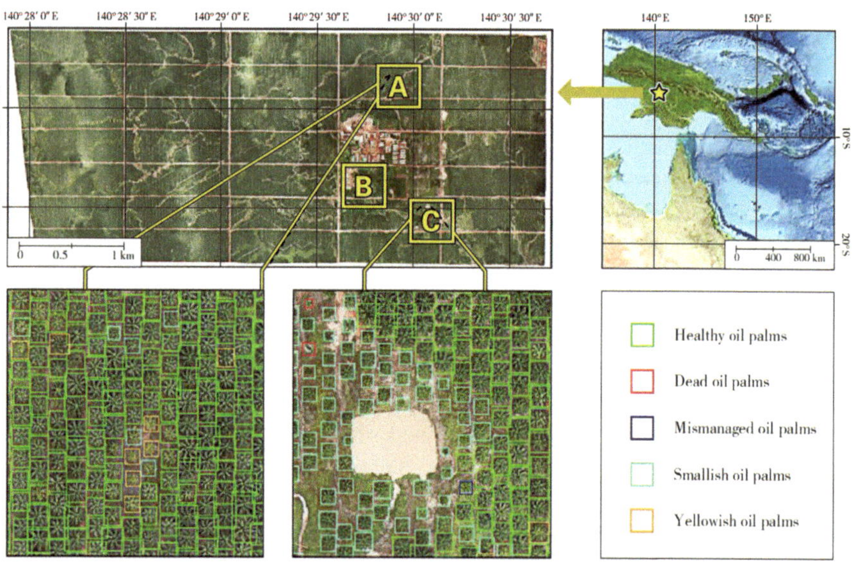

FIGURE 6.1 The study area and three training regions with multi-class oil palm examples.

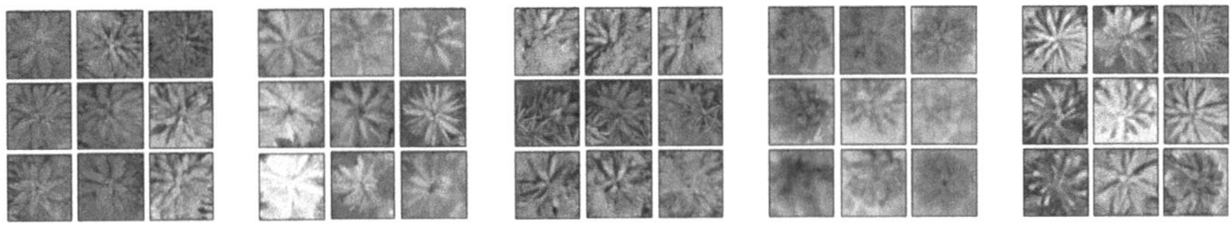

FIGURE 6.2 Examples of different types of oil palms.

(augmented training dataset). It should be noted that our augmented training dataset has 3,000 images and we only select images that have at least two different classes of oil palms. Our test dataset is the whole image area. Table 6.1 lists the number of oil palms in each class for original training dataset, augmented training dataset, and test dataset. We can easily observe that healthy palm has the highest quantity and there exists extremely misbalancing problem among five classes, resulting in great challenges for accurately detecting different classes of oil palms.

6.3 MULTI-CLASS OIL PALM DETECTION (MOPAD)

In this section, we introduce our proposed method, that is, Multi-class Oil PAlm Detection (MOPAD) to count oil palm trees in Papua, Indonesia. This method is inherited from Faster R-CNN (Ren et al., 2015). The framework of MOPAD including a refined pyramid feature (RPF) module, a multi-level region proposal network (RPN), and a hybrid class-balanced loss module is illustrated in Figure 6.3, and each of the module is described below.

6.4 AN RPF MODULE

We integrated deep-level features and shallow-level features to attain more feature characteristic from different level feature maps. Specifically, integrating high-level and low-level feature maps contributes to distinguishing similar classes and detects smallish oil palms. The RPF module involves four steps, including rescaling, integration, refinement, and disintegration.

TABLE 6.1
The Number of Samples of Multi-Class Oil Palms for Original Training Dataset, Augmented Training Dataset and Test Dataset

Index	Original Training Dataset	Augmented Training Dataset	Test Dataset
Healthy palm	3,921	216,743	121,013
Dead palm	9	434	65
Mismanaged palm	40	2,658	406
Smallish palm	1,047	47,411	7,373
Yellowish palm	94	5,724	1,162
Total	5,111	272,970	130,019

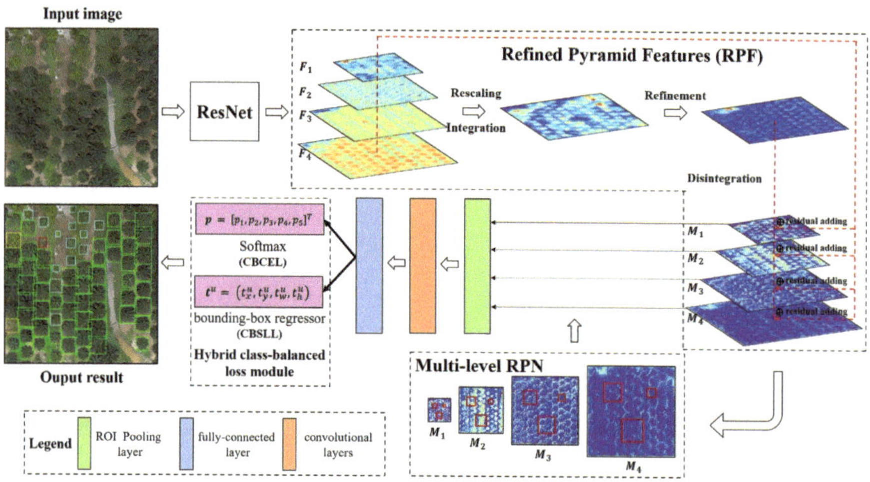

FIGURE 6.3 The flowchart of proposed Multi-class Oil PAlm Detection (MOPAD).

As deep-level features have more semantic meanings, many networks adopt deeper CNN to extract deeper information in images. However, shallow-level features are more content descriptive and have more capacity to detect small size objects. Recently, feature pyramid network (FPN) (Lin et al., 2017) builds high-level semantic feature maps at different scales with lateral connections, achieving significant performance on various object detection tasks. Different from FPN, we focus on integrating deep-level and shallow-level features to realize complementary feature extraction.

Here, we selected Residual Network (ResNet) (He et al., 2016) as our backbone. ResNet formulates the layers as learning residual modules with reference to the layer inputs, which addresses the saturated accuracy and degradation problem caused by deeper networks. In ResNet, we can acquire four different scales of feature map as the upper right of Figure 6.3 shows. For FPN, it makes predictions independently at all four scales. For our approach, inspired by Pang et al. (2019), we integrated four level features and strengthened the integrated feature by refining. Following we introduce four steps of our RPF, including integration, refinement, and disintegration.

ResNet has four levels of feature maps. We denote feature maps at resolution level l as F_l and our multi-level feature maps can be denoted as $\{F_1, F_2, F_3, F_4\}$. First, we rescaled the multi-level feature maps into the same size. For example, if we select F_2 as the intermediate feature map resolution, F_3 and F_4 are interpolated into the same size as F_2, as well as F_1 is max-pooled. Once the multi-level feature maps are resized, the integrated feature maps (F_I) can be calculated as equation (6.1).

$$F_I = \frac{1}{N}\sum_{l_{\min}}^{l_{\max}} F_l \tag{6.1}$$

where N is the number of multi-level feature maps. $l_{\max}$ and $l_{\min}$ are represented as the indexes of the highest and lowest resolution levels, respectively. Here, we obtain the integrated multi-level feature map. Inspired by the non-local module (Wang et al., 2018), we adopt an embedded Gaussian non-local module (denoted by Refinement) to make our integrated feature maps more discriminative. Finally, we use reverse procedure to disintegrate and add the corresponding original feature maps to attain our RPF maps $\{M_1, M_2, M_3, M_4\}$. Notably, each M_i has not only shallow-level semantic features but also deep-level semantic features.

6.5 MULTI-LEVEL RPN

We adopted an end-to-end network to generate region proposals with a wide range of scales and an aspect ratio of $\{1:1\}$ according to the shape of an oil palm tree crown. Following that, we adopted a Region of Interest (RoI) pooling layer to produce the fixed size of feature vectors for all region proposals.

RPN is the main contribution of Faster R-CNN (Ren et al., 2015), sharing full-image convolutional features with the detection network. In Faster R-CNN, RPN slides a dense $n \times n$ spatial window only over the last convolutional feature map. In our MOPAD, we adopted four levels of RPF (i.e., $\{M_1, M_2, M_3, M_4\}$) as the input convolutional feature maps and attached the same spatial window to each level of our RPF. After that, we used two fully connected layers. One is the classification layer and the other is the bounding box regression layer.

In Faster R-CNN, the original default parameters of the anchor size are $\{128^2, 256^2, 512^2\}$ and the aspect ratios are $\{1:2, 1:1, 2:1\}$ for the last feature map of the Faster R-CNN. Instead, in our MOPAD, we assigned an anchor size for each level of RPF. According to the size of oil palms in our image, we defined the anchors to have areas of $\{32^2, 64^2, 128^2, 256^2\}$ pixels on $\{M_1, M_2, M_3, M_4\}$, respectively. As for the aspect ratio, we only assigned $\{1:1\}$ at each level considering the shape of an oil palm tree crown. In addition, our oil palm candidate's generation method in multi-level RPN also accelerates the training and reference process because smaller number of anchors are generated.

6.6 A HYBRID CLASS-BALANCED LOSS MODULE

We first classified the region as foreground (oil palm) or background and localized the oil palm using smooth L_1 loss. Because the number of samples in different classes is extremely imbalanced, we adopted a Class-Balanced Cross-Entropy Loss (CBCEL) and a Class-Balanced Smooth L_1 Loss (CBSLL) to re-weight the fine-grained classification and regression loss according to the number of samples in different classes.

As our training samples are extremely imbalanced among different classes, directly adopting state-of-the-art object detection approaches may have embarrassing results. Many previous researches related to misbalancing problems focus on data augmentation, such as geometrical transformations on images (horizontal flipping, multi-scale strategy, adding noises, etc.) (Ren et al., 2015; Liu et al., 2016), or using random occlusion mask on images or feature maps (Zhu et al., 2015; Wang et al., 2017). However, these methods usually lead to overfitting problem because of generating duplicated images.

On the other hand, imbalanced problem can also be addressed by re-weighting the loss with inverse class frequency or inverse square root of class frequency. Inspired by Cui et al. (2019), we adopted a CBCEL that embed an effective number (EN) for loss calculation instead of directly using the sample number. The EN of samples is the expected number of samples (Cui et al., 2019) and can be calculated as:

$$\mathrm{EN} = \frac{1-\beta^{n_u}}{1-\beta} \tag{6.2}$$

where n_u is the amount of samples in the ground-truth class u and β is a hyperparameter $(\beta \in [0, 1))$. It is noteworthy that $\beta = 0$ corresponds to original loss function and $\beta \to 1$ means re-weighting through inverse class frequency. In other words, our CBCEL is re-weighted by inverse EN. As the number of oil palm class is 5, our predicted outputs from the classifier can be formulated as $\boldsymbol{p} = [p_1, p_2, p_3, p_4, p_5]^T$. Suppose a sample with class label of y, the softmax cross-entropy (CE) loss for this sample can be formulated as:

$$\mathrm{CE}_{\mathrm{softmax}}(\boldsymbol{p}, y) = -\log\left(\frac{\exp(p_u)}{\sum_{i=1}^{C}\exp(p_i)}\right) \tag{6.3}$$

where C is the total volume of classes. Given class u with n_u training samples, our CBCEL can be formulated as:

$$\mathrm{CBCEL}(\boldsymbol{p}, u) = -\frac{1-\beta}{1-\beta^{n_u}}\log\left(\frac{\exp(p_u)}{\sum_{i=1}^{C}\exp(p_i)}\right) \tag{6.4}$$

Moreover, we utilized a CBSLL for our loss function for bounding box regression. Notably, the smooth L_1 Loss is the same as the original Faster RCNN. Our CBSLL can be defined as:

$$\mathrm{CBSLL}(t^u, v) = \frac{1-\beta}{1-\beta^{n_u}} \sum_{i \in \{x, y, w, h\}} \mathrm{smooth}_{L_1}(t_i^u - v_i) \tag{6.5}$$

in which

$$\mathrm{smooth}_{L_1}(x) = \begin{cases} 0.5x^2 & \text{if } |x| < 1 \\ |x| - 0.5 & \text{otherwise} \end{cases} \tag{6.6}$$

where $\boldsymbol{t}^u = (t_x^u, t_y^u, t_w^u, t_h^u)$ is a predicted tuple; $v = (v_x, v_y, v_w, v_h)$ is a tuple of true bounding box regression targets for class u. Therefore, our multi-task loss function can be calculated as equation (6.7) and λ is a trade-off hyperparameter.

$$L(\boldsymbol{p}, u, \boldsymbol{t}^u, v) = \mathrm{CBCEL}(\boldsymbol{p}, u) + \lambda * \mathrm{CBSLL}(\boldsymbol{t}^u, v) \tag{6.7}$$

6.7 LARGE-SCALE OIL PALM TREE DETECTION

We adopted MOPAD described in Section 8.3 to predict the oil palm tree crown using large-scale UAV images of more than 60,000 × 25,000 pixels. As we trained our model using sub-images of 1,024 × 1,024 pixels, we needed to crop the large-scale image into small patches to implement the large-scale image prediction. If we directly crop the large-scale image uniformly into small patches of 1,024 × 1,024 pixels without overlaps, some tree crowns will be separated into two different patches, in which case the tree crown might not be detected or the detected object is incomplete. To solve this problem, we adopted an overlapping partition approach for large-scale UAV image prediction instead of directly split them into several 1,024 × 1,024 sub-images, ensuring every two neighbour patches in the large-scale image have an overlapping width (height) with 200 pixels. By this means, all the detected trees belong to at least one patch. After the oil palm tree detection of each patch, we performed an Intersection-of-Union (IoU)-based merging operation to acquire final detection results. This not only solves the problem of misdetection and wrong detection in large-scale image detection process but also avoids duplicate detection of targets. In addition, we considered the detected healthy palms in which crown sizes are less than 80 × 80 pixels as smallish palms.

6.8 EVALUATION OF MOPAD

We evaluated the experimental results of the proposed MOPAD for multi-class oil palm tree detection. We complemented the experiments based on the MMdetection deep learning framework (Chen et al., 2019), and set the hyperparameter λ (introduced in equation (6.7)) as 1 throughout all the experiments. We trained the model on GeForce RTX 2080 Ti (a graphic card with graphics processing units) and chose SGD as the optimizer with momentum strategy and the number of training epoch is 24. The backbone of MOPAD and other popular object detection methods is ResNet-101 for a fair comparison. The hyperparameter of β (introduced in equation (6.2)) is equal to 0.9999 in the experiments.

The evaluation metric consists of precision, recall, and F1-score. Precision depicts the model's capability of detecting oil palms correctly, while recall describes the model's capability of detecting ground-truth oil palms. F1-score evaluates the overall performance of the model. They can be calculated from the following formulas:

$$\text{precision} = \frac{\text{TP}}{\text{TP} + \text{FP}} \times 100\% \tag{6.8}$$

$$\text{recall} = \frac{\text{TP}}{\text{TP} + \text{FN}} \times 100\% \tag{6.9}$$

$$\text{F1-score} = \frac{2 * \text{precision} * \text{recall}}{\text{precision} + \text{recall}} \times 100\% \tag{6.10}$$

where TP means true positives, denoting the amount of oil palms that are detected correctly; FP means false positives, denoting the amount of others that are detected as oil palms by mistake; FN means false negatives (FN), denoting the amount of ground-truth oil palms that are missing in detection results. When the IOU metric value between the detected palm and an annotated oil palm tree is equal to or greater than 0.5, the oil palm tree will be regarded as correctly detected object.

6.8.1 Oil Palm Tree Detection with MOPAD

MOPAD detected oil palm trees and their types (as described in Section 8.2 and Figure 6.2) at different levels of accuracy as depicted in Table 6.2. The F1-score for oil palm tree detection attains 99.14%, featuring a precision of 98.87% and a recall of 99.41%. Some examples of FP and FN are displayed in Figure 6.4. It is evident that other vegetation and tree shadows are frequently misidentified as oil palm trees. Conversely, FN primarily occur due to the presence of other vegetation, the small size of oil palm tree crowns, and the relatively low quality of UAV images.

As for multi-class oil palm tree detection, the average F1-score for five classes is 70.06%. We can observe that the healthy palm, which has the largest sample number among five classes, reaches the highest F1-score of 99.46%, while the mismanaged palm has the lowest F1-score of 45.96%. The fine-grained oil palm tree detection results of eight selected regions are displayed in Figure 6.5. The confusion matrix for multi-class

TABLE 6.2
Oil Palm Detection Results of MOPAD

Index	Oil Palm	Healthy Palm	Dead Palm	Mismanaged Palm	Smallish Palm	Yellowish Palm
TP	129,252	120,113	36	216	6,744	854
FP	1,483	393	37	318	1,458	566
FN	767	900	29	190	629	308
Precision	98.87%	99.67%	49.32%	40.45%	82.22%	60.14%
Recall	99.41%	99.26%	55.38%	53.20%	91.47%	73.49%
F1-score	99.14%	99.46%	52.17%	45.96%	86.60%	66.15%

FIGURE 6.4 Failure cases for false positives (FP) and false negatives (FN) of oil palm detection.

FIGURE 6.5 Examples of fine-grained oil palm tree detection results of our proposed MOPAD.

oil palm tree detection of the proposed MOPAD is shown in Table 6.3, where the ground truth types are denoted by the top row, and the predicted types are denoted by the left column. Mismanaged palm and yellowish palm have serious confusions with healthy palm. Some examples of the confusions between different classes are shown in Figure 6.6.

The results obtained in this study were compared with the results obtained from other five state-of-the-art object detection methods, including Faster R-CNN (Ren et al., 2015), Cascade R-CNN (Cai and Vasconcelos, 2018), GA Faster R-CNN (Wang et al., 2019), Libra Faster R-CNN (Pang et al., 2019), and Grid R-CNN (Lu et al., 2019). The comparative results (Table 6.4 and Figures 6.7–6.9) demonstrate that MOPAD achieved the highest F1-score for healthy palm, dead palm, mismanaged palm, and yellowish palm. It can be observed that MOPAD demonstrates significantly higher accuracy in fine-grained oil palm detection and fewer confusions among various types of oil palms, resulting in improvements ranging from 15.78% to 3.55% in terms of the average F1-score.

TABLE 6.3
The Confusion Matrix for Multi-Class Oil Palm Detection of the Proposed MOPAD Technique, in Which the Top Row Denotes the Ground Truth Types and the Left Column Denote the Predicted Types

Index	Healthy Palm	Dead Palm	Mismanaged Palm	Smallish Palm	Yellowish Palm
Healthy palm	120,113	3	82	12	258
Dead palm	7	36	2	6	19
Mismanaged palm	166	1	216	22	14
Smallish palm	44	2	86	6,744	5
Yellowish palm	556	4	0	0	854

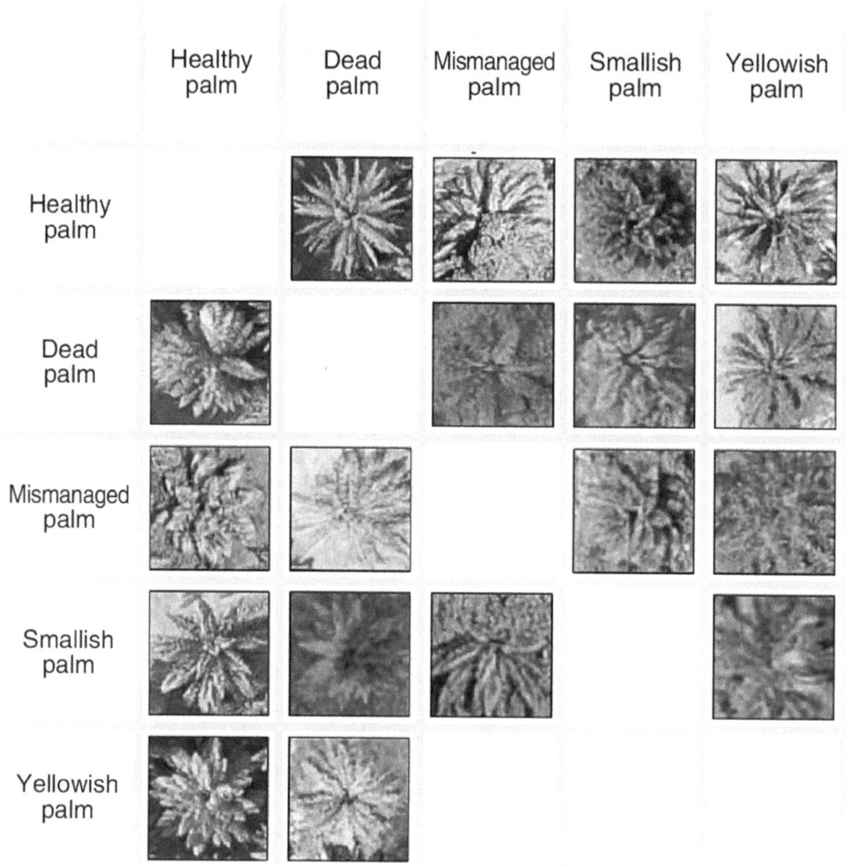

FIGURE 6.6 Failure cases or confusions among different kinds of oil palms, in which the top row denotes the ground truth types and the left column denotes the predicted types.

TABLE 6.4
The F1-Score of Other State-of-the-Art Object Detection Methods

Method	Healthy Palm	Dead Palm	Mismanaged Palm	Smallish Palm	Yellowish Palm	Average F1-Score
Faster R-CNN	98.77%	18.78%	31.14%	73.93%	48.84%	54.29%
Grid R-CNN	98.87%	37.57%	38.85%	87.51%	55.52%	63.66%
Cascade R-CNN	98.79%	44.60%	42.64%	81.23%	57.97%	65.04%
Libra Faster R-CNN	98.37%	42.55%	43.72%	82.31%	59.45%	65.28%
GA Faster R-CNN	98.48%	40.86%	41.58%	87.89%	63.76%	66.52%
MOPAD (ours)	99.46%	52.17%	45.96%	86.60%	66.15%	70.07%

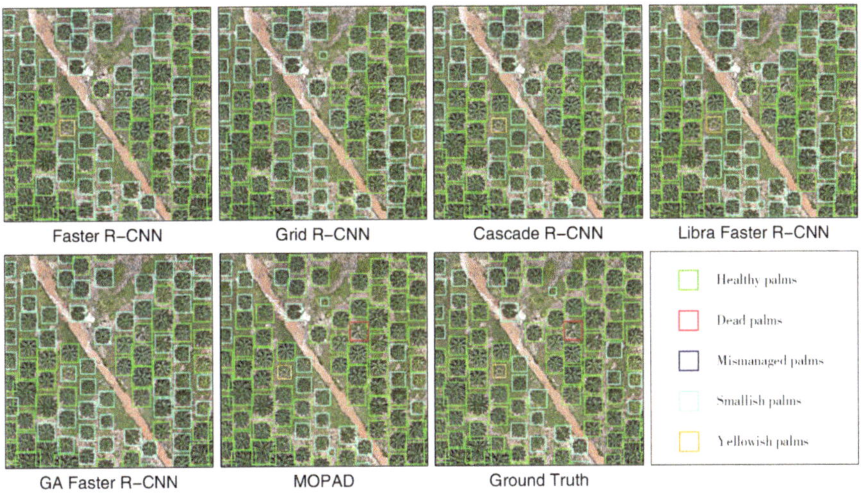

FIGURE 6.7 Multi-class oil palm tree detection results for Region 1.

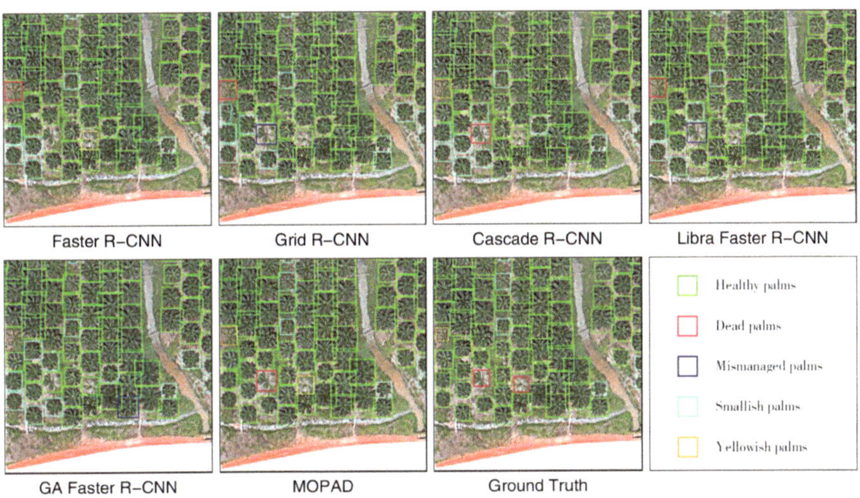

FIGURE 6.8 Multi-class oil palm tree detection results for Region 2.

The performance of MOPAD is influenced by various parameters, including the depth of ResNet, the RPF, and the hybrid class-balanced loss module. Table 6.5 presents the performance of different ResNet depths, namely 18, 34, 50, 101, and 152 layers. Notably, ResNet-101 exhibits the highest average F1-score of 70.06% for multi-class oil palm detection, excelling in detecting healthy and yellowish palms. ResNet-50 achieves the highest accuracy for identifying dead and mismanaged palms, while ResNet-152 performs best in detecting smallish palms with a remarkable accuracy of 86.64%. Consequently, it can be deduced that ResNet-101 is an optimal choice as the backbone for MOPAD.

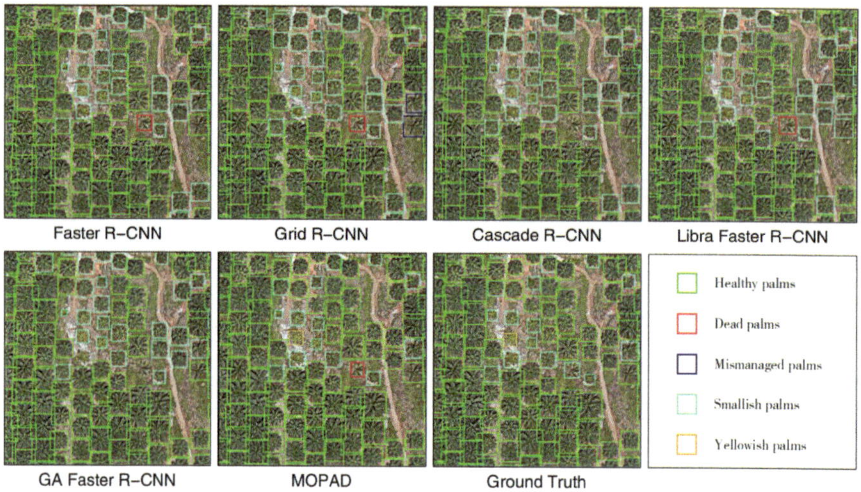

FIGURE 6.9 Multi-class oil palm tree detection results for Region 3.

TABLE 6.5
The F1-Score of Different Depth of ResNet for our MOPAD

Method	Healthy Palm	Dead Palm	Mismanaged Palm	Smallish Palm	Yellowish Palm	Average F1-Score
ResNet-18	97.14%	27.05%	35.18%	80.51%	66.69%	61.31%
ResNet-34	97.49%	44.59%	43.33%	83.64%	53.74%	64.56%
ResNet-50	98.18%	57.38%	46.32%	85.87%	48.78%	68.19%
ResNet-101	99.46%	52.17%	45.96%	86.60%	66.15%	70.07%
ResNet-152	97.22%	40.80%	43.12%	86.64%	65.92%	66.74%

Table 6.6 presents the ablation studies of RPF. Initially, we implemented RPF solely with integration, and it is noteworthy that this naïve feature integration achieved a 2.95% higher average F1-score compared to the corresponding baseline (i.e. Faster R-CNN), without any refinement or additional parameters. This simple procedure ensured that each resolution level received equal information from other level feature maps. Upon incorporating convolutional attention or Gaussian non-local attention (Wang et al., 2018), RPF further enhanced the final results. Specifically, RPF demonstrated a remarkable 5.74% performance gain in terms of average F1-score for multi-class oil palm tree detection when utilizing Gaussian non-local attention. Notably, RPF exhibited a substantial improvement of 14.95% in detecting smallish palms compared to the Baseline (Faster R-CNN), showcasing its efficiency in integrating multi-scale features for accurate smaller tree detection. The performance of these methods in detecting smallish oil palms is visually depicted in Figure 6.10.

It can be seen (Table 6.7) that the hybrid class-balanced loss module yields notable performance gains. We verified the effectiveness of CBCEL and CBSLL separately, which gains 7.66% and 5.03% improvement for multi-class oil palm tree

TABLE 6.6
The F1-Score of Ablation Study of RPF

Method	Healthy Palm	Dead Palm	Mismanaged Palm	Smallish Palm	Yellowish Palm	Average F1-Score
Baseline	98.77%	18.78%	31.14%	73.93%	48.84%	54.29%
+ Integration	98.61%	36.00%	34.02%	79.87%	37.68%	57.24%
+ Refinement (conv)	98.83%	30.43%	39.11%	83.24%	41.66%	58.65%
+ Refinement (non-local) (RPF)	97.66%	31.76%	41.69%	88.88%	40.18%	60.03%

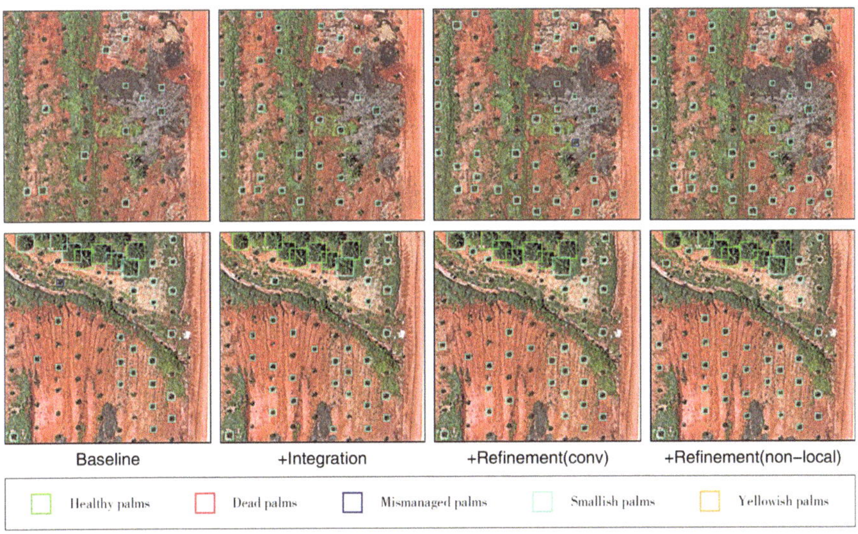

FIGURE 6.10 Comparison of detecting smallish oil palms using different strategies.

detection with respect to average F1-score. Furthermore, combining CBCEL and CBSLL can make a better improvement of 10.70% compared with Baseline + RPF. The hybrid class-balance loss module can effectively increase the detection accuracy by a large margin for low-number of palm classes like dead palms (+20.41%),

TABLE 6.7
The F1-Score of Ablation Study of the Hybrid Class-Balanced Loss Module

Method	Healthy Palm	Dead Palm	Mismanaged Palm	Smallish Palm	Yellowish Palm	Average F1-Score
Baseline + RPF	97.66%	31.76%	41.69%	88.88%	40.18%	60.03%
+ CBCEL	98.82%	49.28%	45.05%	87.27%	62.86%	68.65%
+ CBSLL	98.10%	44.44%	41.44%	86.16%	55.16%	65.06%
+CBCEL+CBSLL (MOPAD)	99.46%	52.17%	45.96%	86.60%	66.15%	70.07%

mismanaged palms (+4.27%), and yellowish palms (+25.97%). Additionally, the performance of healthy palms also has a slight increase with the possible reason that our hybrid class-balanced loss module alleviates the overfitting phenomenon. Although the hybrid class-balanced loss module has a relatively lower performance for smallish palms, it still helps to achieve a favourable improvement for average F1-score of five oil palm classes.

Next, the hyper-parameter β, which controls how fast the EN grows as the number of samples increases is discussed (Cui et al., 2019). According to equation (6.2), we display the F1-scores for multi-class oil palm detection under different $\beta = \{0.9, 0.99, 0.999, 0.9999, 0.99999, 0.999999\}$. If β is close to 1, the weights in the loss module for the class with a low number of samples become larger. It is observed that as β increases, the accuracy of healthy palms consistently increases because of alleviating overfitting phenomenon, while it can negatively affect the detection accuracy for smallish palms. For low-number oil palm classes, mismanaged palms and dead palms achieve highest F1-score when $\beta = 0.999$ and yellowish palms reach best performance when $\beta = 0.9999$. So, $\beta = 0.9999$ was selected for MOPAD as it obtains the highest average F1-score for multi-class oil palm tree detection. The F1-score of multi-class oil palm detection under different β is displayed in Figure 6.11.

MOBAD demonstrates great potential for individual oil palm tree detection and fine-grained monitoring of growing status using visible-band-only UAV images, leading oil palm industries and smallholders to precisely manage plantation and efficiently deal with unhealthy oil palms.

Future studies need to explore and develop more effective and accurate multi-class oil palm tree detection method, and apply them to larger-scale and more complex areas using multi-source and multi-temporal remote sensing images.

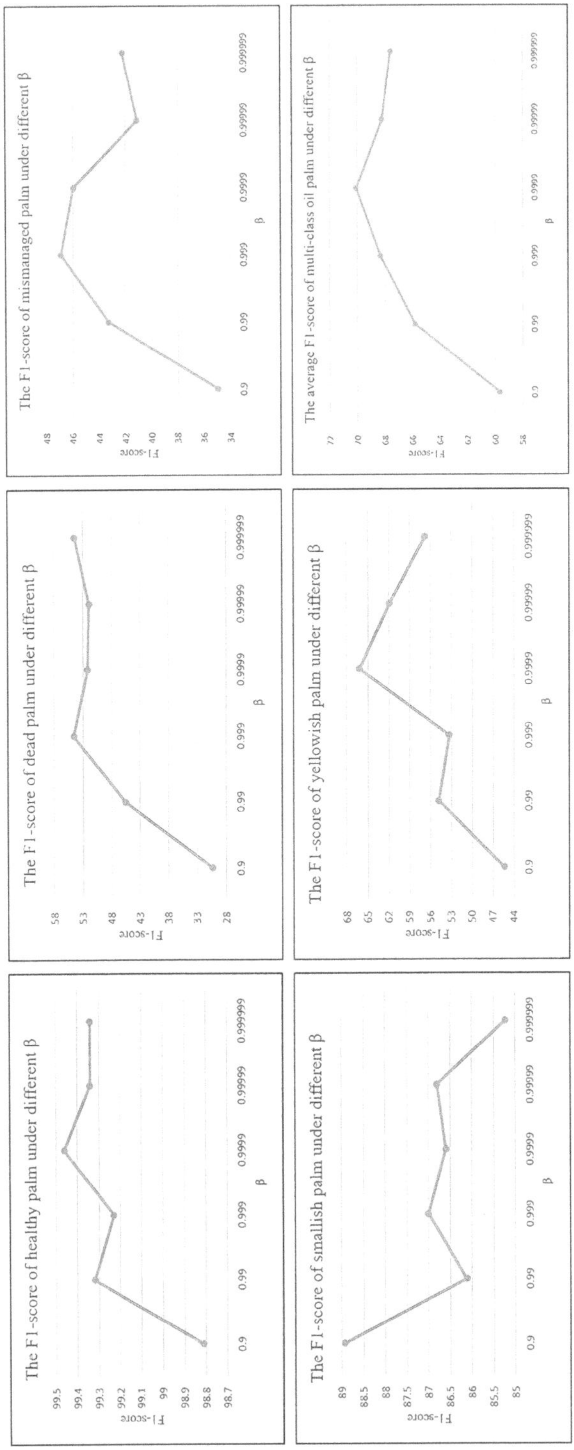

FIGURE 6.11 F1-score of multi-class oil palm detection under different hyper-parameters.

7 Age Classification of Oil Palm Trees

7.1 AGE OF OIL PALM TREES

Age of oil palm trees is an important aspect in oil palm management where it determines the productivity of fresh fruits and yield of oil extracted from the fruits. The yield of fresh fruit bunches (FFB) increases linearly with age from 3 to 7 years which then stabilizes when the trees attain matured phase (8–14 years) and started to decline slowly from 15 years after planting until the end of its life cycle at about 25–30 years (Woittiez et al., 2017). Consequently, age is used as a primary factor in crop management needs including yield forecasting, determining the use of fertilizers and labour force, predicting harvesting time, and planning replanting schedule. Data on age are also used for modelling carbon productivity of oil palm trees (Tan et al., 2013) and for monitoring the progress of small plantation holders so that funds can be allocated for replanting schemes and to ensure a sustainable oil palm plantation as practised by the Roundtable on Sustainable Palm Oil (RSPO).

Typically, the age of palm trees is first recorded when the saplings are transferred from nursery to the plantations. In the field, the age is determined based on the year it was transferred to the plantation. The data on oil palm age are systematically recorded by some large plantation owners; however, they are not shared with the scientific community and researchers to monitor the growth, yield, and carbon production of oil palm plantations. Most of the smallholders do not have the capacity to properly document the age information (McMorrow 2001; Chemura et al., 2015). Consequently, acquiring age information from large or smallholder plantations is rather difficult and it requires technological tools such as remote sensing that can access oil palm plantations and provide age information effectively. Field-based techniques that are widely used to obtain oil palm trees' age are by using measuring tapes and laser rangefinder (Pohl and Loong, 2016; Larjavaara and Muller-Landau, 2013). As one can assume, this technique becomes very laborious, time-consuming, and expensive when large plantations are considered. An alternative method to determine the age of oil palm trees over large plantations but relatively quicker and cheaper is by utilizing remotely sensed data. This chapter presents various promising remotely sensed data and advanced techniques that can estimate oil palm age with high accuracy.

7.2 AGE ESTIMATION USING REMOTE SENSING

Generally, oil palm age is determined using remote sensing by associating the radiance/reflectance/backscattering values of optical and microwave data directly with palm trees' age or by manipulating the reflectance via calculating vegetation indices (McMorrow, 2001; Fitrianto et al., 2018; Thenkabail et al., 2004; Danylo et al., 2021).

DOI: 10.1201/9780429199813-7

The optical and structural properties determine the oil palm canopy reflectance (McMorrow, 1995). Spectral properties of remote sensing data are related to the age of oil palm trees through interaction of radiation with canopy cover and leaf area index (LAI). LAI is defined as the one-sided green leaf area per unit ground surface area. LAI quantifies the amount of foliage in plant canopy and it is the most important factor that affects the total solar radiation intercepted, absorbed, or reflected by plants. In 4-year-old oil palm trees, the total radiation intercepted by the open canopy is only 39%, while 63% is transmitted to the lower layers of the canopy including to the soil surface (Romero et al., 2022). When the trees matured and more leaves are developed and overlapped with neighbouring trees, more radiation can be intercepted by the canopy and less is transmitted (Romero et al., 2022). As much as 95% of the PAR reaching the canopy of oil palm in Africa was intercepted at a high LAI value of more than 5.8 (Squire 1984 in Romero et al., 2022). A steady decrease in blue and red light is noticed as the LAI of oil palm trees increases due to strong chlorophyll absorption for photosynthesis (McMorrow, 1995; Tan et al., 2013). The decrease in the green light results from reduced scattering in the downward direction caused by the crown and a lower level of irradiance at the soil surface (Tan et al., 2013). While, LAI and near-infrared (NIR) radiation exhibit a positive relationship as a result of scattering and weak absorption at cell-wall interfaces within the leaves. This relationship, however, can be confounded by the canopy structure and shadow caused by matured trees. The strong absorption of shortwave infrared (SWIR) radiation by leaf water content causes inverse asymptotic relationship with LAI.

In some cases, the images are enhanced for their texture details before associating with age information (Tan et al., 2013; Hamsa et al., 2019). Another technique used to produce age information from remotely sensed images is by correlating the age with other biophysical properties of palm trees such as tree height, width (diameter at breast height or DBH), and crown size that are measured in the field or extracted from remotely sensed data (Tan et al., 2013; Chemura et al., 2015). These techniques are described below.

7.2.1 Age Estimation Using Spectral Information

The radiance of oil palm trees was found to be increasing swiftly for young trees of less than 4 years old due to their lower canopy cover but exposed bright ground cover or soil surface (Tan et al., 2013; McMorrow, 2001). As the trees grow older, fronds of nearby trees overlap and canopy closure occurs where the radiance started to decrease slowly and stabilizes at the age of 10 (Tan et al., 2013). The strongest correlation between radiance and age was found for the NIR band of Landsat TM (McMorrow, 1995, 2001) and the Disaster Monitoring Constellation 2 (UK DMC-2) (Tan et al., 2013). These researchers demonstrated nonlinear negative associations between age and radiance from the satellite imageries although the strength of the correlation was weaker for the visible bands (McMorrow, 1995, 2001; Tan et al., 2013). The empirical model developed using radiance data from the NIR band was found to be able to predict the age of young (1–10 years) palm trees. Although various spectral indices covering vegetation (Normalized Difference Vegetation Index (NDVI) and Advanced Vegetation Index (AVI)), soil, and shadow indices were correlated with the age, Ibrahim et al. (2001) showed that only young trees of less than 5 years old were able to be identified particularly with the NIR and SWIR bands of Landsat TM data. A similar technique was used by Mohd Zaki et al. (2009) where they identified trees with 8–15 years old with the infrared band from Landsat images although the strength of the correlation between age and NIR reflectance was quite weak in the study (r^2 of 0.37). The reflectance data from all the four bands of IKONOS data (blue, green, red, and NIR) could separate palm trees with 1–3 and 4–5 years old (two classes) in west African oil palms with an accuracy of 75% (Thenkabail et al., 2004).

A latest study on oil palm age classification was conducted by Danylo et al. (2021) using Landsat data and spectral index to determine oil palm plantations' stand age in Indonesia, Malaysia, and Thailand. Time series of Landsat imageries were used to detect the establishment year of palm trees (2–3 years old) using the bare soil index (BSI). BSI is calculated using the blue, red, NIR, and SWIR bands of Landsat data to capture the fraction of oil palm surface that is covered by soil (Diek et al., 2017) (equation 7.1).

$$\mathrm{BSI} = \big((\mathrm{Red} + \mathrm{SWIR}) - (\mathrm{NIR} + \mathrm{Blue})\big) \big/ \big((\mathrm{Red} + \mathrm{SWIR}) + (\mathrm{NIR} + \mathrm{Blue})\big) \quad (7.1)$$

The blue and NIR bands can enhance the presence of palm trees, while the red and SWIR bands are used to highlight the presence of soil/minerals. A higher BSI value in Landsat pixels signifies the prevalence of soil or young oil palm crops, while a lower value indicates a reduction in soil coverage within a pixel as the tree canopy matures with time. A reference BSI value was calculated using the 2017 imagery (established plantation), and the 95th percentile value was considered to be corresponding to the highest BSI value that matches the presence of young trees with 2–3 years since planted. BSI values less than the threshold were considered to be matured trees with canopy closure. For each year since 1984, any new establishment of oil palm trees was determined using the BSI threshold values. This study is significant for monitoring any new establishment of oil palm trees especially in high conservation value areas such as forest, peatland, and riparian zones and for RSPO reporting.

By employing the Landsat-based Detection of Trends in Disturbance and Recovery (LandTrendr) algorithm, Du et al. (2022) generated global maps indicating planting years spanning from 1982 to 2020. Their approach utilized a time-series Landsat archive and the Google Earth Engine (GEE) platform for the analysis of global plantation extent data. The resultant maps offer insights into species types and the most recent planting years of global plantations, presented at a 30-m resolution. Notably, in comparison to the findings reported in Danylo et al. (2021), the oil palm plantation year map within this dataset demonstrated a relatively high F1-score of 78.25%, with a deviation of ±3 years (Du et al., 2022).

As the oil palm trees grow older, their height and LAI increases and canopy becomes denser and closed. Increasing canopy closure will result in the fraction of shadow cast on the ground. There is a positive relationship between canopy closure and the fraction of shadow. Thus, by studying the positive correlation between the fraction of shadow and canopy closure, the age of oil palm trees can be approximated/inferred. The information about shadow casted by oil palm trees has been used to estimate oil palm biomass in a previous study (Morel et al., 2012). In our previous study (Tan et al., 2013), we found a strong relationship between the fraction of shadow and oil palm tree height and thus age ($R^2=0.80$) in an oil palm plantation in Peninsular Malaysia. We manipulated the green, red, and infrared bands from UK DMC-2 satellite data to extract the fraction of shadow by finding spectral endmembers and their shadow abundances in the image, with a value of 0 referring to no shadow and a value of 1 representing full coverage of shadow in a pixel. The shadow index together with several other spectral indices such as the AVI, BSI, and thermal index (TI) was utilized by Fitrianto et al. (2018) to identify percentage of canopy density within each pixel of Landsat 8 Operational land Imager (OLI) and to correlate it with the age of oil palm trees in Bogor, Indonesia. The technique separated young (less than 10 years) and matured (15–20 years) classes but unable to distinguish old trees (>20 years) from the matured class as a consequence of overlapping tree canopies of matured and old trees.

7.2.2 Age Estimation Using Backscatter Information from Radar Data

Similar to optical data, radar images have also been utilized to identify different age of oil palm trees (Tan et al., 2013; Darmawan et al., 2016; Nordin et al., 2022). Previously we (Tan et al., 2013) reported that age separation is possible because matured trees with higher biomass resulted in enhanced backscatter values in the HH polarization ($R^2=0.49$) compared to the HV ($R^2=0.27$) or HH/HV ($R^2=0.26$). Similar findings were obtained by Darmawan et al. (2016) where young oil palm trees aged up to 10 years correlated with the backscattering signal from ALOS PALSAR HH and HV polarization yielded a moderate correlation coefficient ($R^2=0.63$ for HH and 0.42 for HV polarization data). While, Nordin et al. (2022) demonstrated the application of Airborne Synthetic Aperture Radar (AIRSAR) L- and P-band data to classify oil palm trees with different ages at a plantation in the north of Peninsular Malaysia. The L-band HH polarization data could separate palm trees of age 6, 7, 15, and 20 years old, while the P-band HH polarization is effective in detecting 25-years-old palm trees. This study also suggested that de-speckling at different window sizes is important for enhancing the age classification. All these studies proved the use of HH polarization for age detection with Synthetic Aperture Radar (SAR) data. The use of radar data for age classification should be further explored as radar data are very useful for oil palm studies in the tropical environment to overcome the effect of clouds.

7.2.3 Age Estimation Using Textural Information

Similar to size, shape, shadow, tone, colour, pattern, and association, texture is also an important spatial feature and key image interpretation technique to classify oil palm age. In satellite imageries, texture refers to the spatial variability in pixel intensity, that is, colour or tone that creates smooth or rough visual appearance of the images. Incorporating textural information along with the spectral details of the original satellite data can improve the interpretability, thus enable classification of more age classes with enhanced classification accuracy (Tan et al., 2013; Mansor and Sarker, 2015; Kamiran and Sarker, 2014; Vadivelu et al., 2014; Carolita et al., 2015). We (Tan et al., 2013) successfully classified 14 different ages of oil palm trees when texture and fraction of shadow were integrated with the three original UK DMC-2 spectral bands. The overall accuracy of the classification procedure improved from 53% to 56% and 57% when texture measures (mean) from the NIR band and fraction of shadow index were, respectively, added as extra layers in the Random Forest (machine learning) classification technique. Although, a moderate level of overall accuracy was achieved by adding texture and shadow details to the classification, the individual accuracies are low for trees with 10, 11, 19, 21, 22, 24, 25, and 27 years old. This means, matured trees of more than 10 years are not easily discriminated in satellite imageries due to minor textural variability in canopy that is corresponded to LAI that do not develop further (stabilized) after the age of 10 (Corley and Gray, 1976). Consequently, our study effectively classified oil palm trees with 3, 4, 15, 16, 20, and 29 years old with user and producer accuracies of more than 60%.

Incorporation of texture measures into high-spatial and spectral resolution satellite data such as Worldview-2 and SPOT-6 was also attempted to improve the age classification of oil palm trees. These studies used various number of texture measures and scale (different window sizes) but only able to distinguish 2 (Kamiran and Sarker, 2014), 4 (Mansor and Sarker, 2015), and 6 (Carolita et al., 2015) age classes which can be grouped into young (<10 years) and matured (more than 10 years) with an overall accuracy of less than 60%. Nevertheless, a careful selection of texture measures, window sizes, and satellite data could discriminate a large number of oil palm age classes with higher accuracy. We previously experimented eight texture measures (mean, variance, homogeneity, correlation, dissimilarity, entropy, angular second moment, and contrast) and found that a combination of less than two measures at a time produced lower accuracy compared to higher number of textures (Hamsa et al., 2018). The window size for the texture measurement was optimized by considering the size of the age block in the plantation where the maximum window size used was determined based on the smallest block size. This step was crucial to reduce the possibility of overestimating the window size and elucidating the role of texture in the age classification. Each texture set was applied with 39, 29, 19, and 9 window kernel sizes. The classification (maximum likelihood) was run with eight textures and four spectral bands (green, red, NIR, and SWIR) of the SPOT-5 data with 10-m spatial resolution. This technique produced 12 age classes of oil palm trees with an overall classification accuracy of 86% (individual accuracy ranging between 77% and 96%) with all 8 texture measures with 390 m × 390 m on the ground window size. Misclassification mainly occurred between similar ages such as 1–2, 6–7, 8–9, 10–11, 17–18 years old and at the boundary of different ages. To date, this is the best oil palm age classification study produced many age classes with higher accuracy.

7.2.4 Age Estimation Using Biophysical Properties of Oil Palm

Studies exploring the connection between age and biophysical properties of oil palm trees such as tree height, width, crown size, and leaf morphology are limited and warrant further investigation. We found a strong linear association between height of palm trees measured in the field and age ($R^2=0.9$) but a weak correlation with DBH ($R^2=0.02$). We demonstrated that height can be used to predict the age as oil palm trees that continuously grow at a rate of approximately 50 cm/year (Tan et al., 2013).

Besides tree height, tree crown has also been used to estimate the age of oil palm trees. Chemura et al. (2015) determined the age of oil palm trees in Ghana by correlating tree crown projection area that was delineated by classifying the high-spatial resolution Worlview-2 satellite data using object-based image analysis techniques and oil palm age–crown projection area regression model that was developed using field data. This technique could map trees up to the age of 13 years, and for trees more than 13 years old, no significant relationship was found between age and crown projection area. The age identification is highly dependent on the accuracy of the crown projection area delineation in the satellite imageries.

Remote sensing is a useful tool to obtain height data that can be used to estimate the age of oil palm trees at local and regional scale using Light Detection And Ranging (LiDAR) technology on board aircraft or satellites (Rizeei et al., 2018). Since height is directly proportional to the age as mentioned above, Rizeei et al. (2018) proposed a new method of combining high resolution Worldview-3 optical satellite imagery and airborne LiDAR remote sensing data to estimate the age of oil palm trees planted within a university campus in Peninsular Malaysia. The pan-sharpened WorldView-3 image at 0.3 m was classified to delineate the individual crown of oil palm trees which was then integrated with a height model derived from LiDAR data to determine the tree age. The findings showed high accuracy of age classification (85%) where polynomial function was better fit for trees between 4 and 9 years, and exponential function for older trees (22 years). This study demonstrated the importance of high-spatial resolution images for an accurate delineation of oil palm crowns for age prediction using height data from LiDAR remote sensing. The issue of canopy overlapping among the older trees made the age classification of older trees rather difficult despite high-spatial resolution data were used.

Recent studies attempted to derive the height of oil palm trees from airborne (Beese et al., 2022) and spaceborne (Zang et al., 2023) LiDAR are highly potential for estimating age at local to regional scales. Beese et al. (2022) estimated the height of more than 500,000 oil palm trees in Sabah, Malaysia, using point cloud data at 3.3 points/m^2 which was processed to get the normalized point cloud and half a metre resolution canopy height model (CHM). The CHM and aerial photo were used to delineate oil palm crowns manually before a filter algorithm (local maxima) could locate the top of oil palm trees. A window size of 6–10 m (corresponding to the space between trees on the ground of 7–9 m apart) enabled the authors to locate each tree's highest point, that is, the height of oil palm trees.

Satellite-based LiDAR is used to get oil palm height, but usually some additional information such as crown area from satellite images is required to establish a regression model between height extracted from LiDAR and crown area. This method has shortcomings of satellite saturation of optical imageries where very short and very tall trees are not represented well (Potapov et al., 2021). Consequently, some researchers have used age information obtained from satellite images to estimate palm trees height (Zang et al., 2023).

The latest spaceborne LiDAR, that is, the Ice, Cloud, and Land Elevation Satellite-2 (ICESat-2) provides extensive data on canopy height, which presents an opportunity to map the height of oil palm trees at a regional scale. However, there is a challenge in extrapolating the LiDAR data to create maps that are spatially consistent. To address this, Zang et al. (2023) proposed a new method that combines data from ICESat-2 with the stand age of the oil palm trees produced by Du et al. (2022). An empirical model relating the ICESat-2 canopy height data to the age of the oil palm trees (retrieved from Landsat data) was established to estimate the spatial and temporal variations (2001–2020) in oil palm height throughout Peninsular Malaysia at 30-m spatial resolution. The resultant height data showed only a small RMSE of 1.64 m and agreed well with field data with $R^2 = 0.63$. The approach developed in this study based on remote sensing data combining ICESat-2 footprint and stand age from Landsat holds significant potential for producing height and/or age of oil palm trees, leading to improved quantification of palm oil yield and carbon stocks in oil palm–growing regions of the world.

The Global Ecosystem Dynamics Investigation (GEDI) is also collecting height samples globally. GEDI is a full waveform LiDAR instrument mounted on the International Space Station acquiring high-resolution data on surface topography, canopy height, cover, and vertical structure of forests since 2019 (Potapov et al., 2021). The large footprint (1 km) of the data can be enhanced by integrating it with other optical imageries such as Landsat to enable local scale retrieval of canopy height (Potapov et al., 2021) that can be used to estimate oil palm age.

In the following section, we present a case study on detecting the age of oil palm trees using a combination of SAR and Sentinel 2 optical data. As mentioned earlier, age is a significant parameter for explaining the variation in oil palm biomass and yield. Previous studies that attempted to derive biomass information using remotely sensed data have reported that the satellite signal tends to saturate at high biomass levels, which corresponds to oil palm trees that are 10 years and older. When data on oil palm age are unavailable, biomass estimation may be underestimated. Therefore, obtaining accurate age data for oil palm trees is crucial for precise estimates of oil palm biomass. In this case study, we consider three age categories: young (5–11 years old), mid-age (15–23 years old), and mature (26–29 years old) to classify oil palm trees.

7.3 CASE STUDY OF OIL PALM AGE ESTIMATION IN MALAYSIA

For the Malaysian case study, we selected a private oil palm plantation, Genting Kulai Besar, situated in the southern region of Peninsular Malaysia, as depicted in Figure 7.1. This estate spans approximately 1,750 ha and encompasses several administrative divisions within the Kulai district. The trees within this plantation were planted between 1969 and 2017, ranging in age from 1 to 49 years. It is worth noting that the older trees, aged above 30 years, are typically not harvested partly because some of the lands within the plantation have been sold and are slated for conversion into residential and commercial areas in the near future.

Oil palm trees at various stages of growth: intermediate (5–11 years), productive (17–23 years), and mature or non-productive (26–29 years) are showcased in Figure 7.2. The plantation's topography is undulating, with certain areas containing rocks due to nearby quarry sites. In general, the entire plantation consists of mineral soil, along with various land covers such as grass, shrubs, soil, and a lake in the

FIGURE 7.1 The location of Genting Kulai Besar palm oil plantation, in south of Peninsular Malaysia. Different colours and numbers in the blocks are corresponding to the age of oil palm trees as of 2017.

northern part of the plantation. Surrounding the plantation area, there are residential neighbourhoods, highways, and shopping centres.

7.4 CLASSIFICATION OF OIL PALM AGE WITH SENTINEL AND ALOS PALSAR DATA

The study explored the potential of Sentinel-1 (freely available) which has a spatial resolution of 10 m, and ALOS PALSAR-2 data with 25 m, to classify oil palm into distinct age groups. We obtained three scenes of Sentinel 2 Level-1C imagery dated for February, July, and September 2016 from the https://GloVis.usgs.gov/ website. Sentinel 2 offers a temporal resolution of 10 days per single satellite and a combined constellation revisit rate of 5 days. We utilized Sentinel 2 multispectral images with bands 2 (blue), 3 (green), 4 (red), and 8 (NIR).

Additionally, the ALOS PALSAR-2 radar data were acquired from the Japan Aerospace Exploration Agency (JAXA) website (http://www.eorc.jaxa.jp/ALOS/en/palsar_fnf/fnf_index.htm). These radar data, in the fine beam dual-polarization (FBD) mode, offer a spatial resolution of 25 m and include both horizontal transmit–horizontal receive (HH) and horizontal transmit–vertical receive (HV) polarizations.

The digital numbers (DN) from the Sentinel data were transformed into reflectance values using the semi-automatic classification plugin within the QGIS 2.18 software. This plugin also conducted atmospheric correction through the dark object subtraction method. To address any cloud cover in the scenes, masking was applied using the NIR band, followed by image mosaicking as illustrated in Figure 7.3.

FIGURE 7.2 Oil palm trees with different ages in the Genting plantation, Kulai, Malaysia.

Regarding the ALOS PALSAR-2 images obtained from JAXA, they were geometrically projected into the WGS84 projection system. Subsequently, radiometric calibration was performed on the ALOS PALSAR-2 satellite data, involving the conversion of DN to sigma naught. This calibration was executed using the equation proposed by Rosenqvist et al. (2007). Additionally, prior to delivering the data to users, JAXA corrected the SAR data for topographic effects like foreshortening, layover, and shadow.

Given that radar data can be susceptible to speckles, we applied a Lee filter with a window size of 3 × 3 to the image. This filter effectively reduces unwanted noise in the data, utilizing the mean square error technique or the weighted least square estimation (Qui et al., 2004). It ensures the preservation of spatial resolution and features, as highlighted by Medasani and Reddy (2018).

The datasets underwent additional processing to enhance their textural details, a practice supported by previous studies (Section 9.2.3), which have demonstrated improved classification accuracy with the incorporation of texture information alongside the original spectral bands. Various texture measures derived from the

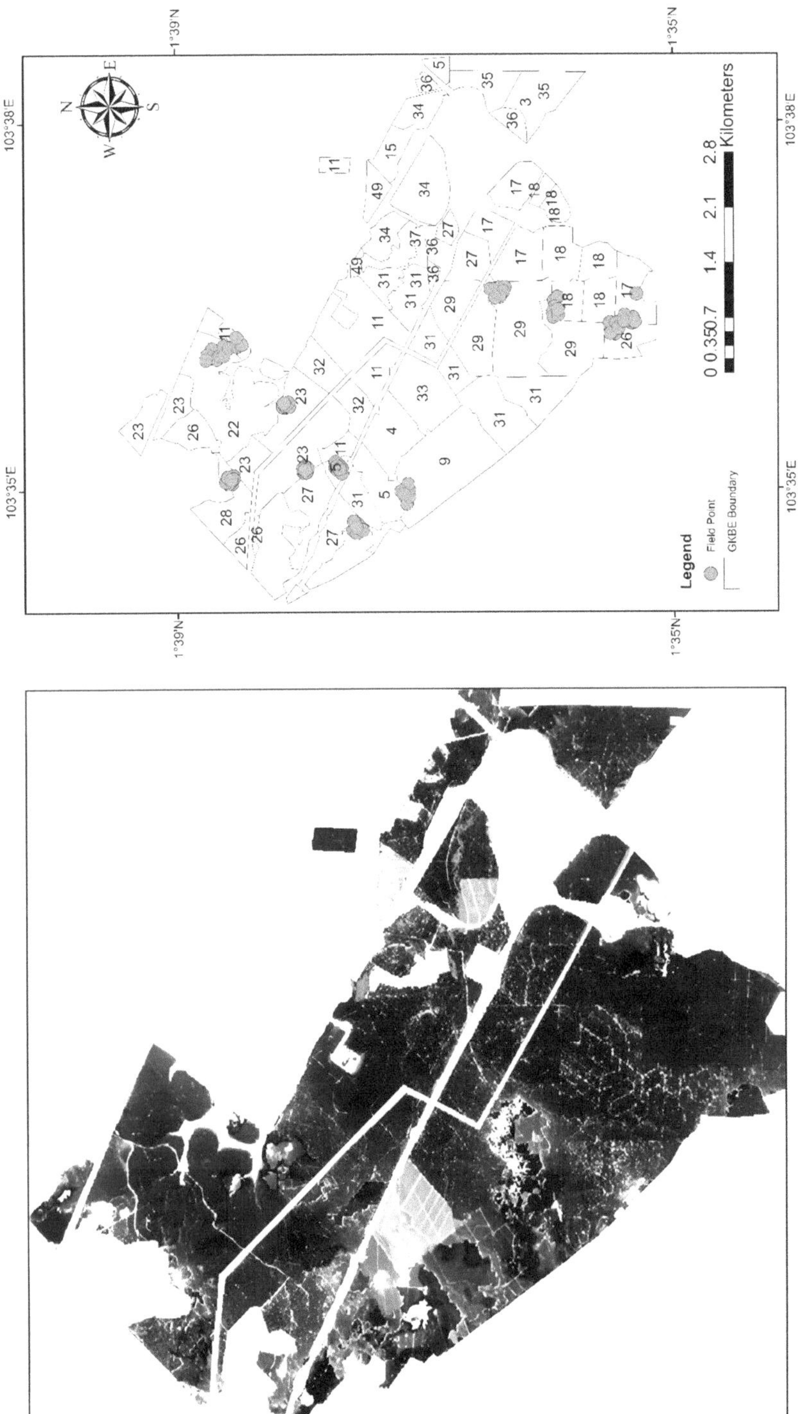

FIGURE 7.3 Sentinel image that was masked out for clouds (left) and the locations of oil palm age recorded in the field (right).

Grey Level Co-occurrence Matrix (GLCM), including mean, variance, homogeneity, contrast, dissimilarity, entropy, second moment, and correlation, were evaluated. A correlation analysis was conducted to assess the relationship between the texture measure types and field-collected oil palm age data.

To determine the optimal window size for differentiation among different age classes, different sizes (3 × 3, 13 × 13, 29 × 29) were experimented with. The selection of the window size was based on the dimensions of the smallest block within the age group, considering intermediate, productive, and mature oil palm categories. In many large plantations, oil palm trees are arranged in a triangular planting style, as depicted in Figure 7.4. The Sentinel 2 data, with its 10-m spatial resolution, can accommodate approximately three trees within a 100 m^2 area. In our study area, the block representing intermediate oil palm trees had the smallest area, measuring 129,021 m^2, which can accommodate roughly 3,870 trees. The largest window size of 29 × 29 (equivalent to 84,100 m^2) effectively covers all the trees within the smallest block without overlapping with other age groups.

The polarization data were carefully manipulated to determine the optimal combination of polarizations for age classification. In addition to evaluating HH, HV, HH minus HV, and HH over HV, we incorporated data from the blue, green, red, and NIR bands of the Sentinel 2 dataset to establish correlations with the age of palm trees gathered in the field (as illustrated in Figure 7.3). We recorded the geographical coordinates of 498 trees in the WGS 84 projection using a GPS Trimble Geo XT 3000 device, ensuring an error margin of less than 100 cm.

The field data collection involved establishing seven plots for each age block and seven trees within each plot. However, due to GPS signal issues and the dense canopy cover in some plots, in two plantation blocks, we were only able to sample up to five plots. Each sampling plot represented an approximate area of 30 m, and the field data collection points' locations are indicated in Figure 7.3.

To determine the most effective combination for capturing age variation, we conducted an analysis involving various combinations of radar polarizations and

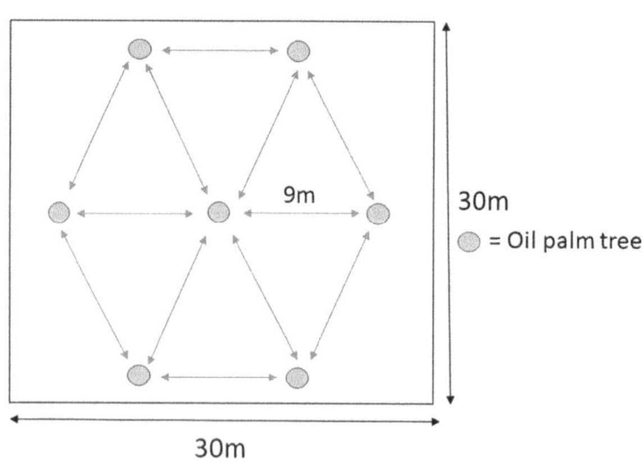

FIGURE 7.4 Triangular planting design of oil palm that is viewed in one pixel.

TABLE 7.1
Various Combinations of Radar Polarizations and Optical Reflectance Bands with and without Texture Measures for Correlating with Age of Oil Palm Trees

Combination	Satellite Data	With/without Texture
1	Sentinel (Red, Green, Blue, NIR) + ALOS PALSAR (HH, HV)	Without texture
2	Sentinel 2 (Red, Green, Blue, NIR) + ALOS PALSAR (HH, HV)	With texture (type: Mean)
3	Sentinel 2 (Red, Green, Blue, NIR) + ALOS PALSAR (HH, HV)	With texture (type: Variance)
4	Sentinel 2 (Red, Green, Blue, NIR) + ALOS PALSAR (HH, HV)	With texture (type: Homogeneity)
5	Sentinel 2 (Red, Green, Blue, NIR) + ALOS PALSAR (HH, HV)	With texture (type: Correlation)
6	Sentinel 2 (Red, Green, Blue, NIR) + ALOS PALSAR (HH, HV)	With texture (type: Dissimilarity)
7	Sentinel 2 (Red, Green, Blue, NIR) + ALOS PALSAR (HH, HV)	With texture (type: Entropy)
8	Sentinel 2 (Red, Green, Blue, NIR) + ALOS PALSAR (HH, HV)	With texture (type Angular Second Moment)
9	Sentinel 2 (Red, Green, Blue, NIR) + ALOS PALSAR (HH, HV)	With texture type: Contrast

optical reflectance bands, both with and without texture measures (as summarized in Table 7.1), correlating them with the age data.

The SVM classification technique was run on the pre-processed Sentinel image to classify the age of oil palm trees and other land uses (built-up area) using training samples (30 polygons for each age class and each polygon contains approximately 100 pixels) collected in the field/obtained from the Google Earth data and the estate administration. The Radial Basis Function (RBF) kernel type of the SVM classifier with a Penalty Parameter value of 100 was used. The penalty parameter functions to trade-off between allowing training errors and forcing rigid margins. The greater the value, the more chances to create more accurate model that may not generalize well. Another parameter to run the classification is the *Pyramid Levels.* A value of 0 was set to process the image only at full resolution.

7.5 OIL PALM AGE MAP FROM SENTINEL DATA

The correlation between backscatter values and oil palm age is shown in Figure 7.5. The results show that generally the correlation between Sentinel 2 and age is higher compared to the correlation between the radar data and age of palm trees in the study area. The green band from Sentinel 2 recorded the highest correlation coefficient ($R^2=0.65$), followed by ALOS HH polarization ($R^2=0.43$), NIR band ($R^2=0.38$),

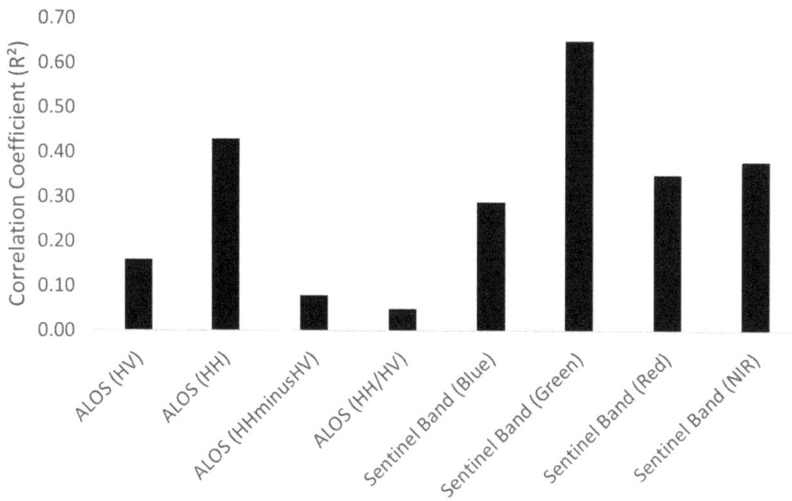

FIGURE 7.5 The correlation coefficient values between age and radar backscatter/Sentinel-2 reflectance values.

red and blue bands. The lowest correlation was recorded by ALOS PALSAR 2 HH over HV, HV, and HH minus HV. Since previous studies also reported that radar data are not significant to improve the age separation of oil palm trees (Section 9.2.2), and when we incorporated the HH band into the SVM classification along with the Sentinel-2 bands, the results were not improved (results not shown); thus, we did not further consider these data for age classification in the study area.

The nonlinear correlational analysis between various texture measures and the age of oil palm trees as depicted in Figure 7.6 reveals that the measures of homogeneity and entropy exhibit the highest R^2 values of 0.61 and 0.63, respectively. In a previous study, we also reported that the reflectance values of SPOT images, processed with entropy and homogeneity texture measures, yielded a high R^2 value when correlated with the age of oil palm trees cultivated in mineral soil in Peninsular Malaysia (Hamsa et al., 2018). Notably, all texture measures displayed the highest correlation with the reflectance from the red band, aligning closely with our prior research employing different satellite data in different study sites.

As suggested by Ouma et al. (2008), heterogeneous texture tends to result in higher entropy values compared to homogeneous ones. In the plantation under examination in this study, the texture of intermediate-aged (5–11 years) palms may significantly differ from that of productive-aged (17–23 years) and mature or non-productive oil palms (26–29 years). Nevertheless, there may be some textural similarities between productive-aged and mature palms. Hence, both homogeneity and entropy texture measures exhibited higher correlations with age. A noteworthy similarity often emerges when tree canopies overlap, typically occurring around the age of 15 years, leading to a strong relation with homogeneity.

The inclusion of the NIR band in texture analysis proves crucial for distinguishing the age of palm trees, as NIR can detect shadows cast by older, taller trees. NIR reflectance tends to saturate as shade increases with tree age. The presence of shade heightens textural variation between oil palm blocks of different ages. Figure 7.6 illustrates a stronger correlation between the NIR band and the age of oil palm trees in the homogeneity texture compared to all other texture measures. The effectiveness of the NIR band in age classification has also been substantiated in various locations by Thenkabail et al. (2004), Chemura et al. (2015), and Tan et al. (2013). Our prior analysis of oil palm age against homogeneity similarly detected a pronounced texture–age connection in the NIR band compared to other bands.

In our study (Hamsa et al., 2018), we obtained higher accuracy for homogeneity (R^2=0.72–0.84) compared to entropy and angular second moment (R^2 0.30–0.76). Initial attempts at SVM classification with all the textures resulted in very low and insignificant outcomes (results not shown), except for the homogeneity texture. Consequently, we chose to employ the homogeneity texture measure for further oil palm age classification using the SVM technique.

The results of classification are tabulated in Table 7.2. Consistent with previous studies, the largest window size (29 × 29) considered in this study produced better overall classification accuracy compared to the smaller window sizes. This finding shows that a large window size provides better discrimination of age classes compared to the small window sizes. This finding is similar to that of our previous study (Hamsa et al., 2018), where a 39 × 39 window size yielded the highest accuracy to discriminate 14 different ages of palm trees. The size of window may vary according to plantations based on their topology, environment, planting style, and site location. The presence of non-oil palm features such as workers' quarters, water bodies, and vacant lands in the plantations may affect the reflectance or backscattering values towards the satellite sensor.

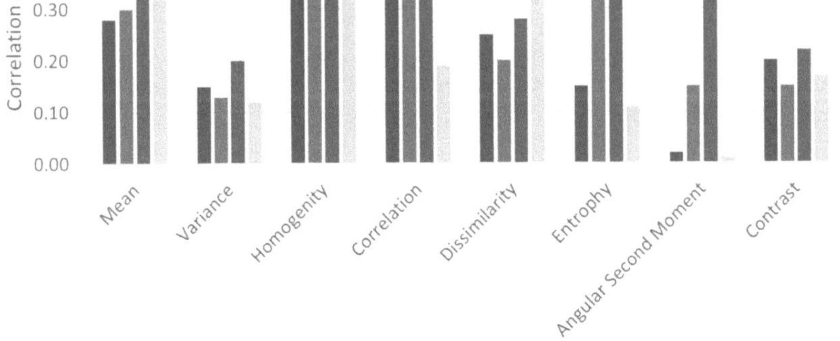

FIGURE 7.6 Determination of correlation coefficient values (R^2) between various texture measures from Sentinel data and age of oil palm trees.

TABLE 7.2
Accuracy of Oil Palm Age Classification Using Sentinel 2 Data with Homogeneity Texture at Different Window Sizes (a) and the Producer and User Accuracies (b)

(a)

Window size	3 × 3	13 × 13	29 × 29
Overall accuracy (%)	61.47	70.92	78.83
Kappa coefficient	0.23	0.49	0.61

(b)

Class	Producer accuracy (%)	User accuracy (%)
Young oil palm	61.88	75.34
Middle age oil palm	70.52	76.53
Matured oil palm	83.88	73.56

In terms of the User accuracy, all three classes produced almost similar accuracies showing that the probability that a pixel labelled as an intermediate, productive, and mature in the Sentinel 2 image is 74%–75% really this class in the field. Whereas, a higher producer accuracy of 84% was achieved for the mature class compared to only 70% and 62% for the productive and intermediate aged classes, respectively. This means, the probability that a mature oil palm tree on the ground is classified as such is 84%. Most of the misclassification were contributed by the productive aged trees that are mixed with mature and intermediate aged oil palm. The confusion of mature with productive age oil palms occurred due to the overlapping of their canopies especially among trees above 20 years. Mature trees are also found to be mixed with intermediate oil palm due to the reduction in canopies. Pruning is normally done for mature oil palm; therefore, the canopy became thin and caused surrounding soil reflectance to become dominant similar to the intermediate (4–10 years old) trees.

The inaccurate categorization of oil palm ages as produced in this study could be due to the large window size used to generate texture information. Although large windows produce reliable measurements within a specific class, they can reduce classification accuracy at the boundaries between different class textures (Ferro and Warner, 2002). As 29 × 29 windows are moved across boundaries, they analysed textures that were a combination of two adjacent textures that represent different ages of palm trees (Coburn and Roberts, 2004). The age blocks map (Figure 7.7) shows intermediate, productive, and mature tree blocks are intermixed and located adjacent to each other, thus may have contributed to a mixed texture detail. Another possible reason for the misclassification could be the canopy similarity although the planting age is different. Various agronomic and topographical issues such as inadequate nutrients, pest infestation, and steep versus low lying areas may affect the physical appearance of the trees. For example, due to nutrient leaching, trees planted in steep areas may have less dense canopy and shorter height, thus different structure compared to trees planted at a same time but in flat areas.

The spatial distribution of different age of palm trees classified in this study is shown in Figure 7.7 and the total area covered by each of the age class was calculated and compared with the statistics, derived from the map provided by the plantation management (Table 7.3).

The estimated total area for intermediate and mature oil palm trees showed only a minor difference compared to the data obtained from the plantation (with a percentage difference of 2% and 9%, respectively). However, the area of productive trees was significantly underestimated due to the misclassification of productive trees as intermediate and mature ones. Additionally, approximately 17 ha of the plantation are occupied by various built-up areas, including management office buildings and workers' settlements. It's worth noting that the total management unit observed in the

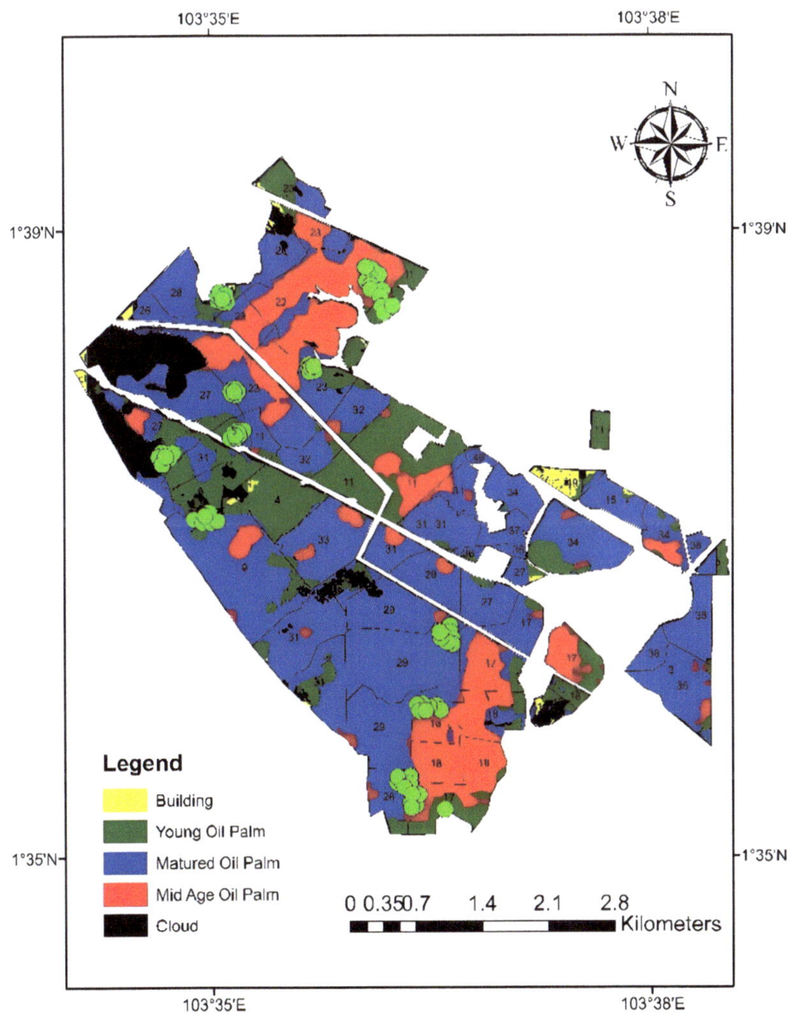

FIGURE 7.7 Age classification map produced using Sentinel 2 satellite data at Genting Kulai Besar plantation.

TABLE 7.3
Comparison of the Oil Palm Area Covering Different Age Groups as Estimated in this Study and that Obtained from Genting Kulai Besar Plantation Management

Oil Palm Age	SVM Area (ha)	Field Area (ha)	%Difference
Young oil palm	351	344	2.01
Middle age oil palm	363	645	55.95
Matured oil palm	1027	936	9.27
Buildings	17	-	
Total	1758.06	1924.67	9.05

satellite image is slightly lower than the statistics provided by the plantation management, partly due to cloud removal from the satellite image.

Age estimation of palm trees is crucial for optimizing plantation management, improving productivity, estimating biomass and carbon stock of palm trees, thus promoting sustainable practices in the palm oil industry. Both optical and radar data have been tested for their potential to predict oil palm age, but optical data are found to be performing better. Among various techniques used, selection of texture analysis with appropriate window size in classifying age is crucial and site specific. The heterogeneity of the site with various features present in the plantation influences the use of remote sensing data for discriminating various age classes. Since height of palm trees are strongly correlated with the age, future studies combining height data provided by spaceborne LiDAR data such as ICESAT and GEDI can be integrated with high-spatial and spectral resolution optical data may further improve the results.

8 Topographic Mapping

8.1 DIGITAL ELEVATION MODEL

Mapping and monitoring the Earth's surface is significant for understanding various environmental and social processes, and it is an inherently three-dimension problem (Elkhrachy, 2018; Yang et al., 2011). So, digital elevation model (DEM) data is an essential supplement for research and observation of Earth, especially the Earth's surface.

DEM is a three-dimensional computer graphics for representing constant terrain using limited elevation data. A DEM can be represented as a raster, which is associated with a two-dimensional coordinate and specific topographic height (Elkhrachy, 2018). The emergence of DEM is a significant addition to the traditional two-dimensional thematic mapping. The term "generated DEM" encompasses both the digital terrain model (DTM) and the Digital Surface Model (DSM). DTM refers to the elevation of the bare ground surface, while DSM refers to the elevation of the Earth's surface with land cover /land use. Hence, it is important to note that the elevation values in a DSM will typically be higher than those in a DTM when measured using the same reference standard. Consequently, researchers and observers should select between DSM and DTM based on their specific research or observation objectives. For instance, DSM may find utility in oil palm tree height detection, while DTM may prove valuable in tasks such as terrain mapping for oil palm plantation and flash flood mapping in oil palm plantations.

The sources of DEM data primarily include digitized topographic maps, field data, and digital aerial photographs or satellite images (Elkhrachy, 2018). Satellites offer an efficient means to generate global digital elevation model (GDEM) data, enabling the observation of Earth's surface characteristics comprehensively. Over the past few decades, numerous satellites have been launched due to significant advancements in Earth observation and sensor technologies. Prominent satellite missions such as Systeme Probatoire d'Observation de la Terre (SPOT), Advanced Land Observing Satellite (ALOS), TanSAR-X/TanDEM-X, European Remote Sensing Satellite (ERS), and Envisat have contributed substantial data volumes for topographical modelling and analysis. Regarding sensor technology, there are four main satellite-based systems for acquiring GDEM data namely Stereo Satellite Imagery, Synthetic Aperture Radar (SAR), Radar Altimetry, and Laser Altimetry.

8.2 SIGNIFICANT DEM PRODUCTS

The open access GDEM datasets are significant in global research and application, and are a great support for understanding and monitoring the Earth and its environment. Some advanced and widely used GDEM datasets include ASTER GDEM, ALOS AW3D30, SRTMGL1, and TanDEM-X 90m DEM.

DOI: 10.1201/9780429199813-8

8.2.1 ASTER GDEM

The Advanced Spaceborne Thermal Emission and Reflection Radiometers (ASTER) Global Digital Elevation Model (GDEM) stands out as one of the most widely utilized GDEM data products. This dataset was jointly released by the Ministry of Economy, Trade, and Industry (METI) of Japan and the United States National Aeronautics and Space Administration (NASA). ASTER is a sensor launched in December 1999 and mounted on NASA's Earth observation satellite, Terra.

The ASTER GDEM was created by utilizing synchronized optical stereo-pair images collected by the ASTER instrument. The initial version, ASTER GDEM v1, was made available in June 2009, boasting a resolution of 30m (1 arc-second). It covers a remarkable range from 83° north latitude to 83° south, encompassing 99% of Earth's landmass (Figure 8.1). In August 2019, the latest iteration, ASTER GDEM v3, was introduced, representing a significant enhancement compared to the initial release. This improvement includes the addition of extra stereo-pair imagery to enhance coverage and mitigate irregular elevation values and water area discrepancies. Alongside ASTER GDEM v3, the ASTER Water Body Dataset (ASTERWBD) was also released, marking the first-ever water body dataset with global coverage. It is worth noting that ASTER GDEM remains unparalleled in its completeness and high-resolution nature, making it available free of charge to all users (Abrams et al., 2015).

8.2.2 ALOS AW3D30

The Japan Aerospace Exploration Agency (JAXA) has introduced the ALOS Global Digital Surface Model (AW3D30), a valuable dataset. This dataset boasts a horizontal resolution of 30m (1 arc-second) and is available free of charge. It covers a substantial range, spanning from 82° north latitude to 84° south (Figure 8.2). AW3D30 was generated from imagery captured by the Panchromatic Remote-sensing Instrument for Stereo Mapping (PRISM) on board the ALOS, a prominent Earth observation satellite mission conducted by Japan.

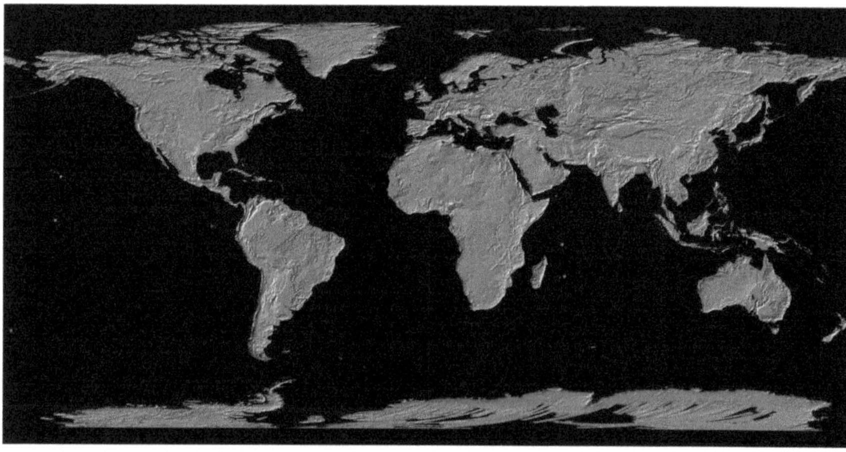

FIGURE 8.1 ASTER GDEM black and white map (NASA, 2004).

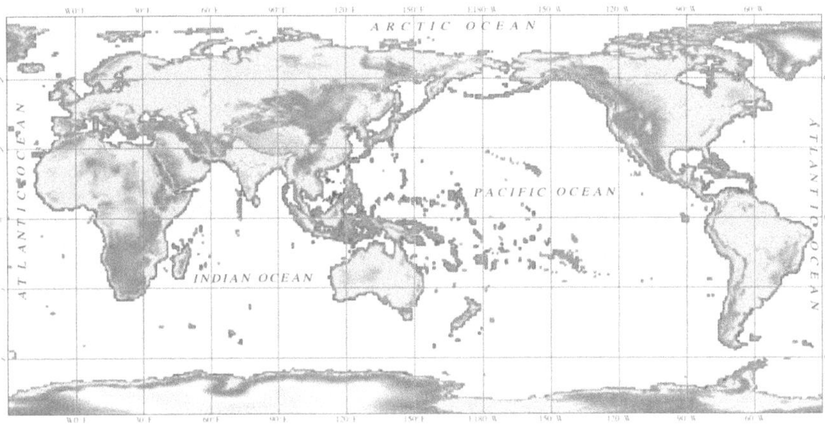

FIGURE 8.2 The coverage of ALOS AW3D30 (ALOS, 2023).

It is important to note that AW3D30 is resampled from AW3D, recognized as the most precise global DSM dataset with a resolution of 5 m. In some regions, even finer resolutions (0.5, 1, 2, and 2.5 m) DSM data are available. AW3D30 made its initial debut in 2016, albeit later than ASTER GDEM. JAXA has maintained a commitment to regularly release advanced versions of AW3D30, and by 2021, it had progressed to version 3.2.

Both AW3D30 and ASTER GDEM rely on optical stereo-pair images for their production. However, AW3D30 has demonstrated itself to be the most accurate quasi-global, medium-resolution land surface model, while ASTER GDEM is known for its higher error rates (Florinsky et al., 2018).

8.2.3 SRTMGL1

The global 1 arc-second Shuttle Radar Topography Mission V3.0 dataset (SRTMGL1) is a collaborative effort between the National Geospatial-Intelligence Agency (NGA) and NASA. The original dataset, SRTM v1, was derived from spaceborne Interferometric Synthetic Aperture Radar (InSAR) data collected during the mission of the Space Shuttle Endeavour, which was launched on February 11, 2000, and operated for 11 days. Initially, this near-global dataset was released at a resolution of 90 m (3 arc-seconds), with higher-resolution data covering the United States at 30 m (1 arc-second). This coverage extended from 60° north latitude to 56° south (Figure 8.3). Subsequently, a refined and aligned dataset, SRTM v2, became available in 2006. The development of SRTM version 3 involved the incorporation of additional data sources, including the ASTER GDEM, GMTED2010 (the Global Multi-resolution Terrain Elevation Data 2010), and USGS NED (National Elevation Dataset).

Starting in September 2014, the US Geological Survey (USGS) began distributing the SRTM plus dataset, which became globally accessible in 2015 (Florinsky et al., 2018). This enhanced SRTM data collection includes GDEM data products at three distinct resolutions: SRTMGL1 (with a resolution of 30 m or 1 arc-second), SRTMGL3 (with a resolution of 90 m or 3 arc-seconds), and SRTMGL30 (with a resolution of 1,000 m or 30 arc-seconds).

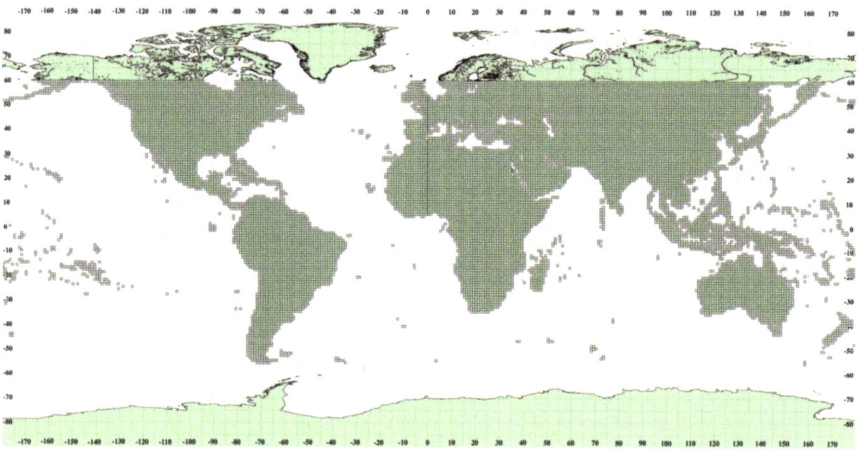

FIGURE 8.3 SRTMGL1 coverage (NASA, 2023).

8.2.4 TanDEM-X 90m DEM

TanDEM-X represents a collaborative Earth observation radar mission between the German Space Agency (DLR, Deutsches Zentrum für Luft- und Raumfahrt) and Airbus Defence & Space. TanDEM-X's high-resolution 12 m (0.4 arc-second) DEM data was generated from imagery captured by a dual-satellite system comprising TerraSAR-X (launched in June 2007) and TerraDEM-X (launched in June 2010). These satellites operated between 2010 and 2015, and the global dataset became publicly available in September 2016.

The primary offering is a truly GDEM with coverage spanning from 84° north latitude to 90° south, encompassing the entire landmass from pole to pole. It is worth noting that initially, this data was provided as a commercial product. Nevertheless, it stands out for its fine pixel spacing and comprehensive global coverage. Additionally, a standard version of the TanDEM-X DEM is accessible with 1 and 3 arc-second pixel spacing, derived from the 0.4 arc-second GDEM as listed in Table 8.1.

TABLE 8.1
TanDEM-X DEM Products Overview (Wessel, 2018)

DEM Product	Independent Pixel Spacing	Absolute Horizontal Accuracy	Absolute Vertical Accuracy	Relative Vertical Accuracy	Coverage
0.4 arc-second	0.4 arc-second (~12 m)	<10 m	<10 m	2 m (slope ≤ 20%) 4 m (slope > 20%)	Global
1 arc-second	1 arc-second (~30 m)	<10 m	<10 m	Not specified	Global
3 arc-second	3 arc-second (~90 m)	<10 m	<10 m	Not specified	Global

The TanDEM-X DEM data is accessible at three distinct resolutions: 0.4, 1, and 3 arc-seconds, which correspond to pixel spacings of 12, 30, and 90 m, respectively (Vassilaki and Stamos, 2020). Importantly, the TanDEM-X DEM product with 90-m resolution at the equator is freely available for users worldwide. Numerous scientists have evaluated the accuracy of the TanDEM-X 90m DEM across diverse terrains. The findings suggest that this DEM dataset exhibits a relatively high degree of accuracy in moderate terrains, such as urban and agricultural areas, but may tend to overestimate elevations in rugged or mixed terrains (Altunel, 2019; Bhardwaj, 2019).

8.3 DEM FOR OIL PALM MAPPING

DEMs serve as invaluable tools for gaining comprehensive insights into the topography and terrain characteristics of oil palm plantations and their surrounding environments. This data is fundamental for a holistic understanding of the landscape, encompassing critical factors such as slope, aspect, and elevation, all of which exert significant influence on various facets of oil palm cultivation. DEMs derived from remote sensing data offer valuable insights for oil palm studies, particularly in terrain analysis, hydrological modelling, site selection, and precision agriculture.

DEMs play a crucial role in the identification of suitable locations for oil palm plantations, enabling assessments of elevation, slope, and soil drainage. This information substantially enhances the prospects for the successful and sustainable cultivation of oil palm. In a recent study conducted by Cheng et al. (2018), the expansion of oil palm in Malaysia was examined using ALOS PALSAR data over a six-year period, employing the Random Forest machine learning technique to classify land use/land cover (LULC) categories, including oil palm, other vegetation, water bodies, and miscellaneous land types.

To ensure the accuracy of LULC classifications, it is imperative to correct for terrain-induced effects, such as topographic relief and slope, which can introduce distortions in the backscattering coefficients of certain land-cover types. DEMs serve as an essential tool for this purpose. Moreover, DEMs facilitate the precise identification of oil palm cultivation areas, as oil palms are typically not planted in regions exceeding 1000 m in elevation due to suitability considerations and reduced palm oil yield (Cheng et al., 2018; Morel et al., 2012). In the study, the SRTM 30-m DEM was employed to filter out oil palm mapping results in areas exceeding 1,000 m in elevation, mitigating terrain-induced confusion. Subsequently, these filtered areas were reclassified as other vegetation types.

The study's findings unequivocally highlight the significance of topography in oil palm cultivation. While the scale of oil palm plantations in Malaysia has substantially increased in the past decade, topographical constraints, such as altitude, slope, and slope direction, have curtailed further expansion into areas characterized by undulating terrain and steep slopes, where oil palm cultivation proves significantly less profitable (Abram et al., 2016).

Analysing suitable elevations for oil palm expansion and the distribution of 2007 oil palm plantations based on 100-m SRTM DEM data revealed key insights. The elevation histogram (Figure 8.4) indicated that oil palm plantations and expansion were concentrated in lower regions (0–300 m), particularly in Peninsular Malaysia,

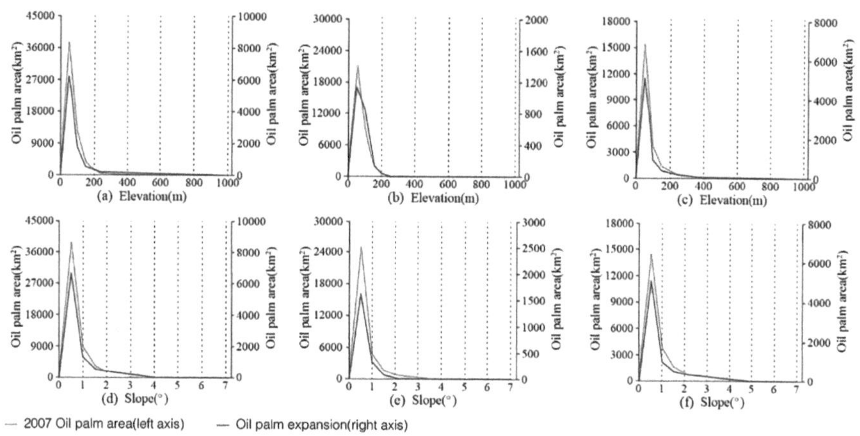

FIGURE 8.4 Frequency histograms of Malaysian oil palm distribution and expansion according to elevation and slope topography: (a) and (d) Malaysia (all 13 states), (b) and (e) Peninsular Malaysia and (c) and (f) Malaysian Borneo.

where oil palm plantations were prevalent below 300 m, with expansion primarily occurring between 0 and 200 m. However, some oil palm plantations were also identified at elevations exceeding 800 m, particularly in Malaysian Borneo. The slope histogram indicated that oil palm plantations typically occupied regions with slopes less than 4°, with expansion occurring in slopes ranging from 0° to 1° in Peninsular Malaysia but increasing to slopes less than 7° in Malaysian Borneo. Nevertheless, it is essential to note that despite the expansion to high-altitude regions (>800 m), most newly planted plantations remained confined to flat areas with slopes less than 7°.

The expansion of oil palm with respect to altitude and slope in two selected Malaysian areas is further illustrated in Figure 8.5. In Peninsular Malaysia (Figure 8.5a and b), where oil palm cultivation primarily occurs at lower altitudes and on flat terrain, the past decade witnessed expansion concentrated in low-altitude, flat terrain regions. In Sarawak, Malaysian Borneo (Figure 8.5c and d), characterized by undulating topography, oil palm expansion extended from low-altitude, flat terrain in the northwest to higher altitudes and rugged terrain in the east, even encompassing regions with elevations exceeding 800 m or slopes ranging from 3° to less than 7°.

8.4 DEM FOR FLOOD MAPPING IN OIL PALM PLANTATIONS

DEMs play a crucial role in the modelling of water flow, runoff, and drainage patterns within oil palm plantations. Precise hydrological assessments are essential for devising efficient irrigation and drainage systems, which are of paramount importance for the successful cultivation of oil palm. Terrain characteristics employed for topographic classification in oil palm plantations encompass attributes like slope, elevation, topographic wetness index, and the modified catchment area.

In a study conducted by Martinez et al. (2021) in an oil palm plantation located in the Colombian Llanos, DEMs sourced from SRTM were employed to model the terrain features and generate maps delineating soil properties. The research findings

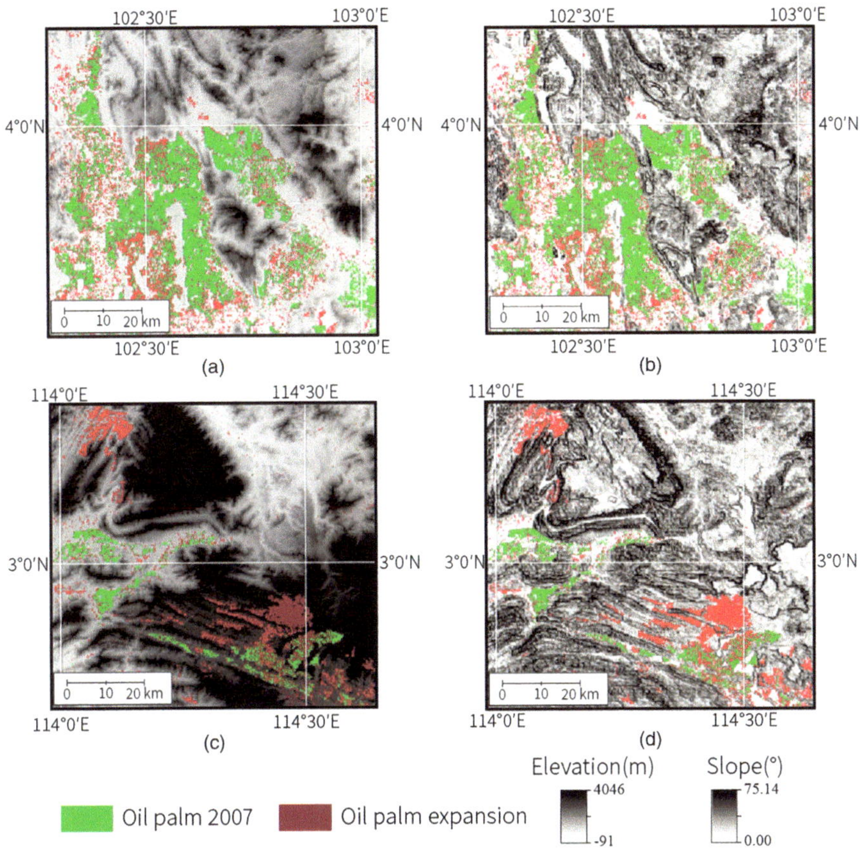

FIGURE 8.5 Elevation (a) and (c) and slope (b) and (d) limitations affecting oil palm distribution in Malaysia. Panels (a) and (b) depict the north of Bandar Pusat Jengka, Pahang, Malaysia, while panels (c) and (d) represent the east of Taman Tinggi, Sarawak, Malaysia.

unveiled a notable impact of the modified catchment area, which quantitatively characterizes the influence of topography on hydrological processes, specifically a grid cell's propensity to either retain or shed soil water. This discovery accentuates the critical role played by soil moisture distribution as a fundamental factor driving variations in yields within this particular agricultural system.

Flooding represents one of the prevalent disasters faced by oil palm plantations (Chong et al., 2017). However, the utilization of remote sensing technology to map flood-affected areas within oil palm plantations is not widespread due to its complexity and associated costs (Azam et al., 2021; Lupascu et al., 2020). Although the utilization of DEMs derived from LiDAR or drones/satellites prior to planting can serve as a valuable reference for monitoring flood-prone regions, this necessitates the collaboration of experts from various disciplines, including hydrology and geology (Chong et al., 2017).

The use of LiDAR data for flood mapping in oil palm plantations is a highly effective approach, as LiDAR provides high-resolution elevation data that can accurately depict the terrain and drainage patterns. GIS/spatial analysis tools are used to calculate flow accumulation, flow direction, and stream networks from the DEM. These parameters enable the understanding on how water flows across the plantation. Hydrological modelling software such as HEC-RAS can simulate flood scenarios based on rainfall data, DEM, and hydrological parameters, soil type, land use, and vegetation cover. Subsequently, flood hazard maps can be prepared based on the flood model results. These maps can indicate areas prone to flooding during different rainfall scenarios. Further, if the flood hazard model/map is integrated with real-time monitoring system that can track rainfall, and river levels, an early warning system can be established to notify plantation managers and workers. Usually, the early warning system is triggered when the rainfall and river levels exceed certain pre-set thresholds values. Based on the hazard maps and the early warning system, mitigation measures can be initiated such as constructing draining system and flood barriers to prevent excessive water to enter the plantation as flood is one of the serious environmental threats in many oil palm plantations in Malaysia and Indonesia. These countries may only see a marginal increase in palm oil output of 3% (as it was speculated in 2023) as floods will affect harvest, logistics, and lead to poorer fruit production (Bloomberg, 2023a).

DEM data plays a pivotal role in the management of oil palm plantations. It is utilized in terrain analysis to identify optimal planting areas, assess the growth status of oil palm trees, and mitigate soil erosion and water management issues (Avtar et al., 2020; Thenkabail et al., 2007; Muhadi et al., 2019). For instance, Muhadi et al. (2016) employed terrestrial laser scanners (TLS) to generate DEMs for the identification of suitable land for oil palm cultivation, aiding governments and farmers in making informed land-use decisions. DEM data can also be leveraged to analyse the ecological environment of oil palm cultivation areas, including changes in forest cover, biodiversity, and wildlife habitats. This information is invaluable for the protection and sustainable management of these ecosystems (Carlson et al., 2012). These applications provide vital support and serve as the foundation for decision-making to ensure the sustainability of the palm oil plantation industry.

Nevertheless, it should be noted that all the benefits described above are largely depending on the accuracy, resolution, and processing methods of DEMs. Low-quality and coarse-resolution DEMs may lead to inaccuracies in terrain analysis and modelling. When using satellite data to obtain DEMs, the dense canopy of oil palm trees can obstruct the view of the ground surface, making it challenging to obtain accurate elevation data. The use of active sensors like LiDAR that can penetrate through the canopies can partially overcome this limitation and provide highly accurate and fine spatial resolution data. However, acquiring LiDAR data requires high cost and technically challenging to process the data. Consequently, the data may not be readily available for all regions of interest and smallholders are not able to employ the data for their operational purposes.

There are several limitations associated with DEM data, including issues with limited resolution, data quality, and access costs. To address these challenges, a multi-source data fusion approach can be employed, combining satellite-derived DEMs, LiDAR data, and ground measurements. This fusion enhances both the accuracy and resolution of the data. Additionally, DEM data must undergo rigorous calibration and filtering processes to mitigate errors and reduce noise effects. GIS and terrain analysis tools should be utilized to process and analyse the data, extracting valuable insights. It is also imperative that DEM data be regularly updated to accurately reflect changes in the ground surface. In summary, DEMs are indispensable tools in oil palm studies, aiding in site selection, terrain analysis, and the mitigation of terrain-induced distortions, ultimately contributing to informed decision-making and sustainable land-use practices in this industry.

9 Productivity and Yield Forecasting of Oil Palm

9.1 SIGNIFICANCE OF OIL PALM YIELD FORECASTING

Oil palm productivity is a measure of the efficiency of palm oil plantations' output per unit area or per palm tree. The productivity of palm oil is typically measured in terms of yield, representing the mass of vegetable oil extracted from the palm trees' fruits. The oil palm tree has a single growing point located at the bottom of its woody stem (Figure 9.1a), from which leaves develop in a dual spiral pattern (Rees, 1964). Each leaf carries a solitary male and female flower inflorescence (Figure 9.1b), that usually develop from 14 to 18 months since planted (Corley and Tinker, 2008) and the trees begin bearing fruits in bunches approximately 3 years after planted (MPOC, https://mpoc.org.my/the-oil-palm-tree/). Each bunch of fresh fruits weighs between 20 and 25 kg and contains 1,000–3,000 individual fruitlets (MPOC). When ripen, the fruits turn from dark purple (Figure 9.1c) to orange-red colour (Figure 9.1d). Each fruitlet (Figure 9.1e) is consisted of two layers; the outer fleshy mesocarp and the inner, hard kernel or seed (Figure 9.1f). Crude palm oil (CPO) is extracted from the mesocarp, while the seed is crushed to produce palm kernel oil (PKO). The extraction of PKO is limited and its production is only about 5% (Carter et al., 2007). The fruit bunches and oil extracted from the fruitlets contribute to palm oil productivity and it is measured as the yield of fruit bunches in tonne per hectare (t/ha) and the palm oil extraction rate (%). The quantity and quality of palm oil yield produced is a major factor affecting food security, thus yield estimation and prediction is important.

Compared to other vegetable oils like soybean, coconut, rapeseed, and sunflower, palm oil stands out as an exceptionally productive crop. It yields large quantities of oil per unit of land area year-round, requiring only 0.26 ha of land to produce one tonne of oil. In contrast, other crops demand nearly 2 ha for the same output (https://www.wwf.org.uk/updates/8-things-know-about-palm-oil).

Despite the theoretical maximum yield potential of palm oil being estimated at 18.5 t/ha/year, the current global average productivity hovers around 3 t/ha/year (Woittiez et al., 2017). This stark contrast highlights the need to comprehensively grasp the environmental and management factors that govern palm oil yield and seek ways to enhance it.

The estimation of palm oil yield serves as the initial step in predicting and forecasting oil palm production, a crucial aspect for making economically sound decisions. These decisions range from replanting in times of low market prices to rescheduling planting when oil prices are high. Moreover, yield estimation aids plantations and smallholders in identifying and resolving issues linked to seeds, pests, and irrigation, ultimately boosting productivity and ensuring food security. Yield forecasting is also critical for operational and financial decision-making. It guides decisions regarding the

DOI: 10.1201/9780429199813-9

FIGURE 9.1 Various parts of oil palm trees.

required workforce, logistics for fruit harvesting, determination of mill facilities, and negotiations of CPO prices. Reliable estimates and forecasts offer significant benefits to agronomists, farmers, product exporters, and policymakers across the agri-food chain.

In order to curb the expansion of oil palm plantations into environmentally sensitive areas such as forests and peatlands, it is imperative to find sustainable methods to increase palm oil productivity. This emphasis on sustainable practices is vital for preserving high-conservation areas.

This chapter provides an overview of geospatial methods to estimate and forecast oil palm yield. Various environmental and management factors that affect oil palm yield are described below. It is important to understand the effect of the factors in order to accurately estimate and forecast oil palm productivity and yield.

9.2 PALM OIL YIELD

Generally, oil palm has an average of 25 years of lifetime although some private plantations in Malaysia have productive trees up to 30 years old. Palm trees are grown from seeds (Figure 9.2) and after a year, the grown seedlings are transferred to the plantations and provided sufficient water, sunlight, nutrients, and protected from pest. In the plantations, generally about 120–150 trees are planted within a hectare in a triangle pattern with 9 m spacing between the trees (Figure 9.2). In terms of yield, the oil palm trees can be categorized into four phases as follows (Woittiez et al., 2017):

i. Yield development (1–3 YAP)
ii. steep increase in yield (4–7 YAP)
iii. Little or no change in yield (8–14 YAP)
iv. Declining yield (15–25 YAP)

In the first phase, the trees are immature and only 2–3 years old after planted in the field. At this stage, the trees have open canopy and started to bear fruits and the yield is being developed. In the second phase, the trees started to produce fruit bunches that can be harvested and this occurs when the trees attain 4–7 years old. At this phase the yield and leaf area increase linearly with the age of the palm trees until 5–6 years old and the maximum leaf area index (LAI) is attained at 10 years (Tan et al., 2013). In this phase,

FIGURE 9.2 Seedlings in a nursery.

after 4 years of planting in the field, trees can produce fruit bunches up to 20 t/ha and on average 10–15 t/ha (Rao et al., 2008). When the trees are in the peak stage of production at the age of 6–7 years old, in some circumstances without any environmental and management constraints, a maximum yield of 60 t/ha can be achieved (Ng et al., 2003). From 8–14 years old, the yield does not increase rapidly but started to stabilize at 25–30 t/ha of fruit bunches and again this happens in plantations that are well taken care of and not constrained by any environmental factors (Donough et al., 2010). The dry matter/carbon production of these mature trees can range between 30 and 35 t/ha (Khoon et al., 2019). In the last phase (15–25 years old), the yield is found to be declining as the trees get older (Jacquemard and Baudouin, 1998).

9.3 FACTORS AFFECTING PALM OIL YIELD

The yield of palm oil is determined by the total number of fruit bunches, their weight, and the fruits' oil content which are influenced by environmental (solar radiation, temperature, rainfall, carbon dioxide (CO_2) concentration), and crop genetic characteristics. While, yield is limited by rainfall, soil, topography, waterlogging condition, and nutrients, it is reduced by weeds, pests, and disease (Woittiez et al., 2017).

9.3.1 Yield-Determining Factors

9.3.1.1 Solar Radiation

Solar radiation provides energy for photosynthesis that is essential for oil palm growth, fruit formation, and oil palm yield. Oil palm grows well in the tropics (10°N–10°S) that receive high mean daily solar radiation exceeding 10 MJ/m^2 or daily minimum sunshine hours of 5.5 hours (Paramananthan, 2003; Van Gelder, 2004; Syahrinudin, 2005; Carr, 2011). A high photosynthesis rate of up to 18–20 μmol CO_2/m^2s occurs at radiation level of 1,100 μmol for oil palm in Ivory Coast (Dufrene and Saugier, 1993), and Malaysia for 12–13-year-old trees (Henson, 1991). However, when radiation gets too high, it may increase the evapotranspiration rate and reduce rainfall which can dry out soil moisture and subsequently lead to yield reduction (Henson, 2000). The amount of radiation captured by the canopy is also depending on the leaf area and their orientation to the angle of radiation where a maximum of 95% radiation is intercepted when the LAI is 6–7 m^2/m^2 (Gerritsma, 1988) and yield may be reduced when the LAI exceeds this value as a consequent of canopy overlapping and competition among palms (Breure, 2010). Any reduction in solar radiation due to haze, dust, and forest fire can affect the productivity and yield of oil palm fruit bunches (Henson, 2000).

9.3.1.2 Temperature

A consistent warm temperature throughout the year of exceeding 24°C but below 32°C is important for optimum yield production (Paramananthan, 2003) while, low air temperature can suppress palm oil yield (Hartley, 1988). It has been shown that heat and drought due to El Nino suppressed the yield of palm oil in Malaysia as much as 9.55 billion USD since 1986 (Khor et al., 2021) and a 4°C increase in air temperature can reduce palm oil production by half (Sarkar et al., 2020). Thus, climate change may pose a threat to the industry if we fail to limit the increase in temperature to 1.5°C by 2030.

9.3.1.3 Carbon Dioxide (CO_2)

Plants that have C3 photosynthetic pathway such as the oil palm can increase their photosynthesis and thereby perhaps the yield under increased levels of atmospheric CO_2 as it has been shown for wheat (Fuhrer, 2003). A twofold increase in the photosynthesis of oil palm seedings was observed when CO_2 concentration was doubled (Ibrahim et al., 2010; Corley and Tinker, 2016); however, similar effect has not been tested for grown-up trees.

9.3.2 Yield-Limiting Factors

9.3.2.1 Rainfall

Oil palm is mostly a rain-fed cropping system where an annual rainfall of 1,800–5,000 mm/year is essential for the growth and optimum yield production (Corley and Tinker, 2016). Yield is reduced if rainfall is less than 1,800 mm/year (Corley and Tinker, 2015). The temporal pattern of rainfall largely affects the growth and yield of palm oil (Fleiss et al., 2017). Insufficient availability of moisture in the soil can affect the formation and quantity of fruits as a consequent of imbalanced male and female flowering ratios (Suharyanti et al., 2020). In Southeast Asia, a reduction in palm oil production up to 30% has been observed as a consequence of drought (Wen and Sidik, 2011). Dry spell also can increase the risk of fire that can degrade and damage the plantations (Fleiss et al., 2017). Even moderate water stress due to vapour pressure and soil water deficits can severely affect palm oil yield by delaying the growth of new leaves and affecting stomatal opening (Henson and Chang, 1990; Henson and Harun, 2005). Increasing air temperature can accelerate soil water loss and further intensify the impact of drought on palm oil yield (Fleiss et al., 2017). On the other hand, extreme rainfall can disrupt fruit harvesting that will reduce the yield.

9.3.2.2 Flooding and Waterlogging

Excessive amount of rainfall received at oil palm plantations is also harmful for trees planted in low-lying areas by inhibiting water and nutrient uptake and photosynthesis of roots and delaying frond opening and fruit production. Consequently, three quarter of young trees were found died and a yield loss of up to 30% occurred in mature oil palm plantations under water-saturated conditions in Peninsular Malaysia (Henson et al., 2008; Abram et al., 2014). Flooding of oil palm plantations due to heavy rainfall in the monsoon season is a serious environmental issue faced by many large plantations in Malaysia and Indonesia. Thus, this factor must be considered when estimating and or forecasting palm oil yield particularly in Southeast Asia. Drought and flood are expected to pose a high risk to the palm oil industry with climate change predictions in the 21st century. Latest studies have developed new indices that can predict the fresh fruit bunch (FFB) production using El Niño events that bring heavy rainfall in regions like Malaysia (Khor et al., 2021). Such yield prediction using rainfall is important to monitor fluctuations in palm oil price due to global supply shortage.

Data on soil moisture can be an important piece of information that can help to identify locations experiencing water stress and thereby take appropriate actions to improve palm oil yield. Remote sensing data can provide soil moisture at plantation/regional scale. The soil moisture and ocean salinity (SMOS) satellite's soil moisture product is generated with a spatial resolution of 27–55 km (Kerr et al., 2015) by using the brightness temperature (TB), physical temperature, and land surface reflectivity or emissivity data. The data have been calibrated for Malaysia and it can be used to assess the soil moisture of oil palm plantations (Kang et al., 2019). Although radar data provide useful soil moisture data, their usage is constraint by coarse spatial resolution. Therefore, these data should be combined with optical data to produce high-spatial resolution soil moisture product (Prakash et al., 2011) or downscaled to generate temporarily continued soil moisture for effective palm oil yield forecasting. Synthetic aperture radar (SAR) with multipolarization data are also useful for retrieving soil moisture information. Shashikant et al. (2023) processed the L-band ALOS PALSAR with HV polarization and LAI of palm trees as the main predictors in the water cloud model to retrieve soil moisture content in oil palm plantation in northern part of Peninsular Malaysia. The water cloud model and neural network (NN) models produced higher correlation and low root mean square errors when compared with the in-situ soil moisture data.

9.3.2.3 Soil

Soil type and soil water availability are important factors affecting the yield of palm oil. The amount of water available in the soil for palm trees' growth and yield production is determined by water received through rainfall, irrigation, and groundwater, and water lost from the ecosystem via evaporation, transpiration and surface water runoff. Palms grown on mineral soil have recorded higher yield (>30 tonnes of fruits/ha) compared to palms grown on peat soil because of the water-logging nature and leaned palm trees in peat soils (Paramananthan, 2013).

9.3.2.4 Topography

Oil palm planted on hilly areas exceeding slope of 10° can reduce the yield as a result of high surface runoff, soil erosion, loss of fertilizers, and insufficient water availability for growth (Paramananthan, 2003; Balasundram et al., 2006). Hamsa et al. (2019) noticed oil palm trees with similar age, but planted at steep areas had shorter stature due to nutrient leaching. Oil palm trees planted in a downhill side in a plantation in Johor, Malaysia, recorded lowest soil moisture levels due to the soil's reduced water-retention capacity (Kang, 2016).

9.3.2.5 Nutrients/Fertilization

Fertilization of oil palm plantations with large amount of potassium, magnesium, nitrogen, phosphorous, and boron is needed to increase the yield. Nitrogen, sulphur, and potassium are essential nutrients to increase the FFBs and oil yields, while phosphorus can increase the number of fruits and fruit bunch weight and yield (Yara, 2023). Large amount of magnesium is usually applied in plantations for frond development, to increase the number of fruit bunches, fruit weight, and to improve the fruit bunches' oil content (Yara, 2023). Calcium usage can ensure the prolonged health of palm trees. Boron is important for root development and also for fruit set. Copper and zinc are also used for palm trees planted on peat soils where zinc promotes new tissue development, and copper, manganese, and iron help to sustain the growth. However, when the nutrients level is exceeding the required quantity, it may exaggerate the palm growth which facilitates pest and disease attacks. These nutrients are supplied through chemical fertilizers and recycled via pruned fronds to counter balance the loss via various processes including fruit harvesting, runoff and leaching due to rainfall (Ng et al., 1999).

9.3.2.6 Plant Physiological Factors

Oil palm breeds that are resistant to extreme weather, pest, and disease and that have high light use efficiency can increase the yield. A new breed that can produce 9.9 t/ha (twice the current production) has been introduced by the Sime Darby plantation company via the genomics-based programme called 'Genome Select' (Murphy et al., 2021) and will plant this variety for all their replanting projects from 2023 (The Edge, Malaysia, 2023).

The number of trees planted per hectare is a crucial factor that determines the yield. A total number of 140–160 trees/ha have been found to be an optimum density for yield in oil palm plantations (Uexküll et al., 2003) because at this density when the trees get matured, there will not be high competition for light. A small percentage (1%–2%) of reduction in plantation yield has been noticed when there is a change of about 10 trees from the optimum number (Uexküll et al., 2003). Planting trees at high density (>160) and later thinning them to 120 when the trees attain 9–16 years after planting on peat soil can maximize the yield as it is practised in Thailand (Palat et al., 2012). At 130–150 trees/ha density and around the age of 10 years, oil palm trees attain highest leaf area (Tan et al., 2014), and this enables larger radiation absorption by the leaves for maximum photosynthesis, that is, net CO_2 assimilation of 64.5 t CO_2/ha year and biomass production of 36.5 t DM/ha year (Noor and Harun, 2004).

9.3.3 Yield-Reducing Factors

9.3.3.1 Pruning

Removing the senescing leaves/fronds is important to optimize resource availability and minimize carbon loss through respiration by deteriorating leaves in oil palm plantations which will increase the yield. In a study conducted by Marcelino and Diaz (2016) on 8-year-old oil palm trees in the Philippines for a period of 2 years, the authors observed significant reduction in terms of the number and weight of the fruit bunches. The optimum number of fronds is between 32 and 40 fronds per tree to produce better yield and any less or non-pruning will reduce the yield (Marcelino and Diaz, 2016).

9.3.3.2 Weeds

Yield is also affected by weeds, pests, and diseases. Although weeds function to protect the soil from erosion, unmanaged weeds can grow wild and compete for nutrients, radiation, and water particularly with young palm trees, thereby reduce the productivity and yield. Leaving some legumes to co-exist with palm trees is advantageous because they can supply nutrients to palm trees by fixing nitrogen (Pardon et al., 2016).

9.3.3.3 Insects

Common insects in oil palm plantations that eat the leaves and reduce the LAI are bagworm, nettle caterpillar, and leaf miner. Rhinoceros beetle, for example, can delay the growth of young trees and occasionally kill them (Manjeri et al., 2014).

Consequently, the yield was found to be reduced up to 50% during the first year of defoliation occurred (Wood et al., 1973). Oil yield is reduced by rats that eat the fruitlets and in mature plantations, up to 240 kg oil per hectare annually is lost due to rats and this is about 5% of loss (Wood and Chung, 2003).

Another major reason for yield lost in oil palm plantations is the pathogenic fungi called *Ganoderma boninense* (Paterson, 2007). In Malaysia and Sumatera, Ganoderma is a serious threat to oil palm industry where this disease caused 80% of mortality in older palm trees (>15 years old) (Idris et al., 2004). Chapter 12 of this book discusses specifically the issue of pest and disease and their detection using geospatial technology. Please refer to Paterson (2007) and Woittiez et al. (2017) for a comprehensive review of factors affecting the yield and productivity of oil palm trees.

The condition of oil palm plantation may be heterogeneous depending on soil and topography, thus incorporating environmental, agronomic, and management factors into the yield estimation, and forecasting models might provide precise outcome rather than considering any one category of the factors (Chapman et al., 2018). Climate and agronomic are dominant factors determining the prediction of oil palm yield (Chapman et al., 2018; Hilal et al., 2021; Ahmed et al., 2021; Khor et al., 2021). The potential yield of oil palm is mostly reduced by water and nutrient limitations where water limitation can reduce the potential yield by one third, while 50% of the potential yield is reduced when oil palm trees are limited by nutrient applications (Woittiez et al., 2017). Further total reduction in yield could happen under severe pest and disease infection (Woittiez et al., 2017). Identifying and correcting these factors can help to increase the yield and most importantly can restrict oil palm expansion into forest and other environmentally sensitive and high-biodiversity areas.

9.4 PALM OIL YIELD ESTIMATION

Typically, palm oil yield is estimated using ground sensing platforms that are hand held, free standing, and mounted on tractor that collect field-scale data such as nutrients content in foliar and soil, and tree biometric data covering the size, height, and leaf density. These data are usually collected on and off and they may be subject to statistical biases (Shamshiri et al., 2018) and consume more time and cost to cover large plantations. Various mathematical models ranging in complexity from simple correlation methods to complex eco-physiological and land surface models that couple vegetation, land surface, and atmosphere through energy, mass, and momentum exchanges have been used to model plant productivities (Pitman, 2003). The models developed for estimating palm oil yield/productivity are described in the following section.

9.4.1 Mechanistic Models for Estimating Palm Oil Yield

The first model developed to simulate the growth and yield of palm oil was the oil palm simulator (OPSIM) by Van Kraalingen in 1985. It was a semi-mechanistic model that considered weather and plant physiological factors such as solar radiation, vapour pressure deficit (VPD), soil water, temperature, LAI, light extinction coefficient, and light saturation of photosynthesis. The model simulated gross photosynthesis of oil palm trees particularly trees aged between 6 and 12 years after field planting. A sensitivity analysis conducted with the model highlighted that any increase in crop photosynthesis resulting from improved husbandry or genotype would lead to a proportional increase in bunch yield without significant influence on vegetative growth. However, the model was criticized for relying heavily on empirical data (Henson, 2000). Subsequent models further improved OPSIM by incorporating more measured data. Dufrene et al. (1990) enhanced the model by integrating measured data of oil palm vegetative parameters. The SIMPALM model, developed by Dufrene et al. (1990), successfully estimated CO_2 intake by oil palm in Africa and Southeast Asia (Lamade and Bouillet, 2005). Henson (2009) gradually improved OPSIM by incorporating additional effects and factors related to oil palm growth and yield, including seasonal yield cycles, VPD, and available soil water. His model is known as OPRODSIM and it also utilized fitted relationships based on measured data from various oil palm studies to better estimate oil palm root turnover, dry matter partitioning, and flower sex ratios. Later studies such as that of Huth et al. (2014) recommended to include a vegetative development component in the model to accurately capture yield and productivity.

Since then, new oil palm models have emerged, such as WaNuLCAS (Van Noordwijk et al., 2011 in Teh and Cheah, 2018a), ECOPALM (Combres et al., 2012), and PALMSIM (Hoffmann et al., 2014). More recent models like APSIM-Oil Palm (Huth et al., 2014), CLM-Palm (Fan et al., 2015), and CLIMEX-Oil Palm (Paterson et al., 2015) are components of larger, more comprehensive modelling frameworks. For instance, APSIM-Oil Palm is one of the sub-models within the APSIM framework, which covers various crops, trees, and pastures.

A more recent model, PySawit, was developed by Teh and Cheah (2018) to simulate oil palm growth and yield. It utilized actual oil palm leaf measurements and a biochemical photosynthesis model to estimate changes in maximum Rubisco capacity rate and intercellular CO_2 concentration for leaf photosynthesis. PySawit also considered microclimate within and under the canopies to account for its impact on yield and productivity. It demonstrated good accuracy in predicting total dry matter, LAI, and trunk height at various planting densities. However, it had limitations in capturing microclimatic changes under very dense canopies, resulting in poor yield simulation in such conditions.

Another recent study by Xu et al. (2021) represented oil palm in a land surface model called ORCHIDEE-MICT. This model simulated oil palm growth, biomass density, and yield, as well as carbon, nutrient, water, and energy fluxes and storage pools. It incorporated oil palm functional traits, phenology, and management practices such as pruning, fruit harvest, and rotation. The model demonstrated good agreement with observations in terms of LAI, biomass density, and fruit yield at various sites. It provided a useful tool for understanding the ecological processes of oil palm growth and assessing the environmental impacts of oil palm plantations.

In summary, various oil palm models have been developed over the years, starting with OPSIM. These models have evolved to incorporate more factors and effects to enhance their accuracy in simulating oil palm growth and yield. Newer models like PySawit and the representation of oil palm in global land surface models offer improved capabilities and insights into oil palm productivity, carbon sequestration, and water dynamics, supporting the sustainable development of the palm oil industry.

9.5 REMOTE SENSING FOR YIELD ESTIMATION

Often modelling exercises focus on simulating detailed processes and functional changes within particular vegetation types, thus they focus on a point or small patch scale and on a limited time. However, by doing so they ignore the spatial heterogeneity in vegetation, the structural pattern, and the possible effects of vegetation redistribution and associated productivity change. Remote sensing techniques on the other hand capture spatial variation and depict the temporal changes in crops' condition to assess and predict the yield. As stated earlier, the FFBs are the yield and they are hidden under the crown. Thus, remotely sensed data cannot be used directly to see and estimate the yield of FFBs. On the other hand, indirect methods are used to observe the trees' healthiness and canopy denseness which are indications of the yield.

Generally, remote sensing data are used for estimating crop yield by (i) deriving biophysical parameters such as LAI which is used in crop models and (ii) correlating LAI and other vegetation indices calculated using spectral information with crop yield via empirical/regression models. The empirical models are then scaled up to estimate crop yield covering large spatial extent. Spectral indices are often used to represent the coverage, density, healthiness, and growth dynamics of vegetation. Healthy leaves generally strongly absorb visible light (i.e., red and blue) and reflect near infra-red radiation and vice versa. This sharp contrast in absorption and reflection is used to calculate various vegetation indices (described in the following section) that can be used to measure crop growth, chlorophyll content, and biomass. High vegetation index values indicate healthy and dense leaves/canopy which is correlated with high yield and vice versa.

9.5.1 Vegetation Indices to Estimate/Predict Palm Oil Yield

The strong correlation between biophysical parameters such as LAI and vegetation indices calculated from remote sensing data and crop yields observed in the field lays a foundation to predict palm oil yield. Balasundram et al. (2013) calculated vegetation indices from high-spatial resolution QuickBird satellite images to estimate palm oil yield covering a period of 12 years. A simple linear regression was constructed between the ratio vegetation index[1] and modified soil-adjusted ratio vegetation index (MSAVI)[2] and yield data (FFBs in t/ha) obtained from 56 management blocks in the field in Peninsular Malaysia. These indices were used to take into account of the confounding effects due to soil (young trees) and canopy shadow (matured trees) that would otherwise result in errors. Empirical models developed using the vegetation indices could estimate the yield well perhaps because yield is connected to LAI and it is reflected in the vegetation indices.

Vegetation indices that reflect water content in oil palm canopy such as the normalized difference moisture index (NDMI)[3] and moisture stress index (MSI)[4] calculated using Landsat Thematic Mapper images were able to simulate oil palm fruit bunches yield in central Costa Rica with satisfactory accuracy with machine learning models (Watson-Hernández et al., 2022). The random forest (RF) model predicted one type of oil palm species (AVROS) well ($R^2 = 0.82$ and RMSE 0.3452 t/ha) while, the NN model predicted multiple species better with $R^2 = 0.82$ and RMSE 0.3452 t/ha.

NN and RF techniques were able to predict palm oil yield using NDMI for lag periods of one and two years. For a long-term lag period (>3 years), the influence of humidity was negligible, but chlorophyll activities represented by vegetation indices such as enhanced vegetation index (EVI),[5] green chlorophyll vegetation index (GCI),[6] and normalized pigment chlorophyll ratio index (NPCRI)[7] showed the best yield relations. The palm oil yield was estimated in this study just based on satellite images alone without any other plant physiological, environmental, or meteorological data from the site but with advanced machine learning techniques. This shows the importance of remote sensing data to get palm oil yield at large spatial extent even when field data are not available.

Advanced machine learning–based statistical techniques are powerful to describe the relationships between various input parameters and yield better. Machine learning algorithms are also time efficient in processing large amount of data especially data generated by Earth observation satellites and capable to describe complex associations between various parameters influencing yield. Nevertheless, machine and deep learning techniques have not been applied widely to predict the yield of palm oil, although their usage is common to predict/forecast crop yields such as maize and grape using vegetation indices derived from satellite images (Aghighi et al., 2018; Arab et al., 2021). Advanced machine learning techniques, however, have high potential to predict palm oil yield (Rashid et al., 2021; Tan et al., 2022).

As described in Section 9.3, the palm oil yield is affected by a combination of plant physiology, environment, weather and climate, and geological and management practices. Thus, besides vegetation indices derived from satellite images, land surface temperature (Landsat 7), rainfall (CHIRPS satellite and ground data), yield, and age (6–28 years) data covering 2005–2018 from 23 blocks and agronomic data covering interpolated leaf nutrients were integrated by Ang et al. (2022) to simulate the yield of palm oil in Peninsular Malaysia. Various statistical models representing a simple linear to more complex machine and deep learning models were attempted to predict the block scale palm oil yield. The GCVI and green normalized difference vegetation index (GNDVI)[8] could explain about 25% of the variation in the palm oil yield. These indices show the influence of plant disease, irrigation, and fertilization on palm oil yield. Any vegetation indices calculated using the green band of satellite images such as GCVI and GNDVI could capture canopy nitrogen content, thus fluctuations in leaf chlorophyll. The deep learning technique (DNN) provided better results in capturing the complex relationship among palm oil yield and meteorology, agronomic, and plant physiological factors. The power of DNN was also proved in other crops' forecasting such as barley, soft wheat and durum wheat (Meroni et al., 2021), and potatoes (Piekutowska et al., 2021). Meroni et al. (2021) particularly used growth-affecting factors (NDVI, precipitation, temperature, and global radiation) derived from satellite data in their forecasting model.

Multisource oil palm growth and yield-affecting factors covering weather and soil moisture (a total of 17 factors) were integrated into several machine learning models by Khan et al. (2022) to predict palm oil yield. Extra Tree and AdaBoost (Adaptive Boosting) models produced higher correlation coefficient values ($R^2=0.61$–0.63) compared to several other tree-based and conventional regression methods such as multiple linear, k nearest neighbour, and several other models (Khan et al., 2022). AdaBoost is a boosting technique in machine learning that automatically assigns higher weights to incorrectly classified instances to solve nonlinear relations among the input parameters. The Extra Tree (extremely randomized trees) technique can improve nonlinear regressions by separating the main data into many subgroups and derives prediction model from each subgroup separately. The final prediction model is then obtained by combining and averaging the models from each subgroup. Results of the study found that soil moisture (root zone wetness), weather (number of rainy days, rain amount), and wind speed were major predictors of palm oil yield in the state of Pahang in Peninsular Malaysia. The study also concluded that machine learning techniques are effective in addressing the complicated relationship among weather, soil moisture, and palm oil yield.

Advanced machine learning–based statistical techniques are powerful to describe the relationships between various input parameters and yield better. Machine learning algorithms are also time efficient in processing large amount of data especially data generated by Earth observation satellites and capable to describe complex associations between various parameters influencing yield. Nevertheless, machine and deep learning techniques have not been applied widely to predict the yield of palm oil, although their usage is common to predict/forecast crop yields such as maize and grape using vegetation indices derived from satellite images (Aghighi et al., 2018; Arab et al., 2021). Advanced machine learning techniques, however, have high potential to predict palm oil yield (Rashid et al., 2021; Tan et al., 2022).

Nevertheless, machine and deep learning techniques can only provide reliable results when large amount of data is fed into the models. Consequently, high computational, memory, and network powers are required to run the models (Liu et al., 2021). Alternatively, cloud computing and edge computing are used to store large amount of data and process them relatively quickly (Chen and Ran, 2019). Edge computing is preferred over cloud computing nowadays due to its sensitivity to time-driven data, enabling super-fast decision-making. It can also be used in remote locations without or with limited connectivity to a centralized location (Chen and Ran, 2019).

The age of oil palm trees is also an important parameter determining the yield. Age was related to yield by a sigmoid nonlinear regression model (Khamiz et al., 2005). Thus, deriving age information from satellite data is important for defining the yield (see Chapter 7) by knowing the total area planted with oil palm.

Multipolarization radar data can be useful to estimate oil palm yield although no studies have been conducted so far. The HH and HV polarizations from Radarsat-2 were manipulated to predict wheat grain yield in Tunisia using empirical statistical models (Barbouchi et al., 2022). Results of the study showed that the ratio of the polarizations produced better results compared to the polarimetric parameters. A combination of the ratio and Alpha (polarimetric parameter) produced best predictor model for wheat grain yields (Barbouchi et al., 2022). Exploration of the application of radar for palm oil yield will be advantageous as oil palm is mainly planted in the tropical region that is frequently covered by clouds. Radar is capable of penetrating clouds and provides useful polarimetric information for estimating palm oil yield.

9.6 PRECISION OIL PALM FARMING/AGRICULTURE

Improving crop quality, increasing productivity, and enhancing profitability by reducing labor time, optimizing resources (such as fertilization and irrigation processes), and minimizing environmental impacts are the primary goals of the palm oil industry. In order to achieve this, high technology and analysis tools are required and this is known as precision farming. Precision farming requires large amount of data acquired using advanced information techniques and processed with innovative analytical methods and communicated to farmers and other decision-makers via fast communication tools (Tan et al., 2022). The data/information produced are used in the decision-making process of sowing seeds, irrigating, fertilizing, applying pesticides, weeding, harvesting, fuel usage, and labour applications. So, precision agriculture is a high-technology system involving information and communication technology and intense capital to grow food in a sustainable manner.

Precision agriculture applies emerging technologies such as geospatial (remote sensing, GIS, and GPS), advanced big data analysis, artificial intelligence (AI), machine learning methods, internet of things (IoT), cloud computing, and communication techniques to enhance the productivity. These technologies are used in precision farming to increase the productivity by optimizing inputs and reducing wastages. GPS technology allows farmers to collect locational data in real time for various applications, including detecting problematic plants, mapping topography and infrastructure in the field, and controlling automated machinery for seed and fertilizer applications, among others. GIS technology is useful for generating 2D or 3D maps and visualizing crop data and environmental conditions in a spatially explicit manner. Remote sensing that produces large amount of digital data (big data) at high-spatial (i.e., sub meter to 5 m), spectral (hyperspectral), and temporal (daily to several days) resolutions are used in precision farming (Huang et al., 2018). These data are stored in cloud computing systems and processed using advanced machine learning techniques to extract information for precision farming such as moisture level, nutrient status, and health condition (Kamilaris et al., 2017; Pavón-Pulido et al., 2017).

9.7 HIGH-RESOLUTION REMOTE SENSING DATA FOR PRECISION OIL PALM

Remote sensing data with high-spatial resolution ranging from submetre to less than 5 m is essential for describing even minor variation in yield across the plantation. One of the early high-spatial resolution satellite data is IKONOS which was launched in the late 1990s. The sensor captures data at blue, green, red, and infrared bands at 4 and 1 m (panchromatic), and the data were used in yield prediction of wheat (Enclona et al., 2004). Its revisit period of 5 days is important for palm oil yield monitoring in the tropical region that has high cloud coverage. Other latest satellites that can be utilized for crop yield monitoring and estimation include GeoEye, Pleiades, Worldview, SkySat, and Superview-1. These satellites not only provide high-spatial resolution data of <5 m but also have daily revisit period and contain visible and near-infrared spectral bands for calculating vegetation indices to estimate yield.

High-spatial resolution remotely sensed data become important in large-scale oil palm plantations to monitor their areal changes, individual tree counts, health condition, and water stress (Putra et al., 2022; Putra and Wijayanto, 2023). Putra and Wijayanto (2023) demonstrated the use of very high-resolution satellite data (Microsoft Bing Maps) with 0.3 m spatial resolution and UAV data for individual oil palm tree mapping in Riau, Indonesia. Advanced machine learning models such as YOLO (You Only Look Once) was used to process the satellite and UAV images for automatically detect palm trees. This object-based deep learning model yielded higher accuracy (F1-score of more than 90%) compared to other threshold methods used in the study. Another newly launched satellite by the European Space Agency in 2015 is the Sentinel series that provide optical and radar data at moderate spatial resolution (10–60 m) but with high revisit period. These data have been used to discriminate oil palm age (Chapter 7) and have the potential for yield estimation (Mohd Najib et al., 2020).

Nevertheless, these high-spatial and temporal resolution satellite images are still expensive and become impossible to be used by small oil palm holders. They are also less preferred by large plantation companies due to clouds obstruction and limited accessibility to the images (not as frequent as they are needed for decision-makings). Alternatively, low-cost remote sensing techniques such as unmanned aerial vehicles (UAV) became popular since the last decades.

UAVs can be flown as it is necessary and more frequent flights can be made during early stage of oil palm growth to monitor their health status, to apply nutrients, to map weeds, and to spray pesticides. The submetre resolution achieved from flying at lower altitudes is essential for various management options. The multispectral, hyperspectral, and thermal sensors mounted on UAV platform can provide lots of data to manage the plantation effectively and to obtain yield-related information such as health condition and fruit bunches count. Sensors mounted on UAV provide very high-spatial resolution (cm) images taken from different angles that can be used to estimate the yield of a single oil palm tree. Areas/locations that show less yield can be supplied with fertilizers, water, or pesticides to increase their yield. Machine vision algorithms can be used to estimate the weight of FFBs and number of fruits in each palm.

UAVs are deployed by large plantation companies such as Sime Darby, Genting Plantations, and Wilmar in Malaysia and Indonesia for their operational purposes, including yield data collection. They offer the flexibility to acquire images as needed, even under cloudy conditions, ensuring cloud-free images, which is crucial for yield monitoring. Multispectral optical camera on UAV has been demonstrated to be effective in detecting nutrient content of oil palm leaves in Indonesia (Santoso and Winarna, 2021). The visible and near-infrared bands were used to calculate vegetation indices such as simple ratio, NDVI, and green NDVI that were then correlated with leaf nutrients data collected in the field via regression analysis. The best model was developed using polynomial multivariant regression, yielding correlation coefficient values between 0.72 and 0.98 for N, P, K, and Mg. These models can be applied to larger plantation areas to upscale the leaf nutrient status and to recommend fertilization requirement for increasing yield/productivity. Algorithms such as the nitrogen fertilizer optimization was used to convert the vegetation indices calculated from remote sensing images to nutrient application rates (Sishodia et al., 2020).

UAVs have been used for identifying and mapping weeds in other crops, such as in rice fields (Huang et al., 2018). Advanced image processing techniques using machine learning approaches were found to be detecting weeds at centimetre scale resolution with high accuracy (Huang et al., 2018; Esposito et al., 2021; Anderegg et al., 2023). A weed pressure index was calculated from drone images and the weed pressure map generated provided spatial distribution of weeds in the field and percentage of weed in each pixel. This information could guide the farmers to employ appropriate spraying techniques which could save time and efforts, manage the plantation efficiently, and most importantly contribute to higher yields (Precisionhawk, 2017).

In addition, thermal, infrared, hyperspectral, and LiDAR sensors are also mounted on UAVs that can provide data for palm oil yield estimation. Husin et al. (2021) determined the level of oil palm fruit bunches ripeness into under ripe, ripe, and over ripe based on the light intensity returned from the FFB to the LiDAR sensor and mapped their spatial distribution and precise locations. Determining the ripeness of the FFB is important to harvest the fruits on time and to avoid any wastage due to late harvest of ripe or over ripe fruits. Thermal infrared sensors mounted on UAVs measure energy emitted from oil palm leaves/canopy and soil surface to estimate their temperature. Information on crop temperature is important to assess crops' water stress, evapotranspiration, and irrigation needs. UAVs are extremely useful to identify areas of palm plantation that are affected by water stress and manage them promptly by irrigating on time.

9.8 ARTIFICIAL INTELLIGENCE (AI)

AI methods play a pivotal role in emulating human intelligence, encompassing thinking and behavioural capabilities, particularly in the management of diverse activities, including agricultural farming (Ampatzidis et al., 2017). One noteworthy application of AI is machine learning, wherein machines such as computers and microcontrollers undergo training to glean insights from extensive datasets. Consequently, they can discern patterns within the data and adapt to new information autonomously, eliminating the need for explicit programming.

In contemporary agriculture, machine learning techniques find practical utility in processing substantial datasets to extract pertinent crop and environmental information. This encompasses aspects like crop yield, soil moisture levels, evapotranspiration, and disease detection (Garcia et al., 2019). These insights are instrumental in automating farm and plantation management tasks, such as precise application of water, fertilizers, and herbicides or pesticides.

The use of autonomous machines and robots is at a preliminary/testing stage in the industry. The loose fruit collecting machine is one example to automate the collection of loose fruits which would otherwise requires large amount of time and manpower (Yusoff, 2019). Some large private plantations, such as Sime Darby, have introduced mechanization to automate the application of fertilizers and harvesting the FFBs (https://www.freemalaysiatoday.com/category/highlight/2022/08/22/getting-more-done-with-fewer-workers/). The plantation company also has adopted geospatial technology, i.e., satellite imageries to detect palms having lower yields.

9.9 INTERNET OF THINGS (IoT)

In the realm of IoT-based precision farming, communications between automated hardware such as sensors, robots, self-directed vehicles in the field and control systems occur through the internet. This interconnectedness enables easy and fast data exchange and farmers gain the ability to remotely monitor their fields without the need for physical presence. Examples of IoT technology that uses wireless sensors, big data, and cloud computing in precision farming are irrigations operated by automated wireless system and smart system to monitor and predict disease and pest (Garcia et al., 2019; Jha et al., 2019; Elijah et al., 2018). Delgado et al. (2017) showcased a prototype of an automated irrigation system for oil palm plantations in Colombia, integrating weather predictions and soil moisture data from a low-cost sensor into a predictive control model for irrigation scheduling. The system employed an Arduino UNO board in conjunction with a Raspberry Pi 2 IoT-based watering systems. The utilization of Arduino UNO boards has gained widespread adoption in agricultural contexts (Garcia et al., 2019). These systems empower farmers to irrigate palm trees selectively, activating watering mechanisms solely when soil moisture levels dip below acceptable thresholds. Implementing IoT for irrigation in the palm oil industry proves effective in water and cost savings, and optimizing yields (Ortega-Reig et al., 2017; Nikkels et al., 2019).

The use of smartphones and mobile applications in precision agriculture can benefit the industry for increasing productivity and profitability. They are used to collect biophysical data of oil palm and environmental data, track the performance of machines, and control the operation of drones. A mobile application for identifying and geolocating nutrient deficiencies in oil palm plantations was developed by Culman et al. (2017). The application enables plantation workers to use mobile phone with Android system to capture images of oil palm trees particularly the leaves and classifies the image into different categories of healthiness based on nutrient deficiencies. The time of occurrence of the issue (timestamp) and GPS coordinate are shown in the screen of the smart phone. Smartphone cameras serve as a valuable tool for capturing images of FFBs and analysing their varying ripeness levels through the application of machine vision algorithms (Septiarini et al., 2021). These algorithms, which are computational techniques enabling cameras to interpret and comprehend visual information from the surrounding environment, were employed for this purpose.

By connecting the mobile phone to internet, all the data can be stored in a cloud platform and the historical data can be visualized and used by plantation owners to manage nutrient deficiencies and predict the yield of oil palm for every tree. When the mobile application is connected with other devices such as sensors monitoring soil moisture, an IoT will be enabled for automation of oil palm management. Mobile applications are used as an important instrument in precision agriculture.

The combination of IoT sensors and satellite imageries enables digital twins (digital replica of a physical object) to remotely monitor oil palm plantations (Peladarinos et al., 2023). This is particularly useful for large or geographically dispersed plantations, allowing for real-time oversight and immediate response to issues. The geographical position of each tree recognized from satellite imageries and with GPS can be used to create a digital twin where the trees' lifecycle can be represented virtually and updated from real-time data. Further physiological features of the trees such as their leaf area and canopy condition can be retrieved from drone images and updated continuously. Other data that also influence the yield, such as weather data and soil information, can be connected to each tree. With simulation models and artificial intelligent, each tree can be monitored for disease and nutrients deficiency and subsequently manage efficiently by supplying with fertilizer, pesticides, and water (Culman et al., 2017).

For a successful operation of IoT in oil palm agriculture, communication technology is of paramount important to send information to the farmers and decision-makers. Newly emerging low-power wide area network technologies (LoRa) are highly preferred in the agriculture sector (Garcia et al., 2019). IoT platforms depending on 5G will be challenging in rural oil palm areas where there is low signal for communication. Alternatively, LoRa technology that is similar to 5G is able to work in remote areas. Ruslan et al. (2021) utilized LoRa module and other components such as pH and moisture sensors to develop a prototype to monitor soil properties in oil palm plantations in Malaysia. Soil and its properties such as temperature, texture, density, porosity, and nutrients content have profound effects on oil palm growth, health, and yield production. Technology to detect and monitor these properties especially the nutrient level continuously is important to ensure the preparation of lands, seeds growth, and health of oil palm trees for high yield.

The integration of geospatial technologies with IoT-based precision farming and AI empowers farmers to analyse the spatial variations within their farms comprehensively. This, in turn, enables them to proactively address issues and enhance crop yield (Koch et al., 2004). The practice of precision can diminish reliance on manual labour, optimize the usage of water, nutrients, and pesticides, and consequently reduce the operational costs. With this, palm oil yield at plantation scale can be increased worldwide more than the current annual rate of ~3 tons of oil per hectare with reduced operational cost. Nevertheless, full automation or implementation of IoT in the palm oil industry will take longer time (Mcconnell, 2019; Pathak et al., 2019). High cost involved in implementing a complete IoT system, demand for high computational power for advanced image processing and analytical techniques such as deep learning and IoT technology, lack of expertise to run and maintain the system, stable and speed internet connection, and resistant to change among farmers are among the factors that hinder the adoption of the technology in palm oil–producing countries (Ahmad Safwan and Zareen, 2019; Shashikant et al., 2023). A solution to these issues lies in the adoption of emerging edge intelligence technology which is still in the early stage of development (Hua et al., 2023).

NOTES

1 Ratio vegetation index = Near infrared/red.
2 Modified soil-adjusted ratio vegetation index = **(2 × NIR + 1 − sqrt ((2 × NIR + 1)2 − 8 × (NIR × R)))**.
3 Normalized difference moisture index = (NIR − SWIR)/(NIR + SWIR).
4 Moisture stress index = SWIR/NIR.

5 **Enhanced vegetation index = G × ((NIR – R)/(NIR + C1 × R – C2 × B + L))**.
6 Green chlorophyll vegetation index = NIR/Green – 1.
7 Normalized pigment chlorophyll ratio index = (Red – Blue)/(Red + Blue).
8 Green normalized difference vegetation index = (NIR – Green)/(NIR + Green).

10 Oil Palm Pest and Disease Detection

10.1 OIL PALM PEST AND DISEASES

Pest and diseases are common destroyers in many crops, and oil palm is not an exception to this. The biggest threat for palm oil yield are pests such as nettle caterpillars and bagworms, and the pathogenic fungi called *Ganoderma boninense*. These pests and disease prevent healthy growth of palm trees at various growth stages, that is, seedling in the nursery, young/immature trees and mature trees, and result in a large decrease of fruit bunch production and palm oil yield or even death of the trees (Chung, 2012). Ganoderma is a major oil palm destroyer in the major palm oil–producing countries like Indonesia and Malaysia. There is no cure for this disease; however with the current treatments, the life time of the trees can be prolonged. Detecting Ganoderma as early as possible can help to restrict their spread, save the trees from dying and thereby reduce the economic loss. It is of paramount important to manage pest and diseases in a timely manner to reduce pesticide usage and to maximize the yield. Geospatial technology has been demonstrated to detect and map the spatial distribution of oil palm diseases and pests for an efficient management (fast and cost effective) of the plantation industry. This chapter describes some common diseases and pests that reduce the yield of palm oil and analyzes the use of remote sensing and other related spatial technologies to manage their outbreak.

Pests are animals that cause harm to the growth and yield of oil palm trees. The common pests that damage oil palm trees in Southeast Asia (SEA) include insects, worms, mites, and mammals like rats (Chung, 2012). In Malaysia, for example, bagworm is an important defoliator and its outbreak was severe in the 1990s and it is still threatening the industry (Wood and Kamarudin, 2019), where *Metisa plana* and *Pteroma pendula* (bagworm species) have reduced palm oil yield up to 33%–47% (Basri, 1993; Mazmira et al., 2011). While, Nettle caterpillar is the main pest destroyer in Indonesia (Harjotedjo et al., 2013). These pests are briefly described, and their symptoms and impact on palm oil yield and management approaches are discussed in Chung (2012).

Diseases in oil palm trees are commonly caused by fungi, bacteria, virus, and parasitic plants (Chung, 2012), and they affect various parts of the trees from roots to fruit bunches. Aderungboye (1977) and Chung (2012) reviewed some of the important pests that affect palm oil yield and they are listed in Table 10.1 and he listed vascular wilt/Fusarium wilt, dry basal rot, basal stem rot (BSR), and Cercospora leaf spot as notorious diseases affecting palm oil yield in Africa. Vascular wilt/Fusarium wilt is also a dangerous disease affecting the industry in Central and South America. Other diseases that also threaten the palm oil industry in the continent are the red ring disease, spear rot/bud rot, and sudden wither. In the Asia and Pacific regions,

DOI: 10.1201/9780429199813-10

TABLE 10.1
Common Pests Found in Oil Palm Plantations in Various Parts of the World

Pests	Africa	Central and South America	Asia and Pacific Region
Leaf miners	✓		
Rhinoceros beetle	✓		✓
Red palm weevil	✓	✓	✓
Leaf eating caterpillars	✓	✓	
Rats	✓		✓
Cutting grass/cane rat	✓		
Weaver birds	✓		
Bunch and stem borer	✓	✓	
Root miners	✓	✓	
Bird (American black vulture)	✓	✓	
Bagworms			✓
Nettle caterpillars			✓
Hairy caterpillars			✓
Long horn grasshopper			✓
Bunch moth			✓
Cockchafer beetles			✓
Termites			✓
Squirrel			✓
Porcupine			✓
Wild pigs			✓
Elephants			✓
Birds (parrots/lorikeet/parakeets)			✓

Source: Chung et al. (2012).

basal and upper stem rot and bunch rot are identified as major diseases affecting oil palm trees.

BSR stands out as the most infamous disease affecting oil palm in the SEA region, home to the two largest oil palm growers and exporters of palm oil. This disease affected a total of about 7.4% of the total plantations in Malaysia and resulted in a loss of palm oil yield worth of approximately RM 1.5 billion (http://sawitsecure.mpob.gov.my/ms/ganoderma-basal-stem-rot/). The infestation is more serious among the smallholders with 9.2% of the total areas surveyed by the Malaysian Palm Oil Board (MPOB) and the reason could be poor management and financial constraint to eradicate the disease. This disease is detected not only in young mature to old palm trees planted in the clay soils along the coastal areas as well as in laterite and peat soil inland (Benjamin and Chee, 1995; Rao, 1990), but the seedlings are also affected. BSR disease is triggered by the fungal pathogen *Ganoderma boninense* which targets the basal stem and roots of the trees. When these roots are infected, the flow of water and nutrients to the foliage and other upper parts of the palms is significantly impeded (Paterson, 2007). The infection will result in unopened young leaves and spears in the crown center and damaged stem which will affect photosynthesis and fruit bunches production (MPOB, 2023). Generally, by the time these symptoms appear on the trees, the Ganoderma fungus has half killed the trees by rotting the basal stem tissue. Meanwhile, the stem base of palm trees that are severely affected by *Ganoderma boninense* will turn to black and have extrusion of gum and formation of fruiting body. These trees will die and collapse eventually. The symptoms of Ganoderma infection at various stages of oil palm growth are listed in Table 10.2, and the severity of the disease is described using four levels of disease severity index that was calculated by MPOB and it is shown in Table 10.3.

BSR is a contagious disease and it can spread to neighboring trees through root contact (Rees et al., 2009) and the symptom of the infection appears on the trunk as well as on the canopy. The infected palms only show symptoms at a later stage, and by the time it is detected, it already spread to neighboring trees. The infected palm trees display symptoms such as closed new fronds, pale green or yellow fronds due to insufficient nutrients, drooping fronds, and smaller canopies as described in Table 10.2. These symptoms are visually identifiable using aerial photos or satellite imageries. Other critical symptoms confirming the infection of BSR are the dry rot and more fruiting bodies at the trees' base especially for advanced stages of the infection. These pests and diseases of oil palm trees have reduced palm oil yield due to reduced productivity or death of the trees. Therefore, using timely and cost-effective measures, the pest and disease effects should be identified and controlled to maximize the productivity and yield of oil palm trees.

Conventionally BSR infection is diagnosed through visual inspection of the trees in the field and chemical analysis that involves the sampling of stem tissues. These methods are time-consuming, labor-intensive, and expensive because laboratory facilities and trained experts are required to perform lab analysis of the samples. This method is not feasible to cover large plantations cost effectively, and importantly it is destructive (non-sustainable) when stem tissues are drilled out from the trees (Noor Azmi et al., 2020). Early detection of the disease is also done using molecular PCR-DNA and GanoSkenTM tomography method that uses transmission or reflection of sound waves for assessing the palms decay and degradation (Idris et al., 2010). This technique can detect the infection even in healthy looking trees and it is non-invasive (Idris et al., 2010). However, the tomography technique is also not practical and costly when large plantation areas are to be monitored frequently. Alternatively, sensors onboard platforms such as aircraft, drone, and satellites that are based on reflection, emission, and backscattering of electromagnetic radiation are able to provide global coverage and continuous measures about the condition of oil palm trees on the Earth. The application of remote sensing technology for detecting pest and diseases affecting the palm oil industry at an early stage is reviewed next.

10.2 REMOTE SENSING FOR DETECTING DISEASE AND PEST OF OIL PALM

Sensors placed onboard various platforms such as on the ground, aircrafts, drones, and satellites can be used to assess palms that are infected with pests/diseases. There are hundreds of satellite remote sensing sensors operating in the optical, radar, and

TABLE 10.2
Symptoms of Ganoderma Infection at Various Stages of Oil Palm Growth

Growth Stages	Symptom Description	Photos
Young plants	Shorter leaves, chlorotic due to insufficient chlorophyll, stunted growth with the spear leaves remain unopened. Can cause mortality	
Immature palms	Similar symptoms as the young plants. Additional symptoms include yellowing of leaves (one sided) patches of different shades/colors of the lower fronts, followed by necrosis. Infected immature palms These symptoms cause death within 6–24 months after the diagnosis. Old leaves turn pale green with some roots rotting for mild infection. For severely infected trees, leaves are completely necrotic. Stem base rotting of >30%	

(*Continued*)

TABLE 10.2 (*Continued*)
Symptoms of Ganoderma Infection at Various Stages of Oil Palm Growth

Growth Stages	Symptom Description	Photos

(*Continued*)

TABLE 10.2 (*Continued*)
Symptoms of Ganoderma Infection at Various Stages of Oil Palm Growth

Growth Stages	Symptom Description	Photos
Mature palms	More unopened spear leaves. Pale green (oldest) leaves which die eventually due to necrotic development. Drooping of dried and dead fronds around the tree. Untreated mature trees will die within 1–2 years from the time of infection.	 

(*Continued*)

TABLE 10.2 (*Continued*)
Symptoms of Ganoderma Infection at Various Stages of Oil Palm Growth

Growth Stages	Symptom Description	Photos
Old trees	Rotted trunk at the base. Yellowed and brittle internal tissue. Mycelium or fruiting body extended through the affected tissue.	

Source: MPOB (2023).

TABLE 10.3
Oil Palm Disease (Ganoderma) Severity Index

Severity Index	Severity Levels	Description
DS 1	Early infection	No foliar symptoms, Fruiting body, mild (<10%) stem decay
DS 2	Moderate infection	Foliar symptoms such as yellowing and collapsed fronds (<50%) and <30% of stem decay Fruiting body
DS 3	Severe infection	Severe (>50%) foliar symptoms and >30% of stem decay Fruiting body
DS 4	Dead	Severe foliar symptoms and Severe stem decay Tree dead and collapsed

Source: MPOB (2023).

light detection and ranging (LiDAR) spectrum of the electromagnetic radiation. These sensors are capable of producing data at various spatial, spectral, temporal, and radiometric resolutions that are suitable to cover small holder to large-scale oil palm plantations (Table 10.4). The revisit period or temporal resolution is an important aspect of remote sensing for disease and pest detection. Images such as Sentinel from the European Space Agency (https://sentinel.esa.int/) are not only having adequate spatial resolution that facilitate identification of individual diseased trees, but they are also produced every 5 days. This is very important for oil palm–growing countries in the tropics that are most of the time covered by clouds. The frequent revisit period can increase the chances of cloud-free images and enable capturing plant status and change detection at 10 m spatial resolution.

Multispectral data are used to calculate various vegetation indices that can discriminate unhealthy/diseased oil palm trees from healthy palms. The hyperspectral sensors with tens to hundreds of narrow bands are extremely useful to detect vegetation phenology and diseases. Active sensors such as radar and LiDAR have gained

TABLE 10.4
Various Satellite Remote Sensing Sensors and Their Spatial Resolution

Very High-Spatial Resolution Sensors (0.5–5 m)	High-Spatial Resolution Sensors (10 m)	Medium-Resolution Sensors (100 m)	Low-Resolution Sensors (>100 m)
Cartosat-2 and 3 series, IKONOS, QuickBird, KOMPSAT series, Pleiades series, SuperView, World-View series	LISS-IV, Sentinel-2, ALOS AVNIR-2, SPOT 5,6,7	(SPOT 1,2,3,4, ASTER, LISS-III, Landsat OLI, AWiFS	MODIS, Suomi NPP VIIRS).

increasing attention in recent years because they provide palm trees structural information and 3D canopy structure that can easily identify the early stage of disease infection.

10.3 DISEASE DETECTION

Based on the fact that infected palm trees generally have unhealthy tree canopy that show distinct spectral properties compared to the healthy canopy (Table 10.2), various studies have used optical data to identify palm trees affected by BSR. Healthy plants generally absorb blue and red lights and reflect more near-infrared reflectance (NIR), while unhealthy plants exhibit relatively higher reflectance in the blue and red bands due to weakened chlorophyll by physiological changes and lower NIR reflectance as a consequence of reduced layers of fronds and increased exposure of background soil (Knipling, 1970). By combining various spectral bands from multispectral optical imageries, valuable vegetation indices can be derived. These indices play a crucial role in detecting plant diseases.

10.3.1 Multispectral Sensors

Vegetation indices calculated from the QuickBird image was used by Santoso et al. (2011) to detect trees infected by BSR in north Sumatera. Reflectance from the red band and various vegetation indices were combined to segment infected and non-infected areas in the plantation. Within the infected areas, different vegetation indices detected the disease with different levels of severity. A range of vegetation indices such as the atmospheric resistant vegetation index (ARVI),[1] normalized difference vegetation index (NDVI), green NDVI (GNDVI), green blue NDVI (GBNDVI),[2] soil adjusted vegetation index,[3] and simple ratio vegetation index were calculated to highlight the vital sign of BSR. The combination of blue, red, and NIR bands is capable of detecting the infected oil palm canopy that have wilting, yellow, and small fronds compared to the healthy trees. The contrast in GBNDVI and ARVI were able to produce higher accuracies in detecting diseased trees with high rate of infection. The high-spatial resolution data enabled separation of infected palm trees from other trees, pruned old fronds, or other man-made objects. Nevertheless, this study is only able to detect late-stage infection which is not very useful for managing the disease.

Consequently, researchers have been striving to enhance the early detection of BSR in palm trees. One notable study by Santoso et al. (2017) employed three distinct machine learning methods, namely Support Vector Machine (SVM), Random Forest (RF), and classification and regression tree models, to classify infected palm trees. Among these methods, RF exhibited the highest overall accuracy of 91% in effectively distinguishing healthy and unhealthy trees. However, when it came to classifying different severity levels of BSR, even the use of very high-spatial resolution Worldview 3 imagery with 8 spectral bands and machine learning algorithms, as demonstrated by Santoso et al. (2019), resulted in only a minimal accuracy ranging from 11% to 52% for the various categories (mildly, moderately, and severely infected trees, and healthy trees).

In an attempt to detect diseased plants, Nuthammachot and Stratoulias (2023) leveraged Sentinel-2 data, a high-spatial and temporal resolution satellite imagery. They employed various vegetation indices calculated from these data and utilized the SVM classification technique. Their study successfully discriminated healthy oil palm with an accuracy of 73% and Kappa coefficient of 0.56, diseased oil palm with an accuracy of 64% and Kappa coefficient of 0.40, and mixed oil palm with an accuracy of 50% and Kappa coefficient of 0.37, using NDVI, RVI, and MSAVI indices, respectively. However, even the adoption of state-of-the-art techniques, such as a very high-spatial resolution remote sensing data captured using a multispectral camera mounted on an unmanned aerial vehicle (UAV) and processed using advanced object-based image analysis (OBIA) by Izzuddin et al. (2021), failed to differentiate between healthy and early-stage infected BSR effectively.

Since the disease is developed on the trunk, it is difficult to use optical remote sensing with multispectral bands for early detection of BSR because optical data can only capture details of the upper canopy. So, based on the canopy changes, the disease is only detectable at the late stage of mortality. Even though these satellites are high in the spatial resolution, their limited number of broad spectral bands (only 5 and 8 bands of the Quickbird and Worldview 3 respectively) are not adequate to detect the physiological changes occurring in the canopy due to the disease infection. Thus, it was challenging to identify mildly affected trees from the healthy palms because of the similar reflectance pattern captured by the multispectral sensors.

10.3.2 Hyperspectral Sensors

Hyperspectral sensors that are mounted on either aircraft, drones, or ground-based tripods have the ability to capture light reflected from palm tree fronds in narrow wavelengths. Thus, they are able to recognize even subtle changes in the oil palm canopy and soil through variations in the reflectance and absorption spectra. Hyperspectral data are therefore suitable for detecting diseases such as BSR at an early stage of infection as a result of physiological changes in the canopy which was not possible with the multispectral images.

Hyperspectral sensors are operating by receiving reflected energy from oil palm canopies. Reflectance from hundreds of the spectral channels produce data cubes that are usually normalized to reduce the data dimensionality. The preprocessed data are then manipulated to extract features by means of classification and other digital image processing techniques. When hyperspectral data with 125 bands within 450–950 nm spectral range was used, the infected seedlings (5 months old) could be separated from the uninfected ones especially in the NIR spectrum at a very early stage (seedlings) without the emergence of any physical symptoms (Noor Azmi et al., 2020). The SVM classification algorithm using the NIR band yielded a classification accuracy of 100%. SVM uses hyperplanes to produce an optimal separation between the two classes (infected and uninfected) and thereby predicted well the ownership of each pixel in the image. This study was conducted using a camera that was positioned horizontally at 2.6 m above the canopy using a tripod in a greenhouse with artificial and constant conditions (Noor Azmi et al., 2020).

Portable hyperspectral instruments were employed in several studies to separate infected palm trees from healthy ones. These studies utilized hyperspectral data captured within 325 and 1,040 nm range and successfully separated severely infected palms from healthy trees using classification and segmentation approaches, respectively (Liaghat et al., 2014). Images captured by the hyperspectral sensors were classified using k-nearest-neighbors (k-NN) (Liaghat et al., 2014) and maximum likelihood classification (Shafri et al., 2011) procedures to identify and separate the classes. Liaghat et al. (2014) particularly successfully separated three stages of infection (slight, moderate, and severe) with an exceptionally high overall classification accuracy of 97% when the second derivative of the hyperspectral reflectance spectra data was used. While, Izzuddin et al. (2017) found the first derivative of the reflectance spectra of hyperspectral images to be more effective in the early detection of BSR. Thus, they leveraged the first derivative of the blue, green, red, and infrared regions to identify additional wavelengths that exhibited noteworthy distinctions between mildly and severely infected oil palm trees. A field spectroradiometer that had 0.5 nm resolution with a spectral range of 350–1,000 nm was used to record the spectral differences among moderately and severely infected four-month-old saplings for a period of six months. A total of 24 narrow bands within the green, red, and infrared spectrum were useful as green and red bands effectively captured chlorophyll changes and the infrared radiation could recognize distorted structure of oil palm fronds due to Ganoderma.

Another successful application of field hyperspectral instrument to detect Ganoderma was demonstrated by Lelong et al. (2010). The Partial Least Square Discriminant Analysis (PLS-DA) performed on the hyperspectral images captured over oil palm canopies using 55 spectral bands enabled the detection of the Ganoderma fungal disease attack from mild to severe stages with a high accuracy of 94%. The technique could even differentiate Ganoderma infection from non-diseased trees that are lack of nutrition. The PLS-DA convolution produced new coordinates that can determine its degree of sickness.

In contrast to prior studies, Izzuddin et al. (2018) provided a comprehensive assessment of the Ganoderma disease across a large plantation area, spanning 25 ha using hyperspectral sensor flown on an aircraft. The study involved the calculation of five vegetation indices, four red edge positions, and four continuum removal techniques to classify the disease into healthy, mildly, and severely infected categories in oil palm trees. Among the vegetation indices tested, the simple ratio index (SRI) and enhanced vegetation index (EVI) successfully detected Ganoderma disease in oil palm, albeit with only moderate accuracy, achieving 44.4% and 40.7%, respectively. Other indices, including NDVI, Vogelmann red edge index, and carotenoid reflectance index, exhibited accuracy levels below 40%.

The red edge region within the electromagnetic spectrum proved highly effective in discerning various infection stages. Positioned approximately between 670 and 760 nm, the red edge light occupies the spectral space between red light and NIR. In this specific region, vegetation exhibits rapid changes in spectral reflectance, with diseased plants showing notably higher reflectance. The distinct spectral signature observed in the red edge region facilitates the precise detection of diseased plants, making it a valuable tool in plant health assessment.

The red edge position technique differentiated healthy trees from severely affected ones and mildly affected from severely affected trees in a plantation in the northern Peninsular Malaysia (Izzuddin et al., 2018). However, it was unable to distinguish between healthy and mildly affected trees. On the other hand, the continuum removal technique showed promise in effectively differentiating mildly infected palms from healthy palms, particularly at the 500 nm absorption region of band depth normalized to area (BDNA), demonstrating statistically significant results.

Nevertheless, a recent study classified hyperspectral images captured by an UAV into four Ganoderma infection stages namely healthy, early and late stages of diseased trees, and dead trees (Kurihara et al., 2022). The images were segmented to get the tree crown details before they were classified into different infection stages using the RF technique. The authors replaced the red band with the red edge band to calculate spectral indices such as normalized reflectance, SRI, and normalized difference spectral index. Incorporating the red edge band into the vegetation indices calculation enables the detection of chlorophyll and nitrogen changes in plants, thereby facilitating the identification of diseases in oil palm trees. Unlike the previous studies, this study could identify the early stage of infection with an average accuracy of 60% (Kurihara et al., 2022). Limited number of spectral bands in the red edge region are sufficient to classify the disease into different infection levels including the early stage.

While hyperspectral remote sensing data have been proven effective in early-stage detection of BSR/Ganoderma disease, recent research by Ahmadi et al. (2022) suggests that combining high-resolution multispectral data (green, red, and NIR bands with a bandwidth of <10 nm) with tree biophysical variables, such as trunk radius, and employing advanced image processing techniques can further enhance the diagnostic capabilities. They employed an artificial neural network with varying numbers of hidden layers to assess different severity levels of infection, along with fine spatial resolution UAV images (<3 cm), resulting in an improved diagnostic accuracy of 73%. This method holds the potential to enable precise and rapid detection of Ganoderma-infected oil palms, offering operational efficiency. Nevertheless, it should be noted that the hyperspectral data captured by cameras mounted on drones often produce noise that can diminish data quality. Denoising the data is an important pre-processing step before the data could be used for detecting Ganoderma. Techniques such as Savitzky-Golay and wavelet spectral denoising were found to improve disease detection in oil palm trees in Malaysia (Izzuddin et al., 2023).

10.3.3 Radar Sensors

Along with the multispectral and hyperspectral data, radar data have also been used by a couple of researchers to explore its potential in early detection of the disease. Radar sensors with relatively larger wavelengths can penetrate clouds and provide oil palm–related information in a spatially continuous manner. Hashim et al. (2021) used ALOS PALSAR-2 HH and HV polarization images to identify and distinguish infected palm trees from non-infected trees using machine learning techniques (multilayer perceptron) and RF. The HH polarization was found to correctly classify the infected and non-infected trees with an accuracy of more than 90%.

Correlation analysis between the radar backscatter and oil palm physiological factors that are associated with Ganoderma disease such as the circumference of the trunk and moisture content of the frond and pinnae was conducted using ALOS PALSAR-2 full polarimetric data (Hashim et al., 2021). The results showed that the backscattering values from HH polarization image decreased rapidly with severely affected trees compared to the mildly affected and healthy trees. Meanwhile, only a slight decrease in the backscattering values was observed in the HV and VV polarization for 8-year-old oil palm trees. An opposite pattern (increasing backscatter values as the disease severity got worsen) was noticed for 16-year palms in the HH, VV, and HV polarizations.

Changes in the backscattering values of the diseased trees are correlated with the structure and moisture content of the trees. Diseased trees are generally dehydrated due to reduced water uptake and increased water loss, thus the moisture content in their tissues of leaves and branches is reduced. Consequently, they are unable to strongly backscatter signal in the radar image compared to the healthy trees. Moreover, as the health of the palm trees become worsen, the structural sturdiness of the stem and fronds may be damaged, leading to a loss of consistency in the radar signal returns. This results in a reduced backscattering response in the HH polarization image that is more sensitive to the vertical structure of oil palm trees canopy and changes in moisture content. HH polarization penetrates the trees canopy parallel to different layers of palm trees, allowing the signal to reach deeper layers of the base of the trunk and disclose changes in the physical appearance of the trees, such as the fungal infections or Ganoderma, which are not readily visible from above the tree crown. These characteristics can be used to monitor diseased palm oil trees using radar remote sensing techniques.

10.3.4 LiDAR Sensors

The remote sensing sensors described above are operating in the visible lights, and infrared radiation in the multispectral or hyperspectral domains of the EMS. These segments of EMR are used for oil palm disease detection due to their high sensitivity to the biophysical and biochemical characteristics of the oil palm canopy. The infected palms can be distinguished from the healthy trees based on the reduced canopy density such as reduced foliage and leaf area. Hence, the canopy structure, geometry, and its morphology/shape of the palm trees are important variables to determine diseased trees. LiDAR offers three-dimensional canopy structure of palm trees that is important to detect diseases such as Ganoderma at an early stage and curb its spread to other trees effectively.

LiDAR utilizes laser light, which is amplified through stimulated emission of radiation, to generate an intense and narrow beam that can be precisely focused into a small spot (https://spaceplace.nasa.gov/laser/en/). A laser is an artificially created light source in which all the light waves have nearly identical wavelengths. Laser scanner is an active remote sensing system that sends out its own source of laser pulses to features/objects on the Earth and receives it back to the scanner to interpret the features' physical size and shape in the form of light intensity (https://spaceplace.nasa.gov/laser/en/). Time taken for laser light to travel back and forth from the sensor/scanner to the objects on the Earth is used to measure the distance of the objects. Laser light is also used to obtain height of the objects through measuring the distances between the objects (highest and lowest points on the ground) and the scanner. The light intensity collected from the objects is point cloud data having XYZ coordinates (3D space) and a very high spatial resolution. These point clouds are processed to generate height, area, diameter, stem volume of the palm trees canopy, and fronts properties (Means et al., 1999). Any minute differences present in the point clouds are valuable to detect changes occurring in the tree structure including changes in the canopy caused by diseases (Means et al., 1999). Laser scanners can be mounted on mobile phones, drone/aero planes, and ground-based tripods such as the terrestrial laser scanning (TLS).

TLS is one of the latest technologies used in precision agriculture which is very useful to detect diseases in oil palm trees. Khairunniza-Bejo and Vong (2014) investigated the use of TLS to detect BSR in one of the palm oil plantations in the south of Peninsular Malaysia. They scanned individual mature palm trees (25 years old) and extracted trees' biophysical properties such as the trunk diameter, height and perimeter, and canopy size and correlated them with healthy and diseased (slightly, moderately, and severely affected) trees. Both the canopy area and the trunk perimeter at 150 cm were found to be correlated well (Pearson correlation coefficient values ranging from −0.53 to −0.57) with different severity levels (slightly, moderately, and severely infected) of BSR as the infection becomes severe, their canopy shrinks to smaller size. For younger trees (age < 10 years), frond numbers, frond angle, and canopy strata at 200 cm from the top were the best canopy parameters to detect unhealthy and healthy trees which were correctly classified with 87% accuracy, while the three severity levels of the disease were accurately classified with 80% accuracy (Husin et al., 2020). Although the TLS technique can acquire data fast by a simple set up and can produce high spatial resolution data with higher accuracy, its usage is limited to a smaller area. This is because scanning is performed over a single tree at a time, not over a group of trees (Khairunniza-Bejo and Vong, 2014; Husin et al., 2020). Future studies should use airborne platform to acquire LiDAR data, but this can be very costly.

Another strategy to manage BSR disease is by linking the geospatial technology with communication technology for a systematic compilation and monitoring of disease data. In a preliminary study, Abdullah and Muhadi (2015) developed a mobile application based on GPS, Wi-Fi/General Packet Radio Services (GPRS) connection, and accelerometer devices on smart phones for BSR data collection in Indonesia. The mobile app would assist workers in the plantations to record the location and condition of diseased trees and update them frequently via mobile applications. This application is connected to a sync-enabled ArcGIS Feature Service that allows the users to view and edit any data uploaded to the system in the absence of an internet connection.

10.4 REMOTE SENSING FOR PEST CONTROL

The application of optical/multispectral remote sensing data in pest control is rather limited because pest outbreak is short lived which needs very high-resolution data to capture their infestation. Nevertheless, the potential of multispectral data to detect bagworm infestation in an oil palm plantation was tested using a field spectroradiometer (Aziz et al., 2012). Severely bagworm-infested palm trees were detected using red and NIR bands that showed considerable increase in reflectance within 570–680nm and a sharp drop in the reflectance between the wavelength range 734 and 1,047nm (Aziz et al., 2012). Izzuddin et al. (2021) acquired multispectral images (covering green, red, red edge, and NIR bands) from an UAV over eight-year-old oil palm area. They found the red edge; red and green bands could visually differentiate bagworm-infested oil palm from the healthy trees based on leaves defoliation of approximately 10%–13% (MPOB, 2023).

The thermal infrared bands are also applied for pest detection based on the fact that higher temperature is conducive for the development and survival of bagworm species such as *Pteroma pendula* (25°C–30°C) and *Metisa plana* (30°C–35°C) until optimum temperature is reached (Muharam et al., 2017). Typically ground temperature data are obtained from weather stations that are sparse over a region, thus spatially continuous data are not available. Currently thermal bands that can be used to calculate land surface temperature are provided by satellites/sensors such as the Landsat 9 (thermal infrared data at 100m), MODIS (land surface temperature product at 1km), AVHRR sensor (temperature data at 1km), ASTER (thermal data at 90m), and Sentinel 3 (1km) (Table 10.5). Although these satellite-based remote sensing tools are useful for pest detection covering large plantations, the revisit period of these satellites is several days which may not be efficient to track the short-lived bagworm of about 10days to several months (MPOB, 2023). The SEVIRI onboard the MSG-2 geostationary satellite is capable of offering data at every 15 minutes,

TABLE 10.5
Thermal Sensors Onboard Various Satellites

Satellites/Sensors	Spectral Resolution (μm)	Spatial Resolution (m)	Temporal Resolution (days)
Landsat 8/9-thermal infrared sensors	10.6–12.51	100m	8 days when Landsat 8 and 9 are combined
Advanced Spaceborne Thermal Emission and Reflection Radiometer (ASTER)	8.125–11.65	90m	16days
National Oceanography and Atmospheric Administration's (NOAA) Advanced Very High Resolution Radiometer (AVHRR)	3 bands (3.55–12.5)	1,000m	1day
Moderate Resolution Imaging Spectroradiometer (MODIS) on TERRA and AQUA platforms	16 bands (3.660–14.385)		1–2days
Spinning Enhanced Visible and Infrared Imager (SEVIRI) onboard Meteosat Second Generation (MSG) satellite	8 bands (3.9–13.4)	3,000m	15 minutes

but the spatial resolution is too coarse to capture any spatial variation in the infestation, and this satellite does not cover the Asia region frequently. Alternatively, UAVs equipped with thermal sensors can be to flown as frequent as it is necessary to monitor the bagworm infestations.

10.5 FUTURE OUTLOOK

The primary threats to the oil palm industry are pests like nettle caterpillars and bagworms, along with the pathogenic fungus, *Ganoderma boninense*. These infestations hinder healthy palm tree growth at different stages, leading to significant reductions in fruit bunch production and palm oil yield. Identifying outbreaks and preventing their spread throughout the entire plantation are of utmost importance. Manual diagnosis involving visual inspection of the affected palms' canopy, trunk, and roots is susceptible to human bias, labor-intensive, and impractical for large-scale operations. Laboratory analysis, though providing insights, has limited processing capacity and involves non-sustainable practice.

Remote sensing technology offers a sustainable way to detect and monitor pests and diseases across extensive areas continuously. Hyperspectral sensors operating at various regions in the electromagnetic spectrum are effective in early detection of infected palms; however, they are optical sensors, thus affected by cloud cover. Radar systems, capable of penetrating clouds, may be viable alternatives for disease detection, but their application in palm tree research is still relatively unexplored. LiDAR technology is promising but faces challenges due to high operational costs, especially for large-scale scanning. Mobile laser scanning mounted on tractors could offer a faster solution, though certain technical aspects must be addressed.

Combining hyperspectral images and LiDAR data provides powerful tools to detect diseases, particularly when coupled with advanced image processing techniques like machine learning and deep learning (Haw et al., 2023). While these techniques require substantial computing power and extensive field data, they hold great potential for such research. Miniature hyperspectral sensors designed for commercial use, such as Micro- and Nano-Hyperspec, HySpex VNIR, and FireflEYE, can be mounted on aerial platforms like drones to monitor disease outbreaks promptly.

Laser scanners mounted on drones for airborne scanning present an opportunity to extract essential palm parameters and analyze them using advanced machine/deep learning techniques to diagnose oil palm diseases early. Additionally, recently available LiDAR data from NASA's Global Ecosystem Dynamics and Investigation (GEDI) mission can aid in studying oil palm structures and their relationship with diseases at regional scale.

Incorporating remote sensing technology into pest and disease detection and monitoring strategies is essential for the palm oil industry's effective management. The choice of data (multispectral, hyperspectral, radar, or LiDAR) depends on the financial capabilities of the companies. Large plantation companies can deploy UAVs equipped with hyperspectral or LiDAR data, while smallholders may face financial constraints, making their oil palms more vulnerable to disease and pest attacks.

NOTES

1 Atmospheric resistant vegetation index (ARVI) = [NIR – (2R – B)]/[NIR + (2R – B)].
2 Green blue normalized difference vegetation index = [NIR – (G + R)]/[NIR + G + R].
3 Soil adjusted vegetation index = [(NIR – R)/(NIR + R + L)] x (1 + *L*) Where $L = 0.5$ for medium density vegetation and it represents the soil background adjustment factor.

11 Climate Suitability for Oil Palm Expansion

11.1 OIL PALM EXPANSION INTO UNSUITABLE REGIONS

Rising temperatures in the 21st century will open up new areas for oil palm cultivation, although the overall suitable land area for this purpose will shrink. The future holds challenges such as more frequent droughts and flooding, which will likely lower oil palm yields. However, some of these losses may be offset by elevated temperatures and increased carbon dioxide levels, which can enhance plant productivity (Habib-ur-Rahman et al., 2022). To sustain oil palm cultivation without causing additional harm to forests and natural habitats, it is crucial to gain a better understanding of where suitable locations for growing oil palm will be in the future.

Palm oil stands as the most widely produced vegetable oil globally, fuelling the rapid growth of the palm oil industry in tropical regions. Oil palm is a high-yielding oil crop, generating a versatile product, palm oil, with applications spanning edible oil, lubricants, food additives, biofuels, and more. Over the past few decades, oil palm cultivation has experienced unprecedented expansion, establishing itself as one of the fastest-growing oil crops.

Nonetheless, the rapid expansion of oil palm plantations, encroaching upon tropical forests, cultivated land, and peatlands, has precipitated severe ecological and environmental challenges, including biodiversity loss and ecosystem imbalances (Sarzynski et al., 2020). Moreover, the oil palm cultivation areas in Malaysia and Indonesia are either approaching or have exceeded their sustainable limits, emphasizing the urgent need to identify suitable land for future oil palm cultivation. The exploration of suitable land for oil palm cultivation plays a pivotal role in promoting the sustainable development of this industry while averting environmental damage resulting from inappropriate agricultural practices. However, the unique geographical environments across the world's regions, coupled with geographic isolation, give rise to distinct habitat characteristics for oil palm.

To support the expansion of oil palm plantations worldwide, an efficient evaluation process is imperative for identifying suitable land. Consequently, this chapter focuses on the proper assessment of global oil palm suitability and provides recommendations for areas suitable for oil palm cultivation. To initiate this assessment, first we monitored the expansion of oil palm plantations from 1993 to 2019, utilizing global datasets on oil palm distribution generated by Du et al. (2022) using the Landsat archive at 30-m spatial resolution. The sources of oil palm expansion's land cover were determined for the period spanning 1993–2019. This identification process involved overlaying the global oil palm maps (Du et al., 2022) with the European Space Agency Climate Change Initiative (ESA CCI) global land cover datasets. The ESA CCI has generated annual global land cover datasets from 1992 to 2019,

DOI: 10.1201/9780429199813-11

featuring a spatial resolution of 300 m (ESA, 2017). These datasets encompass a total of 37 distinct land cover classes, which we subsequently reclassified into ten broader categories. These categories include cropland, forest, grassland, shrubland, wetland, water, tundra, impervious surface, bare land, and snow/ice.

We further assessed the extent of oil palm encroachment into peatlands and protected areas, focusing on those designated as categories I and II within the World Database of Protected Areas (WDPA) spanning from 1981 to 2019. Subsequently, our efforts shifted toward the development of comprehensive global-scale maps, outlining lands suitable for oil palm cultivation. To facilitate this, we established a foundational global oil palm sample database, enabling a thorough analysis of oil palm growth conditions across varied regions. Our assessment also took into account the environmental constraints linked to oil palm cultivation, addressing these constraints at the global, continental, and national levels, respectively.

Our analysis draws upon a comprehensive literature review and the extraction of key biophysical characteristics from the oil palm sample database, encompassing factors such as climate, soil quality, and terrain (Table 11.1). These data are employed to determine minimum, optimal, and maximum suitability values for assessing oil palm suitability. Subsequently, we apply filters to exclude areas where future oil palm expansion is nearly impossible, such as protected areas and urban zones, while also preserving land with special value in biodiversity-rich regions. Finally, we conduct an analysis of both past and current land cover within the identified oil palm suitable areas.

TABLE 11.1
The Restrictions for Oil Palm Suitable Map

	Criterion	Unit/Description	Original Spatial Resolution	Dataset Used
Climate	Annual precipitation	mm/m²	30 arc seconds	WorldClim (Hijmans et al., 2005)
	Number of dry months	Monthly precipitation <100 mm/m²		
	Average annual temperature	°C		
	Min annual temperature	°C		
	Max annual temperature	°C		
	Wind	m/s		
	Daily solar radiation	MJ/m²		
	Water vapor pressure	KPa		
Soil	Predominant soil texture type	Soil texture classification	30 arc seconds	HWSD (Nachtergaele et al., 2012)
	Water content of soil	cm		
	Effective soil depth	cm		
	Soil drainage			
Topography	Slope	Sloping degrees	30 arc seconds	NASA SRTM (NASA, 2010)
	Elevation	m		

11.2 OIL PALM SAMPLE DATABASE

Building upon our previous study (Cheng et al., 2017), we utilized an equal-area hexagonal sampling approach, implemented using the Discrete Global Grid Generation (DGGRID) software developed by Sahr et al. (2003), to establish the sample database (Figure 11.1). Based on the global oil palm sample database, the environmental

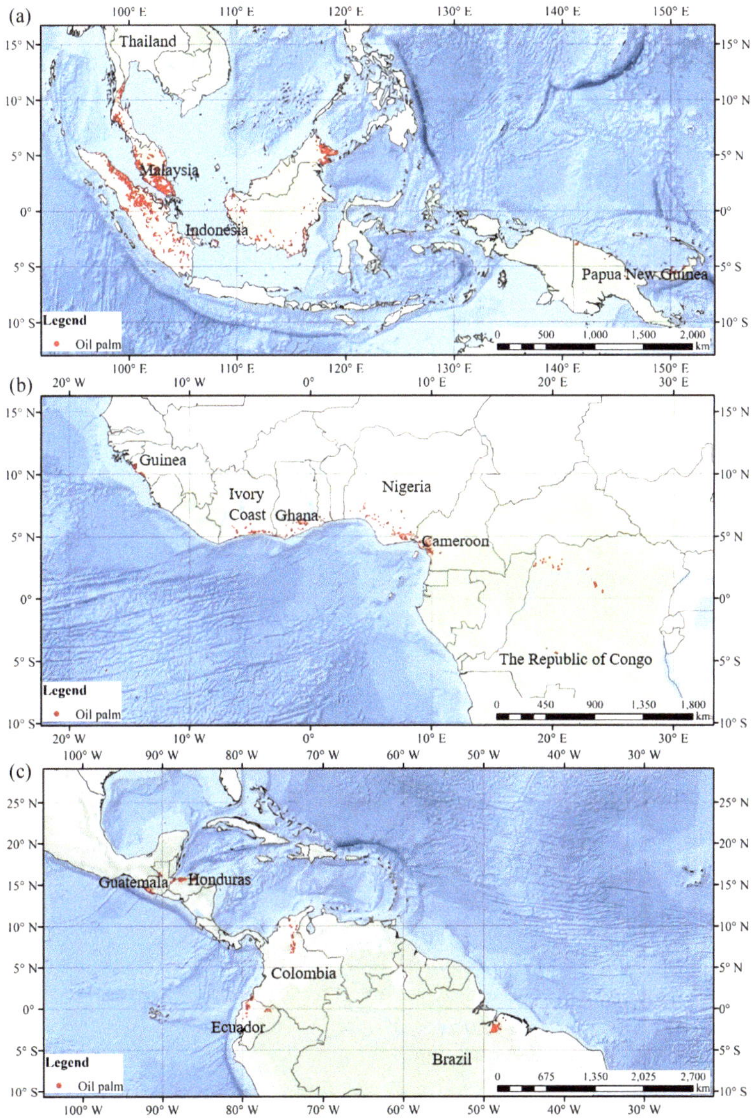

FIGURE 11.1 The distribution of oil palm samples database: (a) Indonesia, Malaysia, Thailand, and Papua New Guinea; (b) Nigeria, Ghana, Guinea, Cameroon, Congo, and Ivory Coast; (c) Brazil, Colombia, Ecuador, Guatemala, and Honduras.

factor thresholds that restrict oil palm planting were extracted, including climate, soil, and terrain factors. These thresholds are more representative of the real environmental conditions for regional oil palm cultivation.

11.2.1 Different Restrictions for Oil Palm Suitability Map

Climate, soil, and terrain are fundamental factors influencing the growth and yield of oil palm. While it is possible to enhance these factors through practices such as fertilization and irrigation, they can also escalate costs. Therefore, when determining suitable areas for oil palm plantation, it is essential to consider a broad range of bio-physical characteristics. To achieve this, we conducted a detailed analysis of oil palm distribution characteristics for 15 countries, which collectively account for 97.0% of global oil palm production. Table 11.1 provides the dataset used to analyze these oil palm distribution characteristics. Our analysis drew upon data from the WorldClim dataset (Hijmans et al., 2005), the Harmonized World Soil Database (HWSD) (Nachtergaele et al., 2010), and the Shuttle Radar Topography Mission (SRTM) digital elevation model (DEM) data.

11.3 Characteristics of Oil Palm Plantation Distribution

Five climatic factors play a pivotal role in oil palm cultivation: temperature, precipitation, daily solar radiation (DSR), wind, and water vapor pressure (WVP) (Goh, 2000). When examining temperature requirements across various regions, it becomes evident that oil palm thrives in high-temperature conditions, with mean temperatures exceeding 19°C (Table 11.2). Additionally, the average temperature during the coldest month of the year was observed to be 14.6°C in Indonesia and 13.7°C in Papua New Guinea, which is slightly lower than the previously reported 15°C (Corley and Tinker, 2008). In the cases of Colombia and Guinea, the minimum annual rainfall falls below 1,000 mm, and the dry season (defined as the number of months receiving less than 100 mm of precipitation) extends for more than 6 months. Notably, the availability of abundant solar radiation is of utmost importance for oil palm growth and production (Hoffmann et al., 2014), with oil palm being cultivated in areas receiving more than 11.8 MJ/m^2 day, as indicated in Table 11.2.

Soil and terrain play a vital role in the assessment of suitable areas for oil palm plantation, significantly impacting potential yield and suitability. Notably, oil palm exhibits a relatively modest set of requirements concerning the chemical and physical properties of the soil (Pirker et al., 2016). It can tolerate varying levels of soil acidity, different soil depths, and diverse water content levels, with soil pH typically exceeding 4.0 in both topsoil and substrate, as detailed in Table 11.3.

Terrain features also exert a substantial influence on the distribution of oil palm plantations. Therefore, we extracted soil and terrain features suitable for oil palm cultivation based on the oil palm sample database (Table 11.3). Previous research has suggested that oil palm cultivation predominantly occurs in regions with elevations below 1,000 m and slopes measuring less than 8° (Cheng et al., 2018). However, as highlighted in Table 11.3, oil palm cultivation has been observed in regions such as

TABLE 11.2
Suitable Climatic Factors for Oil Palm Cultivation in Different Countries

	Temperature (°C)			Precipitation		ET	DSR (Daily Solar Radiation)	Wind	WBC (Water Budget Condition)	WVP (Water Vapor Pressure)
	Min (°C)	Max (°C)	Mean (°C)	Annual Rainfall (mm)	Dry Season	(mm)	(MJ/m²)	m/s	(mm)	(kPa)
Brazil	21.1~22.1	30.8~32.2	25.9~27	2255~2594	3~6	940~1,500	15.3~16.2	1.2~1.8	755~1,543	2.7~2.8
Ecuador	18.6~22.1	27.1~30.6	22.8~26	1,397~3,645	0~8	88~1,310	11.8~13.5	0.9~1.9	515~3,517	2.4~2.8
Colombia	21.8~23.5	28.1~34.1	25.4~28.5	970~3,065	0~6	470~1,500	13.3~19.4	1.1~3.9	−330~2,544	2.7~2.9
Guatemala	19~23	29.4~34.4	24.5~28.1	1,422~3,067	1~6	542~1,500	18.2~19.8	1.4~2.6	−62~1,711	2.3~2.8
Honduras	19.4~23	29.5~32.5	24.4~27.3	1338~2888	0~6	538~1,500	18.8~20.0	1.5~3.5	−98~2,152	2.4~2.8
Guinea	17.4~24.6	31.8~36.5	24.5~30	2,110~3,507	3~7	316~1,403	15.9~19.1	1.4~2.9	706~3,076	2.2~2.7
Ghana	20.7~23.7	29.1~31.8	25.3~27.3	878~2,129	3~9	504~1,500	15.1~17.0	1.1~2.2	−114~872	2.6~2.9
Congo	18.9~20.6	29.1~30.8	24.2~25.5	1,625~1,898	0~4	521~1,423	14.6~16.6	0.9~1.3	317~1,106	2.4~2.6
Cameroon	19~23.2	26.8~31.2	22.9~27.2	2,107~3,962	0~3	343~1,161	13.7~14.4	1.0~2.3	1,175~3,221	2.4~2.9
Ivory Coast	21.7~23.1	29.2~31.5	25.6~27.1	1,272~2,024	4~6	350~1,322	16.5~18.0	1.1~2.3	162~1,366	2.6~2.9
Nigeria	20~23.1	29.6~32.3	24.8~27.4	1,205~2,985	3~6	248~1,285	14.1~15.8	1.3~1.9	−2~2,602	2.3~2.9
Thailand	21.2~23.2	29.9~32	25.5~27.5	1,600~3,452	3~5	704~1,500	17.4~19.1	1.1~2.3	363~1,808	2.6~2.9
Malaysia	19.4~23.4	29~32.9	24.2~27.5	1,901~3,746	0~3	76~1,500	15.5~17.8	1.1~2.2	401~3,010	2.6~3.0
Indonesia	14.6~24.1	24.6~32	19.6~27.4	1,491~3,966	0~5	196~1,500	15.1~18.1	0.8~2.3	259~3,058	1.9~3.0
Papua New Guinea	13.7~23.6	25.1~32.2	19.4~27	1,795~5,517	0~3	797~1,500	15.4~19.6	1.2~3.2	859~2,900	1.8~2.9

TABLE 11.3
Suitable Soil and Terrain Factors for Oil Palm Cultivation in Different Countries

	Depth	Texture (%)			PH		AWC	C (Machine Carbon)		Topography	
	(cm)	Sand (%)	Silt (%)	Clay (%)	Top (−log(H+))	Substrate (−log(H+))	(mm)	Top (%)	Substrate (%)	DEM (m)	Slope (°)
Brazil	100	46~80	3~17	17~37	3.8~5.2	3.8~4.8	150	1.0~1.6	0.4~0.6	8~79	0.03~0.49
Ecuador	100	8~54	20~57	15~59	4.3~7.0	4.1~7.8	150	0.5~3.1	0.2~2.5	15~570	0.02~3.82
Colombia	100	13~94	4~39	2~54	4.5~5.6	4.6~6.7	150~100	0.4~3.9	0.2~1.7	8~161	0.02~3.22
Guatemala	100	24~55	24~45	14~37	5.5~8.0	5.4~8.1	150	0.5~7.3	0.2~2.9	10~643	0.03~1.44
Honduras	100	24~44	27~33	23~49	5.1~7.0	5.1~7.3	150	0.7~2.6	0.3~1.0	3~541	0.01~9.10
Guinea	10~100	8~77	9~34	14~58	4.8~7.6	4.2~5.0	150~0	0.6~2.6	0.3~3.5	1~400	0.02~0.86
Ghana	10~100	16~89	5~41	6~53	4.6~7.7	4.8~7.9	150~15	0.4~1.6	0.2~0.6	1~227	0.05~1.59
Cameroon	10~100	34~79	7~48	14~26	4.8~7.7	4.8~7.9	150~15	0.6~7.4	0.3~3.3	6~524	0.05~5.97
Congo	100	29~86	1~13	13~58	4.1~5.1	4.3~5.2	150	0.7~1.2	0.2~0.5	352~594	0.09~2.71
Ivory Coast	100	26~89	5~39	6~41	4.4~6.2	4.5~5.7	150~50	0.4~1.3	0.2~0.4	7~174	0.01~1.22
Nigeria	100	26~82	7~40	8~49	4.8~6.2	4.8~6.3	150~50	0.6~1.6	0.3~0.5	15~451	0.01~1.71
Papua New Guinea	100	11~47	29~39	19~55	5~6.4	5.1~7.0	150	0.8~3.4	0.3~1.5	4~1551	0.03~5.70
Thailand	100	8~83	11~43	6~58	4.4~6.6	4.2~6.7	150~50	0.5~30.7	0.3~34.8	4~272	0.04~8.32
Malaysia	10~100	8~83	11~34	6~59	4.4~8.1	4.2~8.2	150~25	0.5~30.7	0.3~34.8	2~1078	0.01~5.82
Indonesia	100	17~90	6~48	3~55	4.4~6.4	4.5~6.4	150~50	0.4~30.7	0.4~34.8	1~1315	0.01~6.71

Indonesia, Malaysia, and Papua New Guinea, where elevations exceed 1,000 m, as well as in areas like Honduras and Thailand, characterized by slopes greater than 8°.

11.4 LAND POTENTIALLY AVAILABLE FOR OIL PALM EXPANSION

In this section, we address two key restrictions when mapping oil palm suitability: land that cannot be converted for other purposes and land with high biodiversity value. The datasets employed for this analysis are outlined in Table 11.4. Initially, we generated a suitability map for oil palm by considering country, continent, and global values, as derived from Tables 11.2 and 11.3. Subsequently, we applied a filter to exclude areas where oil palm expansion is nearly impossible, utilizing a filter mask comprised of land that cannot be converted for other uses (as indicated by global land cover products released by the ESA CCI) or land designated as protected (based on data from the WDPA released by the United Nations Environment Programme (UNEP) and the International Union for Conservation of Nature (IUCN)). Finally, we excluded areas with high-carbon stock forests, defined as those with above-ground biomass equal to or greater than 100 t/ha (Avitabile et al., 2016). The remaining land represents the areas we recommend for future oil palm plantation expansion.

As depicted in Figure 11.2, a substantial amount of land remains available for oil palm cultivation, even after excluding areas designated as unsuitable due to their inability to be converted to other uses (such as urban, water, and wetland) or areas protected for conservation. Much of the land in Southeast Asian countries falls under the 'country' category, with Malaysia and Indonesia being notable examples. This highlights why Southeast Asia has witnessed the most significant expansion of oil palm plantations. Additionally, the regions in South America, Central America, and Africa hold significant potential for oil palm cultivation, as they align with the 'continent' value category. Furthermore, as previously analyzed, the Congo Basin, Brazil, and the coastal areas of Central Africa emerge as particularly suitable regions for future oil palm plantation expansion.

However, considering convertible or protected land is not comprehensive, it is necessary to consider environmental condition and support sustainable development. Therefore, we consider the impact of planting oil palm on carbon stocks. The global remaining areas suitable for oil palm cultivation: 298.2 M ha of global

TABLE 11.4
Additional Constraints on Oil Palm Suitability Mapping

	Criterion	Unit/Description	Original Spatial Resolution	Dataset Used
WDPA	Land already under use	Protected areas	30 arc seconds	IUCN and UNEP (2015)
Biomass	Above Ground Biomass (AGB) map	t/ha	30 arc seconds	Avitabile et al. (2016)
Land cover	Land not possible to convert	Wetland-water-urban	300 m	ESA (2017)

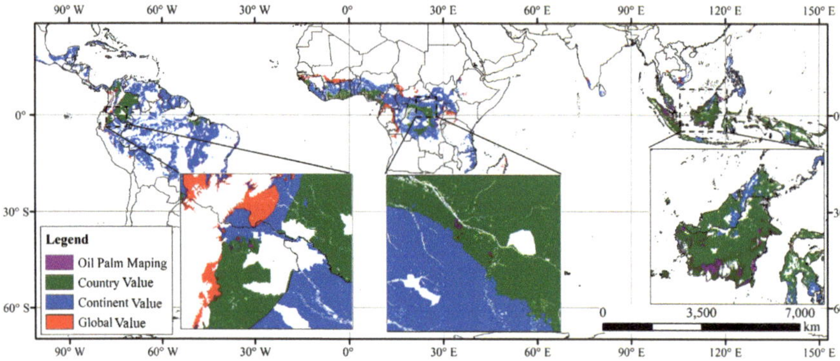

FIGURE 11.2 Oil palm suitability map generated based on different value criteria, excluding land that cannot be converted to other uses or is protected.

value, 289.34 M ha of continent value, and 73.25 M ha of country value separately (Table 11.5). As shown in Figure 11.3 and Table 11.5, there are fewer areas remaining suitable for oil palm cultivation in Southeast Asia than another region. South America and the coastal areas of Central African countries are the suggested regions for future oil palm expansion, such as Colombia, Brazil, Nigeria, Ghana, Ivory Coast, and Congo. However, the remaining areas of Brazil that of global value and continent value are much greater than the area that of country value. So, the regional suitability and outliers of restrictions should be considered more.

When compute suitable map for oil palm using country value, Colombia, Nigeria, Ghana, and Indonesia are suggested for the expansion of oil palm plantation. However, the strict oil palm suitability areas of Malaysia and Ecuador are negative value, demonstrating that the oil palm in Malaysia and Ecuador has already exceeded its sustainable planting area and encroached the high-carbon stock forests or protected areas (Table 11.5). Furthermore, Cameroon, Guinea, Papua New Guinea, Honduras, and Guatemala are not suitable for drastic oil palm expansion, but moderate planting is recommended.

11.5 LAND COVER CHANGE IN SUITABLE AREAS FOR OIL PALM

As depicted in Figure 11.4, the land cover in areas recommended for oil palm cultivation has undergone significant transformations since 1992 (Figure 11.4a). Over the past few decades, there has been a substantial increase in agricultural land, accompanied by a notable decline in forested and shrubland areas. This transformation is particularly evident in countries such as Indonesia, Malaysia, and Thailand, where nearly all forests have been converted into agricultural land, primarily for oil palm cultivation. This phenomenon reflects the rapid growth of the agriculture sector in these regions. Moreover, while there has been an increase in forested areas in countries like Nigeria, Ghana, Ivory Coast, and Congo, it is worth noting that much of this growth is due to the conversion of shrublands. As illustrated in Figure 11.4b, the suitable land for future expansion of oil palm cultivation in most countries now

TABLE 11.5
Remaining Oil Palm Suitability Areas

Country	Bio-physical Suitability (M ha)	Global Var (M ha)	Continent Var (M ha)	Country Var (M ha)	Oil Palm Map (M ha)	Strict Oil Palm Suitability Area 1 (M ha)	Strict Oil Palm Suitability Area 2 (M ha)	Strict Oil Palm Suitability Area 3 (M ha)
Brazil	287.52	121.41	120.05	1.38	0.19	119.86	121.22	1.19
Nigeria	44.54	36.38	36.36	11.89	2.86	33.5	33.52	9.03
Congo	96.28	33.56	31.88	2.46	0.06	31.82	33.5	2.4
Colombia	90.01	25.48	24.27	18.05	0.24	24.03	25.24	17.81
Indonesia	160.38	24.02	23.90	19.75	14.4	9.5	9.62	5.35
Ivory Coast	26.78	17.73	17.33	4.37	0.66	16.67	17.07	3.71
Ghana	31.64	16.58	14.97	6.45	0.39	14.58	16.19	6.06
Cameroon	33.75	9.19	7.65	0.30	0.27	7.38	8.92	0.03
Thailand	7.69	2.84	2.37	2.24	0.77	1.6	2.07	1.47
Honduras	5.77	2.83	2.79	1.14	0.20	2.59	2.63	0.94
Guatemala	5.41	2.62	2.43	1.06	0.10	2.33	2.52	0.96
Malaysia	31.97	2.49	2.48	2.33	6.76	−4.28	−4.27	−4.43
Papua New Guinea	31.80	1.63	1.59	0.97	0.18	1.41	1.45	0.79
Guinea	5.82	1.15	1.06	0.85	0.35	0.71	0.8	0.5
Ecuador	4.69	0.29	0.21	0.01	0.19	0.02	0.1	−0.18

Strict oil palm suitability area 1 = Global Var area-Oil Palm, Strict oil palm suitability area 2 = Continent Var area-Oil Palm, Strict oil palm suitability area 3 = Country Var area-Oil Palm.

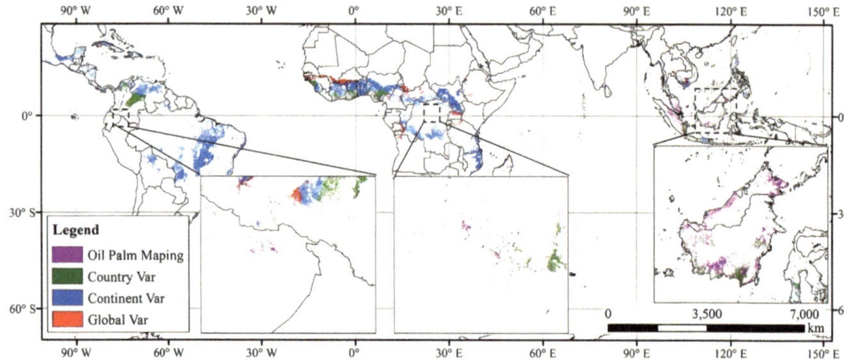

FIGURE 11.3 Oil palm suitable map computed by different value excluded high carbon stock forests that above-ground biomass greater or equal to 100 t/ha areas (Avitabile et al., 2016), and compare with global oil palm plantation mapping (Cheng et al., 2018). The regions into three oil palm focus areas (from right to left): Kalimantan Island, Congo Basin, and Amazon region.

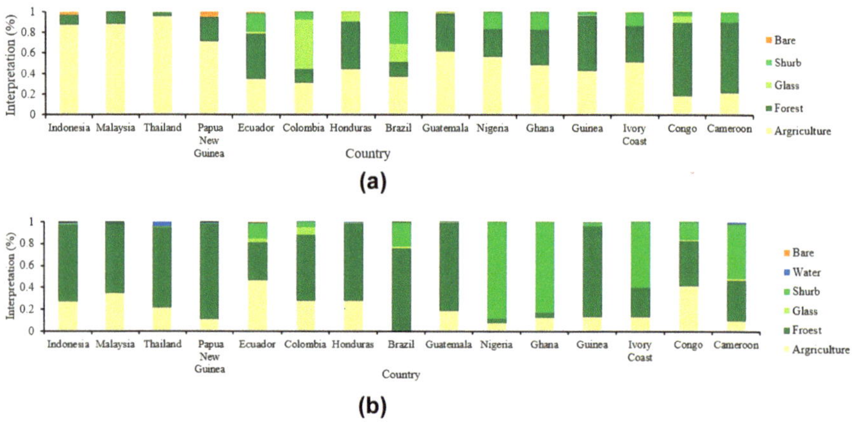

FIGURE 11.4 (a) Land cover in suitable areas in 1992 and (b) current situation of land cover in suitable areas.

predominantly comprises agricultural land, a practical and environmentally acceptable choice.

Human activities, particularly agriculture, pose one of the gravest threats to wildlife and biodiversity. Despite our exclusion of protected areas from our recommended oil palm–planting zones, it is important to note that many existing oil palm plantations have encroached upon these protected areas, as illustrated in Figure 11.5.

Across 15 countries, a total of 157 strict nature reserves, wilderness areas, or national parks have been impacted by encroachments from oil palm plantations. Malaysia bears the highest encroachment burden, with 64 such instances, followed by Thailand with 49 and Indonesia with 23. Collectively, these encroachments amount to 146,528 ha of oil palm plantations within strict nature reserves, wilderness

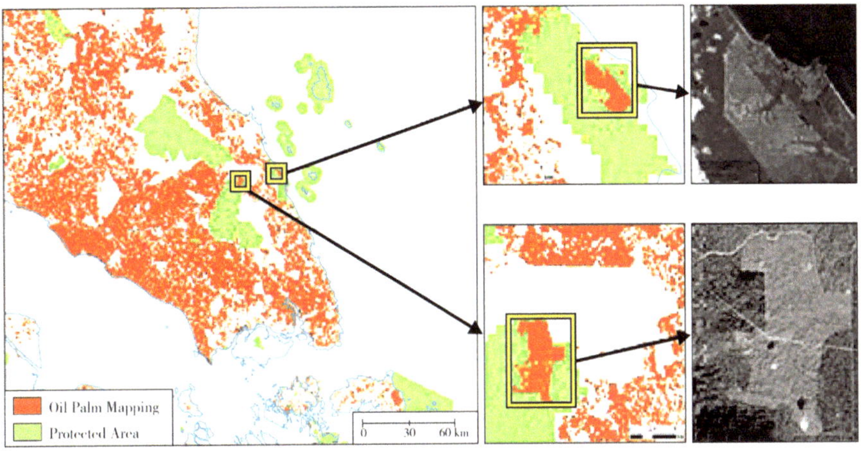

FIGURE 11.5 Oil palm plantations encroached upon protected areas in the region of Johor, Malaysia. Two protected areas namely Endau-Kota Tinggi (a wildlife reserve) and Ulu Sedili (a forest reserve) have been affected: as depicted from top to bottom.

areas, or national parks, constituting 0.56% of the total oil palm plantation area. Notably, Indonesia has experienced the most severe protected land encroachments, with 75,388 ha, accounting for 51.45% of the global total, while Malaysia follows with 23,005 ha (15.7%).

Through the analysis of a dynamic global oil palm samples database, we identified suitable conditions for oil palm cultivation on a global, continental, and country scale. Our findings reveal substantial areas worldwide available for potential oil palm expansion, totaling 298.2 million ha at the global level, 289.34 million ha at the continental level, and 73.25 million ha at the country level. Notably, South America and the coastal regions of Central African countries emerge as promising areas for future oil palm expansion, including Colombia, Brazil, Nigeria, Ghana, Ivory Coast, and Congo. However, challenges are evident, as some regions, such as Malaysia and Ecuador, have already exceeded their sustainable planting areas, encroaching on high-carbon stock forests and protected areas. Additionally, Cameroon, Guinea, Papua New Guinea, Honduras, and Guatemala are recommended for moderate rather than drastic oil palm expansion. The importance of avoiding outliers and considering regional suitability cannot be overstated, especially in cases like Brazil. Addressing existing oil palm plantations encroaching on protected areas is a pressing issue requiring concerted efforts at the national level. Moreover, the role of human interference with land and policy factors remains crucial and warrants further consideration in the context of future oil palm plantation expansion.

11.6 FLOOD IN OIL PALM PLANTATION

The sustainable oil palm map we developed did not exclude peatlands. However, encroachment on peatlands by oil palm plantations is an important cause of flood and water logging (Merten et al., 2020). Oil palm plantations and their expansion increase the intensity and frequency of flood risk (Khatun et al., 2017; Sumarga et al., 2016). The primary causes are subsidence due to peat decomposition and shrinkage caused by drainage in plantations and compaction during land clearing (Carlson and Garrett, 2018; Hooijer et al., 2015). Oil palm planted in peatlands requires drainage, which will cause the soil to sink by 3–5cm/year (Hooijer et al., 2012; Wösten et al., 2008). All of these will affect the ability of the surface to absorb precipitation, which directly leads to flood and waterlogging. Merten et al. (2020) found that the groundwater level of oil palm plantations is fluctuant and has lower water storage than forests, which makes oil palm plantations more vulnerable to flooding. Runoff in areas surrounding plantations will be affected, causing social conflicts, especially if surface runoff is diverted to residential areas.

Flood and water logging have always been important issues in the management of oil palm plantations, which can directly affect palm oil production and even cause tree death (Anzum and Naeem, 2021; Koon and Kun, 2006). However, waterlogging in plantations is not considered a serious problem because it does not pose a threat to the public, but there are still a small number of stakeholders who have experienced flooding (Middelberg et al., 2019). Remote sensing is not widely used for detecting floods within the plantations. 3D mapping before planting oil palm is an important means of assessing flood risk, but it is a complex, time-consuming, and costly process (Azam et al., 2021; Chong et al., 2017; Lupascu et al., 2020). Monitoring plantation water levels and soil quality using remote sensing technology is worth exploring but has not yet been implemented (Anzum and Naeem, 2021; Chong et al., 2017). A proper drainage system can effectively reduce the frequency of flooding, and research in plantations can help to develop effective solutions to reduce the damages of flooding (Anzum and Naeem, 2021; Lupascu et al., 2020). Soil protection and rational planning of land use are essential steps to avoid flooding, while planting flood-tolerant crops suitable for degraded peat in mature oil palm plantations can reduce flood impacts (Middelberg et al., 2019; Sumarga et al., 2016).

12 Carbon Cycle in Oil Palm Plantation

12.1 CARBON

Carbon serves as the fundamental building block of life on Earth, essential for the creation of crucial molecules such as proteins and DNA in all living organisms. Carbon is found in the atmosphere as carbon dioxide (CO_2) gas, in the ocean water as a dissolved gas, in the rocks as carbonate minerals containing calcium or magnesium and as organic carbon stored in plants (plant biomass) and soils. These are reservoirs and the carbon atoms are continuously travel from one reservoir to the other, a process known as the carbon cycle. CO_2 from the atmosphere is absorbed by plants on the Earth and ocean through carbon fixation process called photosynthesis. The absorbed carbon is then released back into the atmosphere when organisms respire or die and decay, when fossil fuels are burned, and through other mechanisms such as volcano eruption, forest fire, and land/soil disturbance. Carbon is vital to regulate the global temperature by warming up colder regions and provide food and energy to sustain all life in the universe.

The significance of plants in terrestrial ecosystems lies in their fundamental role in the global carbon cycle. Carbon stored within plants is relocated to other organisms such as herbivores and omnivores through ingestion processes. Research findings have revealed that terrestrial vegetation has the capacity to sequester as much as 183 Pg C annually (Sha et al., 2022), making it a crucial player in the global carbon cycle and climate change dynamics. Given that oil palm plantations cover an extensive area of 19 million ha (Murphy et al., 2021), their impact on the global carbon cycle can be substantial, leading to concerns about the palm oil industry's alleged contribution to CO_2 emissions (Cooper et al., 2020). Therefore, it is crucial to quantify the potential of oil palm in sequestrating and/or emitting CO_2 for making well-informed decisions regarding sustainable land use, climate change mitigation, and biodiversity conservation. This chapter explores the carbon cycle in oil palm plantations, by examining carbon inputs, storage, outputs, and the potential for carbon sequestration. The chapter highlights the role of geospatial technology to quantify carbon uptake and storage by palm trees and presents a case study in one of the oil palm plantations in Malaysia.

12.2 CARBON CYCLE OF OIL PALM PLANTATIONS

Oil palm trees take up CO_2 from the atmosphere to produce organic matter, which forms the basis of biomass. CO_2 from the atmosphere is absorbed by the fronds for the growth of fronds, stems, roots, and trunk (Lamade and Bouillet, 2005). Additionally, carbon inputs also come from the decomposition of organic matter,

DOI: 10.1201/9780429199813-12

such as fallen leaves, fruits, and plant residues, which release carbon back into the soil. The absorbed carbon is stored in aboveground and belowground components of the oil palm ecosystem (Figure 12.1). Trunks, fronds, and fruits form the aboveground components, and these biomasses can accumulate significant carbon stocks over its lifetime. While carbon is stored belowground in the extensive root systems of oil palm trees, as well as in the soil. The carbon stored in the soil is crucial for soil health and fertility, contributing to the overall stability of the ecosystem.

As oil palm trees mature and grow taller, their carbon stocks increase, a trend influenced by factors such as seed type or varieties, which can impact the growth rate of oil palm trees, averaging between 25 and 40 cm/year (Setyamidjaja, 2006). The photosynthetic activity of oil palm trees varies depending on the number of fronds in the crown. Young palm trees can produce more than 20 fronds/year, with fronds growing to lengths of 5–8 m, and by the ninth year, the canopy captures over 80% of the available light and produces maximum biomass (Lamade and Bouillet, 2005). The trunk of an oil palm plant begins to form after approximately 3 years of plantation, with the apex reaching its full diameter (forming an inverted cone) and intermodal elongation taking place (Verheye, 2010). Carbon accumulation in oil palm trees generally continues as the trees age (Figure 12.2). As the tree matures, it develops more leaves and fronds, and the trunk also increases in height leading to an increase in biomass. Lignification that occurs in older trees makes the trunk and branches denser and heavier, thereby increase in biomass (Corley and Tinker, 2016). The relationship between carbon stock and the age of oil palm is illustrated in Figure 12.2.

Oil palm trees as a living organism respire and release part of the stored carbon in the form of CO_2 as a by-product. Mature trees exhibit the highest release of CO_2 through respiration, emitting approximately 200 kg C/year under optimum ecological conditions (Lamade and Bouillet, 2005). This respiration process occurs in various parts of the palm, including leaves, fronds, and roots, releasing CO_2 into

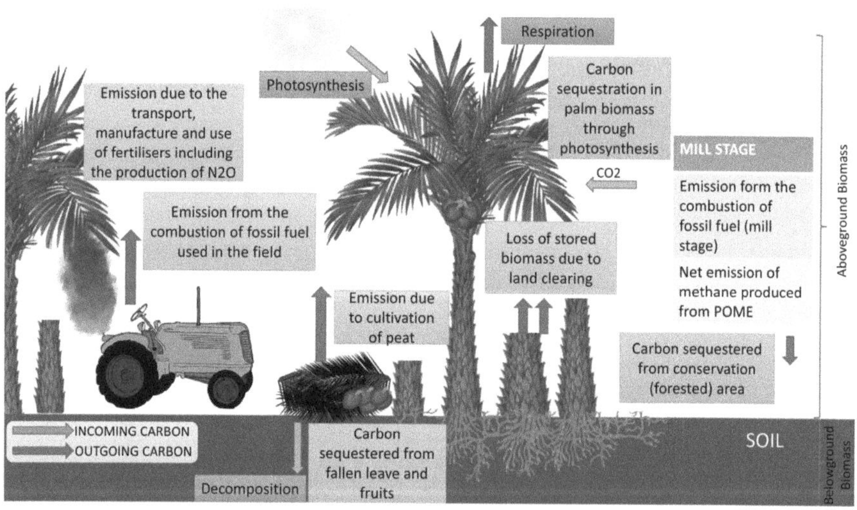

FIGURE 12.1 Carbon cycle in oil palm plantations.

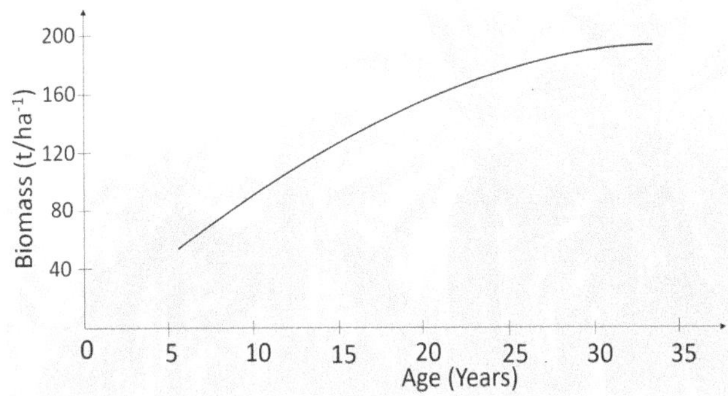

FIGURE 12.2 Changes in biomass of oil palm trees according to age (Tan et al., 2014; Corley and Tinker, 2003; Syahrunudin, 2005).

the atmosphere (Lamade and Bouillet, 2005). Carbon loss also occurs from the ecosystem via decomposition of dead plant materials. Most important carbon emission contributed by palm oil industry is through the conversion of natural ecosystems such as forests and peatlands into oil palm plantations (Figure 12.2).

The conversion of tropical forests into oil palm plantations not only releases massive amounts of stored carbon but also significantly reduces the planet's ability to sequester CO_2. A comprehensive study by Carlson et al. (2013) reveals the alarming carbon loss associated with palm oil-driven deforestation, emphasizing the urgency of addressing this issue. Oil palm cultivation often takes place on peatlands, where drainage for plantation establishment leads to the release of substantial amounts of CO_2 and other greenhouse gases. These emissions result from the decomposition of peat and the oxidation of organic matter. Hooijer et al. (2010) demonstrated the considerable emissions linked to peatland degradation and emphasized the importance of sustainable peatland management to mitigate CO_2 releases.

Other CO_2 emissions by the industry include fertilization, utilization of machines and transportations in the plantations, and palm oil mill effluents (POME) (Putro, 2022). The intensive use of fertilizers and agrochemicals in oil palm plantations contributes to CO_2 emissions through various pathways. Additionally, soil disturbance during plantation establishment and maintenance can lead to increased CO_2 emissions from soil respiration.

A previous study has provided valuable insights into the carbon dynamics of oil palm plantations, reporting that oil palm sequesters approximately 161,000 kg CO_2/ha/year through gross assimilation, while releasing about 96,500 kg CO_2/ha/year back to the atmosphere through respiration. This results in a net carbon storage of 64,500 kg CO_2/ha/year (MPOC, 2007). Considering an estimated economical lifespan of 25 years for oil palm trees, it is projected to sequester a net total of 1,612,500 kg CO_2/ha over its lifespan. Overall, the carbon cycle and dynamics of oil palm plantations are influenced by a combination of factors related to tree growth, age, environmental conditions, management practices, and land use/cover disturbance. Understanding these processes is vital for assessing the carbon sequestration potential of oil palm plantations and implementing effective strategies to enhance carbon storage while sustainably managing this critical agricultural industry. The following section presents frequently employed techniques for deriving assessments of the carbon cycle in oil palm systems.

12.3 MEASURING CARBON CYCLE IN OIL PALM PLANTATION

Various methods are used to obtain estimates of carbon sinks and sources in oil palm plantations at scales ranging from leaf, plot, and plantation.

12.3.1 Infrared Gas Analyser

The photosynthesis rates (CO_2 uptake in μmol/m^2 second) in plants at a leaf scale can be measured effectively using an infrared gas analyser. This instrument records the absorption of CO_2 molecules in the air that passes over a leaf surface at infrared radiation (Baskauf and Eickmeier, 1994). The instrument is also able to measure CO_2 concentration and humidity that enters the leaf chamber and the intensity of photosynthetically active radiation and CO_2 concentration that exit the chamber (Baskauf and Eickmeier, 1994). In this way, the net uptake of CO_2 can be determined. However, for oil palm it is difficult to measure the photosynthesis rate using the infrared gas analyser due to the aerenchyma-like spongy mesophyll and corrugated structure of the leaves which make it difficult to seal the leaves in the infrared gas analyser chamber (Apichatmeta et al., 2017). Consequently, not many studies have quantified the photosynthesis rate of oil palm trees. Apichatmeta et al. (2017) employed blue-diode pulse amplitude modulation fluorometry (PAM), a fluorescence-based technique, to assess the light-dependent photosynthetic reactions. Unlike infrared gas analysers, which measure CO_2 gas exchange, PAM focuses on quantifying the actual photosynthetic processes. Using this approach, Apichatmeta et al. (2017) conducted an investigation into photosynthesis among various growth stages, including seedlings (<6 months old), young (1–2 years), and mature (>5 years) palm trees within the gardens of Prince Songkla University, Phuket campus, Thailand. They reported a maximum quantum yield of photosynthesis was similar for both young and mature trees, but the rate of photosynthesis under optimum sunlight during the midday were two times higher for mature trees (≈600 μmol photon/m^2 second) compared to the young. Oil palm trees growing in tropical regions strategically optimize photosynthesis under high sun light both in mature palms and in young trees and seedlings.

12.3.2 Eddy Covariance-Based Flux Tower

Measuring the absorption (photosynthesis) and release (respiration) of CO_2 by palm trees on an ecosystem level can be accomplished through a collection of automated instruments positioned on towers and employed via eddy covariance (EC) methodologies (Baldocchi, 2003). The EC technique operates by measuring the CO_2 exchange (CO_2 flux) between the atmosphere and the oil palm canopy interface. CO_2 is exchanged within the atmosphere–canopy boundary by eddies (turbulent air motions) that grow vertically below, within, and above the canopy (Figure 12.3). The

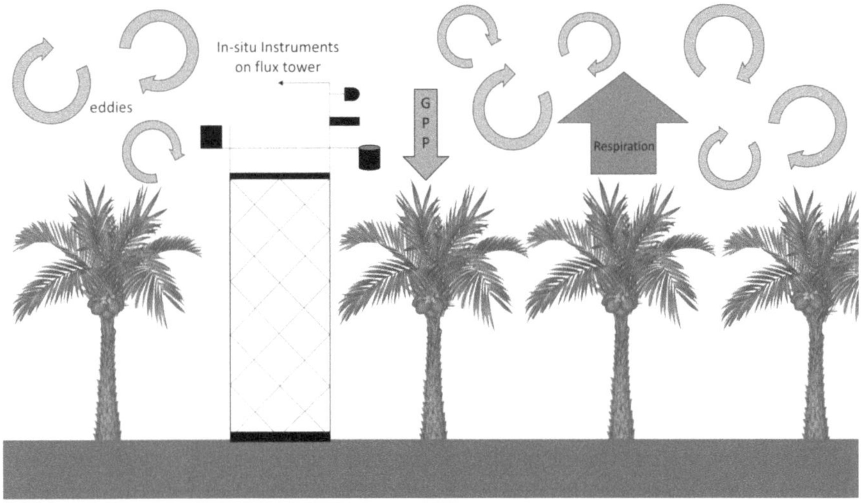

FIGURE 12.3 Operation of eddy covariance-based flux towers over oil palm ecosystem. Redrawn from Tan et al. (2012).

EC method is more effective in capturing CO_2 through photosynthesis and CO_2 losses from plants and soil (via respiration) concurrently. It is thus able to provide estimates of net ecosystem exchange (NEE) that quantify the overall carbon exchange between an ecosystem and the atmosphere per unit ground area. This serves as a key indicator of the ecosystem's capacity to sequester or emit carbon. The EC technique obtains the net CO_2 exchange (NEE) of the ecosystem using the mean air density, mean vertical wind velocity, and mixing ratio of CO_2 (Baldocchi, 2003). The NEE value can be negative if the ecosystem sequesters more CO_2 than it emits to the atmosphere and vice versa. This technique provides carbon flux data at high-temporal resolution that can define whether palm trees are a sink or source for carbon.

The NEE is partitioned into gross primary production (GPP) and ecosystem respiration by establishing a correlation between night-time NEE (which is assumed to primarily represent ecosystem respiration) and air temperature. This correlation is then employed to estimate the contribution of ecosystem respiration to daytime NEE, with the remaining portion being attributed to GPP (Reichstein et al., 2005). The amount of CO_2 absorbed by oil palm leaves (GPP) is estimated by deducting respiration from plants, roots, and soils from NEE at approximately 1 km^2 footprint (flux tower scale). The net primary productivity (NPP) is derived by subtracting the autotrophic respiration from GPP (Turner et al., 2003). Respiration is estimated at night when the CO_2 concentration by respiration is dominant and the values during the day are adjusted as a function of air temperature (Henson, 1993). Other variables measured on the tower include wind speed and direction, radiation, temperature, humidity, precipitation, and canopy and air temperature (Henson, 1993). This micro meteorological technique can provide reliable long-term (entire life cycle) ecosystem scale CO_2 flux that is used for CO_2 balance assessment despite its high operational cost (Henson, 1993; Henson and Harun, 2005). CO_2 flux estimates of NEE, GPP, and ecosystem respiration (Reco) are generally used to assess the carbon balance of oil palm ecosystem.

The EC-based techniques were first used by Henson (1993) to study the diurnal changes in CO_2 flux within 9-year-old palm trees in a coastal region in Peninsular Malaysia. The study reported solar radiation and vapour pressure deficit to be the dominant drivers of CO_2 flux. In a later study, Henson and Harun (2005) collected CO_2 flux data measured from an 8.7-m flux tower instrumented within three years old oil palm trees in the northern state of Kedah, Malaysia. These studies reported a fourfold higher absorption of CO_2 in the wet (−4.31 μmol CO_2/m^2 second) compared to the dry season. Subsequent studies investigated the effect of rainfall on CO_2 exchange between young palm trees and the atmosphere (Haniff et al., 2004; Henson and Harun, 2005).

An extended 18-month EC-based study was conducted by Haniff et al. (2016) at a plot covered by 15-year-old palm trees in Peninsular Malaysia. Notably, the study revealed pronounced seasonal fluctuations in the monthly averaged CO_2 fluxes throughout the observation period. With a monthly average CO_2 flux of −3.5 μmol CO_2/m^2 second, the findings indicated that the mature oil palm ecosystem within an inland mineral soil region in Peninsular Malaysia acted as a CO_2 sink. Young palm oil trees are, however, found to be releasing more CO_2 than their potential to accumulate carbon. Recently, Meijide et al. (2020) investigated the net carbon flux of an oil palm plantation on an ecosystem scale in Jambi, Sumatra. They found the young plantation (1 year old) acted as a carbon source (1,012 ± 51 gC/m^2year), while the mature plantation (12 years old) acted as a carbon sink (−754 ± 38 gC/m^2year). A mature plantation in Riau Province further supported these findings, with mature oil palm acting as a carbon sink assimilating 36–40 t CO_2/ha/year (Septiwibowo et al., 2019).

The carbon dynamics of oil palm ecosystems are influenced by soil CO_2 storage, particularly in peat soil. Disturbing peat for planting palm trees can release stored CO_2. Kiew et al. (2020) discovered higher CO_2 emissions (mean annual net emission of 36.4 Mg CO_2/ha/year) from oil palm on peat soil compared to mineral soil studies (Hirano et al., 2012; Meijide et al., 2020). Similarly, McCalmont et al. (2021) monitored carbon exchange from newly planted and mature oil palm plantations converted from peatland in Sarawak, Malaysia. Young plantations exhibited eight times higher net CO_2 emissions (137.8 Mg CO_2/ha/year) than mature ones, emphasizing the role of plantation age and biomass accumulation. Overall, as oil palm plantations mature, they tend to exhibit increased CO_2 uptake and reduced soil CO_2 losses. However, even after 12 years, plantations on peat soil remained a carbon source, and the regaining of lost carbon within a typical plantation rotation remains a challenge. Additional EC investigations on peatland plantations, particularly during the initial conversion years when the highest net emissions are anticipated, should be undertaken. These findings emphasize the complex interplay between plantation age, soil type, and carbon dynamics in oil palm ecosystems. The conclusions drawn from prior research emphasize the necessity of conserving tropical peatlands to leverage their potential in addressing global warming and climate change.

12.3.3 Biometric Method

A biometric method involves the estimation of oil palm biomass (dry weight) by analysing the organic materials present in different living and deceased components, including the trunk, frond, leaf, fruit, and aboveground litter of the oil palm, as well as the belowground weight of its roots (Corley and Tinker, 2008) (Table 12.1). The weight or combined mass of the various components of oil palm trees is called biomass and approximately 50% of the biomass is carbon where oil palm accumulates carbon in biomass over time (Corley and Tinker, 2008; Houghton et al., 2009). Another 50% is consisting of hemicellulose, cellulose, lignin, water, and trace amounts of extractives and minerals (Corley and Tinker, 2008). The biomass density is expressed as milligrams or tonnes per hectare. Biomass primarily consists of carbon compounds, which are derived from atmospheric CO_2 through photosynthesis and released back to the atmosphere via decomposition and combustion activities (Figure 12.1). Thus, estimating oil palm biomass can provide a snapshot of CO_2 that resides in the ecosystem and monitor its dynamics as a consequence of land-use changes.

The biometric method to obtain oil palm biomass involves several processes, such as cutting down the trees, separating each component of palm trees, including fronds, leaflets, fruit bunches, flowers, slabs of stems, and roots, drying them, and weighing their dry weight. This method is destructive but can produce highly reliable outcomes, as each part of the oil palm is precisely measured to derive carbon stock estimates based on the dry weight. Kiyono et al. (2015) calculated both the aboveground biomass (AGB) and belowground biomass (BGB) of mature (12 years old) palm trees in Sarawak, Malaysia, using the biometric technique. They pruned and weighted the aboveground components of flowers and fruits of the trees and felled the stems and cut into logs to calculate their volume. The roots were dug and weighed at every 0.3 m depth to get the BGB. Lastly, all of the samples' dry/fresh weight ratios were determined, and the samples were oven-dried at 75°C or until they reached a constant weight to obtain the palm trees' dry biomass.

In another study, Sunaryathy et al. (2015) estimated the AGB of palm trees by felling a total of nine trees, categorized them into three age groups (1–3, 4–10, and 11–20 years). The process involved measuring the diameter at breast height (DBH) before and after felling, obtaining height measurements with and without fronds post-harvest, and segregating fronds, leaflets, fruit bunches, and flowers from the felled trees. The tree stems were sliced and analysed for volume and dry weight calculation, using the Newton formula to account for varying stem sizes. The method involved measuring weight and wet volume of smaller cubes through submersion in paraffin and water, and subsequently obtaining constant dry weight through heating and desiccation. Other tree components were similarly dried in an oven at 105°C ± 3°C or until a constant weight is achieved to determine their dry weight. For the three age groups, the AGB was found to be 0.75, 22.17, and 105.41 t/ha, respectively, with mature trees exhibiting the highest stem biomass (62%). The study also compared the biomass distribution among different components. In younger trees (<3 years), fronds constituted a larger portion of AGB, while stem biomass gradually increased with tree maturity.

TABLE 12.1
Estimation of Aboveground (AGB) and Belowground (BGB) Biomass/Carbon of Oil Palm Trees Using Biometric Method

Source	Age (years)	Site	AGB/BGB	Components
Ng et al. (1968)	11	Pen. Malaysia (mineral soil)	AGB	Trunk
Khalid et al. (1999a,b)	23 (10 trees)	Kluang, Johor, Peninsular Malaysia	AGB	Leaflets, Rachis (including petiole), Spears, Growing point or 'cabbage', Trunk, Frond bases, Inflorescences.
Nordin et al. (2002)	25	Tapah, Perak, Peninsular Malaysia	AGB	-
Corley et al. (1971)	1.5–27.5 (38 per age class)	Pen. Malaysia (mineral soil)	AGB	Trunk
Corley and Tinker (2003)	1.5–27.5	Mineral soil (Malaysia, Indonesia, Africa)		
Henson (2003)	2–12	Mineral soil (Peninsular Malaysia)	AGB	Canopy
Tenkabail et al. (2004)	1–5 years (1–3 and 4–5) 7 trees	Benin, West Africa	AGB	Above ground core, leaf (dead and live), rachis (dead and live), flower, and palm nut.
Syahrinudin (2005)	3–30 (3, 10, 20, 30) (3 trees per age class)	Mineral soil (Sumatera)	AGB, BGB	AGB (Spear, leaf, rachis, frond base, trunk) BGB (trunk base, roots) Undergrowth, non-leaf litter, leaf litter
Aholoukpè et al. (2013)	(25 trees)	Benin, West Africa	Frond biomass	Petiole, rachis, leaflets, fragment (along the rachis)
Khoon et al. (2019)	21 (10 trees)	Lambir, Miri, Sarawak (mineral soil)	AGB, BGB	Trunk, Frond (Leaflets, rachis, petiole, frond base, spears, cabbage, inflorescence, fruit bunch, epiphytes), and roots
Lewis et al. (2020)	3, 8, 12 years (9 trees)	Peat soil, Sarawak	AGB	Frond (rachis, petiole, and leaflet)
Migolet et al. (2020)	35 years (18 trees)	Congo basin, west Africa ferralitic Cambisol-dominant soil	AGB	Fruits, leaflets, petioles, rachises, and stems

Earlier research by Yulianti (2010) and Hartley (1967) supported the dominance of stem biomass, which varied between 57% and 69% for mature trees aged 10–17 years, while fronds and leaflets contributed smaller proportions. This trend was consistent across different soil types, highlighting the regularity of biomass distribution in oil palm trees. Younger trees exhibited higher frond biomass due to lower stem density and higher water content, resulting in less biomass compared to older, less water-rich stems. This indicated that the stem's lower water content in older trees contributed to their higher biomass content (Tsoumis, 1991).

In the Congo Basin of the West Equatorial Africa, Migolet et al. (2020) estimated the biomass of various oil palm trees components (fruits, leaflets, petioles, rachises, and stems) in a mature plantation dominated by trees aged 35 years old using the destructive sampling (Table 12.1). The samples were oven dried at 105°C (except for the leaflets at 65°C) to obtain the dry biomass. The results showed the stem contained 73% of the total AGB, while other components such as rachises, petiole, fruits and leaflets contained 13%, 8%, and 6% of the AGB, respectively. The AGB of trees planted on peat also demonstrated similar outcome where Lewis et al. (2020) harvested nine palm trees covering three trees for each age classes: immature (3 years after planting), young-mature (8 years), and mature (12 years) and quantified the AGB in Sarawak, Malaysia. Different parts (fronds, frond base, rachis, petiole, leaflets, trunk, frond base, inflorescences, fruits, spear, cabbage, and epiphyte) of the trees were sectioned and oven dried at 105°C or until a constant mass was reached. Results of the study confirmed that most of the total dry weight (46%–78%) of the palm trees are contained in the trunk and frond base of the 8–12 years old trees. On the other hand, young trees stock most of its biomass in the fronds. This study also revealed that palm grown in the peat and mineral soil show no difference in terms of the trunk and total frond dry weight.

12.3.4 Allometry Equations

Destructive measurements to get oil palm biomass are essential, however, they are non-sustainable as trees need to be cut down, high cost, laborious, and time-consuming. Alternatively, allometric equations which are developed by relating harvested field data (i.e. biomass) to non-destructive or semi destructive measurement of biophysical variables such as age, height of trunk, and DBH (Corley and Tinker, 2003; Henson, 2003; Syahrinudin, 2005) are invaluable. This enables biomass estimation straight forward without the need for cutting down the trees and importantly they enable extrapolation/upscaling of AGB estimates over a large spatial extent such as the block scale, plantation, or even regional scale.

Since, biomass calculation is very site-specific, as it depends on the topology, plantation techniques, growth rate, and so on, different allometric equations are suitable to be implemented at different locations (Table 12.2). The allometric equation formulated by Corley and Tinker (2003) was initially designed for use in Indonesia, Malaysia, and Nigeria. The equation provides a quantitative approach to estimate the biomass of a palm tree, in kilograms per tree incorporating both structural and developmental aspects such as tree's height, cross-sectional area of the trunk, and a factor related to its age. While, the equations of Henson (2003) and Syahrinudin (2005)

TABLE 12.2
Allometric Equations for Estimating Oil Palm Aboveground Biomass (AGB)

Source	Parameters	Models
Corley and Tinker (2003)- kg/tree	Height, radius, age (1.5–27.5 years)	Biomass (kg/tree) $= \pi(0.777r)^2 h [(0.0076x + 0.083)/10]$
Henson (2003)- tonne/ha	Age (2–12 years)	Biomass (t/ha) $= -0.00020823x^4 + 0.000153744x^3 - 0.011636x^2 + 7.3219x - 6.3934$
Syahrinudin (2005)-tonne/ha	Age (3, 10, 20 and 30 years)	Biomass (t/ha) $= 68.2 \ln(x) - 36.7$
Aholoukpè et al. (2013)	(25 trees) Length	$DW_{frond} = 1.147 + 2.135 \times DW_{rachis}$ $DW_{rachis} = 1.133 \times L_{rachis}\ DW_{fragment}/0.30$
Khalid (1999)	Height (age 23 years)	AGB $= 725 + 197H$
Germer and Sauerborn (2008)	Age	AGB $= 18.95 \times$ YAP05
Carlson et al. (2012)	Age	AGB $= 1.526(5.97 \times$ YAP0.62)
Lewis et al. (2020)	Age (3, 8, 12 years)	AGB $= 6.389$ YAP$\times - 17.59$
Migolet et al. (2020)	35 years	$[B_{stem} + B_{leaf}] = [a_1 + b_1\ In(DBH^2\ H_T) + a_2 + b_2\ In(DBH^2\ H_T N_F)$

represent the estimated biomass of a palm tree stand or plantation, measured in metric tons per hectare (t/ha). The terms in Henson's (2003) create a complex equation that estimates biomass based on the age of the tree stand in Malaysia. While, the equation developed by Syahrinudin (2005) uses the natural logarithm (ln) of the age to estimate biomass. The specific constants (68.2 and –36.7) determine the shape and magnitude of the relationship between age and biomass specifically in Sumatera, Indonesia. These three equations are widely used by many researchers around the world to get the biomass of oil palm trees. More details about other allometry equations and the parameters used in the equations are provided in Table 12.2.

An allometric equation for oil palm AGB estimation on tropical peat soil is defined by Lewis et al. (2020) based on destructive approach of nine palms supported by non-destructive analysis of another 902 palms. This equation was built on the palm trees of year 3, 8, and 12, while no palms more than 12 years was harvested, which shows the uncertainty of this equation applied on older oil palm trees. When considering the dry weight of fronds, significant factors included the linear density of the frond rachis and the cross-sectional area of the petiole, collectively accounting for 76%–82% of the observed variation in dry weight. This finding contradicts to the finding of Aholoukpè et al. (2013) who concluded rachis linear density was a better predictor compared to petiole cross-sectional area. It should be noted however that measuring the rachis linear density from the dry weight of a rachis fragment in the field is difficult.

For the palm dry weight and total palm biomass, trunk volume was the best predictor. Instead, when trunk height was used to estimate palm dry weight in mature and young mature palms, a large underestimation was observed by Thenkabail et al. (2004) but low underestimation of 16% was observed by Dewi et al. (2009). Singh et al. (2018) reported higher crown biomass (65%) in younger palm (1–3 years), while the trunk with old frond bases biomass showed a larger percentage (68%) in the older oil palm aged 4–11 years in northeast India. Total AGB was highly correlated with the trunk height ($r=0.985$), total height ($r=0.994$), and age ($r=0.973$). On an average, portioning of biomass and carbon stock was in the order: AGB > BGB > standing litter biomass > deadwood biomass > understory biomass. An 11-year oil palm plantation accumulated 111.96 Mg/ha biomass with a carbon density of 49.90 Mg C/ha and could sequester 3.70 Mg C/ha/year in 10 years after planting in Mizoram, Northeast India.

Based on the available allometric equations developed up to date, age of oil palm is the most significant and could be the single parameter for biomass estimation. Some allometric equations use additional parameters such as height and DBH to improve the allometric model, but only Khalid (1999a), Aholoukpè et al. (2013), and Migolet et al. (2020) generated the allometric models without the age information. Height can be associated with age as the height increases linearly/positively (significantly) with the age of palm trees, but only weakly and negatively with DBH (Tan et al., 2013). In Sabah, height was found to be a reliable predictor of AGB for trees aged 2–24 years, yielding a root mean square error (RMSE) of just 18% (Alometri et al., 2016). Integrating a wide range of parameters, the Corley and Tinker equation is employed to compute biomass at the field level (Tan et al., 2014). The resulting value is then extrapolated across all trees to calculate the actual biomass of oil palms, a process further influenced by the planting density (Kho and Jepsen, 2015).

The belowground components of palm trees such as the roots and soil contain a large pool of carbon. Quantification of soil carbon storage and losses indicates that the increase of root biomass offers short-term carbon sink, while the soil respiration, soil surface CO_2 flux, fertilization, and initial soil carbon losses due to land conversion could turn oil palm BGB into carbon source in a long term (Danielsen et al., 2009).

All the three methods described above to get carbon flux, biomass, and carbon content of palm trees have limitations due to the insufficient network of flux towers covering palm oil plantations, its nature of being laborious, expensive, and non-sustainable practice of the destructive biometric methods. Instead, to obtain a large-scale carbon absorption, stock, and emission of oil palm, mathematical modelling and remote sensing-based models become very effective. These methods are discussed next.

12.3.5 Modelling

Mathematical models play a crucial role in understanding and predicting oil palm productivity across different time and space scales. These models employ a set of equations to describe and simulate the complex interactions that influence oil palm growth and yield. Developed based on biophysical and environmental factors, these equations offer a powerful tool for conducting sensitivity analyses by altering physiological and environmental parameters. A significant application of mathematical models lies in their capacity to simulate oil palm biomass yields. These models, rooted in physiological processes and phenological characteristics such as flowering and rotation dynamics, allow for the predictions of carbon storage pools, biomass density, and growth patterns. Tan et al. (2012) explored the adaptability of existing ecosystem models to estimate oil palm primary production, enabling efficient estimation without the extensive data collection required by field studies.

The land surface models (LSM) have been harnessed to simulate the intricate processes within the oil palm ecosystem. Models like the community land model-palm (CLM-Palm) and ORCHIDEE-MICT-OP were developed to capture the intricate interplay between carbon, water, and energy cycles. CLM-Palm, for instance, successfully predicted variations in NPP, yield, and leaf growth. The integration of sub-canopy structures and phenology in CLM-Palm enables the exploration of land-use changes from natural ecosystems to oil palm plantations, shedding light on their impacts on carbon, water, energy cycles, and regional climate.

Critical to accurate simulation is the incorporation of oil palm's functional traits, encompassing morphological, physical, and physiological characteristics. Xu et al. (2021) addressed this challenge by integrating oil palm-specific morphology, phenology, and harvest processes into the ORCHIDEE-MICT global LSM. Their refined model considered age-specific parameterizations for photosynthesis, autotrophic respiration, and carbon allocation, resulting in successful reproduction of leaf area index, biomass density, and fruit yield across various observation sites.

While mathematical models offer a comprehensive means of explicit carbon process representation, they can become intricate when applied to fine temporal and spatial scales. Models often concentrate on detailed processes within specific vegetation types, sometimes neglecting the spatial variability, structural patterns, and possible consequences of oil palm redistribution on productivity. To account for spatial variation in oil palm productivity and biomass, spatially explicit models are needed, tracking temporal dynamics in vegetation. Remote sensing techniques provide a solution, translating spectral properties of oil palm into valuable data on carbon flux, productivity, and biomass, enabling a more comprehensive understanding of oil palm growth and its environmental impacts.

12.4 REMOTE SENSING

The estimation of oil palm productivity and biomass through remote sensing techniques has garnered attention, albeit being relatively limited compared to other vegetation types. While numerous studies have focused on tropical forests and various vegetation types, remote sensing offers a valuable approach for assessing oil palm biomass and productivity (the intake of carbon), critical parameters for monitoring and managing the oil palm ecosystem. This methodology involves analysing radiometric properties of images and vegetation indices to derive essential information. Optical sensors, such as the advanced very high-resolution radiometer (AVHRR), have been pivotal since the 1980s, providing data on vegetation characteristics including normalized difference vegetation index (NDVI) and LAI which are calculated from reflectance measurements in the red and near-infrared channels, offering insights into vegetation health and density.

The utility of NDVI in terrestrial productivity estimation was initially demonstrated by Tucker et al. (1983), who linearly correlated NDVI with grass biomass and extended this relationship to predict NPP in Africa. NDVI's temporal variations were associated with climate and biome types, revealing its potential in reflecting environmental changes. Subsequently, remote sensing models incorporating NDVI and environmental constraints were developed for estimating vegetation biomass at larger scales. The light use efficiency (LUE) logic, proposed by Monteith (1972), became foundational in quantifying plant GPP and NPP.

12.4.1 Spectral Information for Oil Palm Biomass Estimation

With the deployment of high-spatial resolution sensors such as IKONOS and SPOT-5, remote sensing capabilities expanded, allowing for enhanced biomass estimation. Optical data and vegetation indices, including NDVI, were utilized to estimate oil palm biomass in various regions. Nonlinear regression techniques, such as those employed by Thenkabail et al. (2004) and Singh et al. (2014), provided satisfactory results achieving accuracies ranging from 64% to 72% when correlating optical data with biomass obtained from destructive method or using allometric equations for oil palm trees in the savannah eco-regions of West Africa and Borneo, respectively. Notably, spectral signatures exhibited distinguished correlation with biomass, particularly in cases with less density trees, such as the 1–4-year-old palm trees in West Africa (Thenkabail et al., 2004).

However, limitations in optical-based methods arise when dealing with high biomass, elevated LAI, and complex canopy structures. Signal saturation in the context of optical remote sensing data refers to a situation where the measured signal or response from a sensor reaches a maximum value, beyond which further increases in the property being measured (such as vegetation biomass) do not result in proportional increases in the measured signal. This saturation point can lead to inaccuracies, that is, underestimation in estimating biomass, as the sensor is no longer able to distinguish between different levels of biomass due to its limited dynamic range.

To address this challenge, Asari et al. (2017) implemented a model combining reflectance data from bands 3 (red) and 5 (shortwave infrared) extracted from Landsat Thematic Mapper data with vegetation indices calculated from the bands like infrared index (IRI), enhanced vegetation index (EVI), and wetness from tasselled cap (WT). They successfully correlated the indices (log-log) with oil palm AGB, achieving a moderate prediction capability of approximately 62% variability in AGB and an RMSE value of 3.68 t/ha in the Selangor state of Peninsular Malaysia. The AGB for trees aged 6–23 years was determined through field-based measurements, utilizing the allometric equations developed by Corley and Tinker (2003) as well as Khalid et al. (1999a,b). Notably, the IRI exhibited a significant correlation with oil palm biomass due to the increased infrared reflectance from oil palm canopies, while healthy vegetation absorbed shortwave infrared. This index demonstrated a strong association with oil palm age, as noted by McMorrow (2001). Conversely, EVI exhibited potential in estimating wood volume, thus proving valuable in biomass estimation, as highlighted by Ni-Meister et al. (2010). When applying the satellite-derived data to calculate mean AGB for the palm trees, the average value computed was 58.50 t/ha, with a range spanning from 35 to 175 t/ha. Notably, higher AGB values were observed in correlation with older trees. This contrasts with the average AGB values obtained from the allometry equations proposed by Khalid et al. (1999a,b) (47.19 t/ha) and Corley and Tinker (2003) (40.77 t/ha), indicating a tendency for the satellite-based approach to yield comparatively higher AGB estimates.

Another effective spectral approach for estimating AGB in young oil palm trees (<10 years old) is the normalized difference fraction index (NDFI). NDFI leverages spectral information from optical data and is calculated using the fractional components of green vegetation (GV), non-photosynthetically active vegetation (NPV), shade, and soil within each pixel of Landsat ETM+ images, achieved through spectral mixture analysis (SMA) (Morel et al., 2012). Remarkably, NPV and shade components were found to closely align with oil palm biomass (ranging from 78% to 80%). This strong correlation may be attributed to the relatively simple canopy structure of young trees, which exhibit increased shade fractions as canopy closure intensifies. Consequently, an empirical equation for estimating AGB in young oil palm trees was developed using shade fraction imagery. Employing this technique, AGB values were successfully retrieved for oil palm pixels in the study area near Sandakan, Sabah, spanning 2000–2008, resulting in a sequestration estimate of 53.4 Teragrams (Tg) of CO_2 equivalent. This study emphasizes the potential of medium-resolution satellite data, particularly when combined with appropriate techniques like SMA, to offer reliable estimates of oil palm biomass, while synthetic aperture radar (SAR) data struggled to distinguish AGB in oil palm plantations (Morel et al., 2011).

In our previous study, we estimated GPP and NPP (biomass sink) of a private oil palm plantation in Peninsular Malaysia using LAI calculated from spectral data (NIR band of the UK DMC 2 satellite). GPP calculations require the fraction of photosynthetically active radiation (fPAR), which was determined through an empirical model linking radiance from UK-DMC 2's NIR with field-measured LAI. The upscaled LAI was converted to upscaled fPAR using Beer's law. Biomass, a function of age, changes gradually as oil palm matures, and NPP is related to biomass age-wise. The inter-annual biomass changes provide insight into carbon absorption, with annual NPP representing 45% of biomass. NPP estimation for oil palm (ages 1–25 years) involved allometric equations from Corley and Tinker (2003), Henson et al. (2012), Syahrinudin (2005), and data from the Malaysian Palm Oil Board (MPOB). Field-collected biometric data, encompassing DBH, trunk height, and age, were employed to compute biomass values.

This study (Tan et al., 2014) yielded GPP and NPP estimates for 2010 of approximately 0.09 Pg C/year (~0.33 Pg CO_2/year) and ~0.03 Pg C/year (~0.11 Pg CO_2/year) respectively. These outcomes suggest that Peninsular Malaysia's oil palm plantations could significantly contribute to global carbon sequestration. However, determining whether the oil palm ecosystem acted as a carbon sink or source in 2010 requires consideration of other carbon emission components, including ecosystem respiration, land-use preparation, fertilization, POME, and transportation. Although challenges remain in discerning the overall carbon balance, these findings offer valuable insights for oil palm management in anticipating the ecosystem's response to potential climate change impacts. Future studies should aim to estimate these emission components comprehensively to enable a comprehensive carbon cycle assessment for oil palm.

12.4.2 Texture Analysis for Oil Palm Biomass Estimation

Texture analysis has emerged as a promising solution to address the saturation challenges associated with optical data. Unlike spectral reflectance, texture information holds greater significance, particularly in enhancing biomass estimation within intricate stand structures (Lu, 2005; Sarker and Nichol, 2011). The underlying premise of texture analysis is that frequency signatures strongly correlate with the grain size of the canopy. Singh et al. (2014) conducted a study in Sabah, Malaysia, using SPOT 5 data where they correlated the reflectance values from all four bands (green, red, near-infrared, and shortwave infrared) with field-based AGB data to develop biomass estimation models for various forest types and young oil palm (5–8 years old).

While a robust relationship was identified between field-based AGB and SPOT-derived AGB, the latter tended to underestimate AGB values. Interestingly, weak correlations were found between vegetation indices such as NDVI and SAVI and AGB. However, reflectance values from red band demonstrated a strong association with field-based AGB values ($R^2=0.85$, $p<0.01$). So, texture details were extracted from the red band of SPOT-5 data using various measures from the grey level co-occurrence matrix (GLCM), including mean, standard deviation, contrast, and entropy. These texture measures proved to be informative about forest stand parameters and biomass dynamics. However, the results for oil palm plantations (5–8 years old) were not statistically significant.

When advanced texture analysis techniques were utilized, Migolet and Goita (2020) effectively estimated AGB of mature oil palm trees (>30 years old) in the Congo Basin, Central Africa. Their methodology harnessed the two-dimensional fast Fourier transform (FFT) and principal component analysis (PCA), collectively known as the Fourier transform texture ordination (FOTO) method. The Fourier transform converted radiation signal intensity into sinusoidal waves of distinct spatial frequencies. Concurrently, PCA performed textural ordination of the r-spectrum values derived from the FFT, facilitating the description of canopy characteristics. By focusing on the first three principal components, the variability in canopy attributes—such as canopy size, openness, and health status—was effectively characterized in satellite images. The resulting factor scores were used as textural indices (FOTO indices), categorizing palm canopies based on factors like age, canopy structure, and health status. To derive textural information for canopy structure, the infrared bands of PlanetScope and FORMOSAT-2 satellite images were processed. This information was then correlated with in situ AGB estimates obtained through laboratory and biometric data (tree height and DBH), employing allometric equations. Employing multiple linear regressions (MLRs) and multivariate adaptive spline regressions, they achieved high accuracy with low RMSE values <10%.

The study yielded AGB values ranging from 36 to 72 t/ha for the study area characterized by homogenous palm stands. It highlighted the suitability of the FOTO method for homogeneous environments, where textural variations primarily stemmed from topographic features (such as hilly versus flat areas). The spatial resolution of satellite images (2–3 m) emerged as a key determinant in effectively delineating palm crowns and subsequently enabling accurate AGB estimation. Notably, the choice of window size for texture analysis, discussed in Chapter 9, played a crucial role.

Optimal window sizing was essential, as too small a window might overlook textural variations in a homogenous surface, while overly large windows could underrepresent canopies while capturing broader landscape features.

12.4.3 Radar Data for Oil Palm Biomass Estimation

Optical imageries are proved to be effective in estimating oil palm biomass through upper canopy information, yet the bulk of biomass resides in palm tree trunks. To comprehensively estimate volume and AGB, satellite data capable of penetrating deeper into the canopy and revealing structural details are valuable. Radar-based remote sensing, particularly SAR, has emerged as a promising alternative. SAR operates in the microwave frequencies and can penetrate dense canopies, providing structural information about vegetation, including biomass of branches, stem, and vegetation density.

SAR is an active system, emits microwave pulses, and captures their reflections. When microwave radiation encounters vegetation, it interacts with various components such as leaves, branches, stems, and trunks. The interaction is influenced by factors like polarization, frequency, and incidence angle, leading to varying scattering mechanisms. This interaction forms the basis for biomass estimation using different frequency bands. SAR data, collected in K, X, C, L, and P bands, provide diverse AGB information (Table 12.3 and Figure 12.4). Longer wavelengths, such

TABLE 12.3
Different Bands of Microwave Data that Are Useful for Oil Palm Biomass Studies

Band and Wavelength	Biomass Estimation	Example of Satellite/ Sensors
X (~3.1 cm)	Interacts with the leaf canopy surface, so suitable to get biomass of tree crown	TerraSAR-X, TanDEM-X, COSMO-SkyMed Polarimetric interferometric SAR (Pol-InSAR)-
C (~5.6 cm)	Passes through foliage and scattered as it encounters small branches and underlying elements	Sentinel-1, Radarsat-2, RCM
S-band (~9.4 cm)	Passes through foliage and scattered as it encounters larger branches and underlying elements. Can result in volume scattering due to distribution of light across the entire tree	NovaSAR-S and NISAR
L (~23.5 cm)	Greater ability to penetrate through tree crown, main branches and tree trunk	ALOS -2 PALSAR-2, SAOCOM-1, NISAR-L
P (~69.0 cm)	Passes through foliage and interacts with tree trunk and trunk-ground interface, thus enables below canopy biomass retrieval. Sensitive to highly dense and structurally complex canopy	BIOMASS

Source: Sinha et al. (2015).

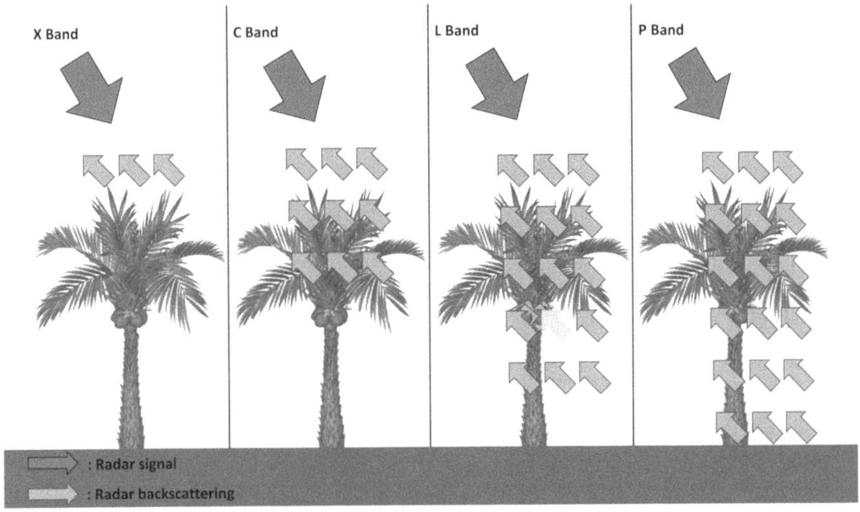

FIGURE 12.4 Scattering characteristics of different bands of radar with oil palm.

as L and P bands, can penetrate through foliage and interact with larger structures, resulting in a positive correlation with AGB. On the other hand, shorter wavelengths, that is, X- and C-bands are particularly sensitive to sparsely vegetated areas with low biomass or canopy biomass (Yadav et al., 2021). Both C- and L-bands are commonly employed for assessing biomass of individual crowns or entire canopies. The L-band radar data are also capable of penetrating clouds, thus essential for cloud-covered tropical regions. Changes in vegetation structure, which are often related to changes in biomass, can be detected by SAR even in areas with high vegetation cover where optical sensors might encounter saturation due to dense canopies.

SAR sensors offer valuable insights into vegetation structure and composition by utilizing different polarizations during data acquisition. These sensors emit radar waves with varying polarization orientations, enabling multiple perspectives of the same scene. The four primary polarizations used are HH, HV, VV, and VH. HH polarization interacts predominantly with surface features, making it sensitive to canopy roughness and understory vegetation, aiding in biomass estimation. HV polarization provides insights into volume scattering within the canopy, revealing the distribution and orientation of vegetation components. VV and VH polarizations are essential for probing interactions with large vegetation elements like tree trunks and stems, contributing significantly to biomass estimation in palm trees.

Polarimetric information from SAR data is employed to estimate oil palm biomass. When leaves within a dense vegetation canopy align perpendicularly to the radar illumination, they can directly reflect the signal back to the SAR sensor in a co-polarization manner, resulting in brighter appearances in C-band SAR images (Figure 12.4). Co-polarization, whether HH or VV, is particularly useful for studying biomass of tree stems when they are perpendicular to the ground surface, generating enhanced signals due to double bounce scattering. As biomass increases, backscatter values also increase due to higher moisture content in the scatterers (Nizalapur et al., 2010).

In three-dimensional features like forests or oil palms, volume scattering, especially at L-band, occurs on twigs and branches within canopies, leading to unpredictable backscattered signals in both co-polarization and cross-polarization. Despite this, HV polarization at L-band has shown a robust correlation with oil palm AGB (Shashikant et al., 2012). However, precipitation can diminish HV polarization's accuracy in predicting AGB. In areas like Sabah, Malaysia, where oil palm tree leaves' scattering obscures AGB information using HV polarization (Morel et al., 2011), dry period images or P band data that can penetrate raindrops may be more suitable for AGB estimation.

SAR data offer advantages over optical data in providing biomass information even at high levels, with saturation levels varying by bands. Specifically, C-band data saturates at around 60–70 t/ha (Nizalapur et al., 2010), while L- and P-bands saturate only at 100 t/ha for complex tropical forests, potentially reaching 250 t/ha depending on tree species (Sinha et al., 2015). Saturation levels are influenced not only by wavelength but also by polarization, vegetation structure, and ground conditions. For instance, PALSAR L-band HV polarization saturates at approximately 200 t/ha in Malaysian mangrove forests (Hamdan et al., 2014). Combining data from different bands and polarizations significantly enhances biomass estimation, with L-band data from the ALOS satellite being a preferred source for this purpose.

12.4.4 LiDAR Data for Oil Palm Biomass Estimation

LiDAR technology offers another avenue for accurate biomass estimation by providing detailed canopy height information. Airborne LiDAR, mounted on unmanned aerial vehicles, has been successful in deriving top of canopy height (TCH) and AGB estimates for oil palm trees (Fawcett et al., 2019; Beese et al., 2022; Nunes et al., 2017). Nunes et al. (2017) developed equations to estimate aboveground carbon stock in oil palm plantations consisting of oil palm trees (8 and 14 years old) in Sabah by correlating LiDAR-derived height data with palms' canopy height and cover (as predictors) with carbon stock. The LiDAR point clouds were classified in the study into ground and non-ground returns, with a digital elevation model (DEM) generated from the ground returns to obtain terrain elevation. Subtracting terrain elevation from non-ground points produced a canopy height model (CHM). Area-based approaches proved to be more precise in predicting carbon stock within oil palm plantations compared to the tree-centric approach. In plantations consisting of mature trees, the overlapping crowns often led to challenges in individual tree differentiation, causing inaccuracies in tree counts, and therefore carbon stock (Nunes et al., 2017).

While airborne LiDAR offers promising oil palm canopy height results, its coverage is limited and operational costs are high (Fawcett et al., 2019; Beese et al., 2022). Alternatively, spaceborne LiDAR like NASA's Ice, Cloud, and Land Elevation Satellite-2 (ICESat-2) holds promise for extensive biomass mapping as there is a strong correlation between height data derived from LiDAR and AGB (Nunes et al., 2017). ICESat-2 employs photon-counting technology to provide 11-m wide along-track samples and superior horizontal geolocation accuracy (approximately 3 m) of its footprint heights for canopy height estimation (Neuenschwander et al., 2021). Although LiDAR's biomass saturation level reaches 1,000 t/ha, exceeding that of passive or radar sensors (Yavasli, 2012), it should be noted that LiDAR data analysis necessitates expertise and time.

The next section showcases an oil palm biomass case study, utilizing a synergistic approach involving multi-band and multi-polarization radar data.

12.5 CASE STUDY

The objective of this study was to estimate AGB of oil palm trees based on their age in a plantation located in Johor, Peninsular Malaysia. The analysis utilized radar data from ALOS PALSAR 2 and Sentinel 1 satellites. The oil palm population was categorized into three distinct age groups, as detailed in Chapter 9, to facilitate AGB calculations. This age-based estimation approach was adopted to mitigate the limitations posed by optical data saturation, particularly in cases of high biomass levels.

For the assessment of oil palm AGB in a private plantation (Genting Plantation—Figure 7.1) belonging to one of Malaysia's largest oil palm companies, L-band, HH and HV polarizations from ALOS PALSAR 2 and C-band VH and VV polarizations from Sentinel 1 were harnessed. The L-band data was particularly valuable for capturing information related to oil palm trunk biomass, while the C-band data facilitated the estimation of upper canopy biomass.

Biometric data encompassing DBH, trunk height (h), and age were gathered for oil palm trees ranging from 5 to 29 years in age within the plantation. The trunk height was measured using a Nikon Forestry Pro Laser Rangefinder, while DBH measurements were taken at a height of 1.3 m above ground level. These recorded parameters were employed to calculate in situ AGB utilizing allometric equations derived from the works of Corley and Tinker (2003), Henson et al. (2012), and Syahrinudin (2005) as depicted in Table 12.2.

To enhance the ALOS PALSAR 2 data, a preprocessing step involving speckle filtering through the Lee filter was carried out. Additionally, terrain correction was applied to mitigate the impact of foreshortening, layover, and shadow effects in radar data. For Sentinel 1 imagery, terrain effects were corrected to minimize the influence of SAR geometry. Recognizing the potential of texture information to enhance AGB estimation, texture details from both ALOS PALSAR 2 and Sentinel 1 were improved using the GLCM measure.

Various polarization combinations (HH + HV, HH − HV, HH/HV, HH × HV, [(HH + HV)/2], and $\sqrt{}$HH + HV), with and without texture enhancement, were correlated with in situ biomass using an MLR technique, drawing from methodologies outlined by Zheng et al. (2004). The outcome of this analysis was the production of optimal AGB estimates for the studied oil palm population.

12.5.1 Oil Palm Above Ground Biomass of Genting Plantation, Malaysia

Oil palm AGB calculated using the three allometric equations are listed in Table 12.4. We have previously observed that the allometric formula developed by Syahrinudin (2005) represents the complete biomass in relation to age (Tan et al., 2014). Importantly, there is a significant alignment between the results of Henson (2003) and Corley and Tinker (2003) for trees aged less than 18 years, as well as between the

TABLE 12.4

Aboveground Biomass (AGB) of Oil Palm Trees at Various Ages in the Genting Plantation Estimated Using the Allometric Formulas Established by Corley and Tinker (2003), Henson et al. (2012), and Syahrinudin (2005)

Age (years)	Height (m)	DBH (m)	Radius (cm)	Trunk Biomass (kg/tree)	Canopy Biomass (kg/tree)	Fruit Biomass (kg/tree)	Total AGB (kg/tree)	Total AGB (t/ha) Corley and Tinker (2003)	Total AGB (t/ha) Henson et al. (2012)	AGB + Root Biomass Syahrinudin's (t/ha)	AGB: Syahrinudin (2005) (t/ha)
5	1.38	0.89	44.56	62.89	39.23	107.89	210.01	30.03	29.81	73.06	68.76
9	2.2	0.76	38.04	91.41	86.33	154.05	331.78	47.45	57.31	113.15	105.40
11	5.4	0.75	37.56	240.73	102.63	169.81	513.16	73.38	69.90	126.84	117.37
17	6.6	0.83	41.38	454.85	122.52	204.00	781.37	111.74	98.08	156.53	141.89
18	8.6	0.72	35.97	463.85	121.60	208.49	793.94	113.53	100.67	160.42	144.93
22	13	0.61	30.40	570.07	105.86	224.25	900.19	128.73	101.91	174.11	155.17
23	9.8	0.70	35.01	587.47	98.91	227.74	914.12	130.72	99.45	177.14	157.34
26	7.4	0.79	39.47	613.56	70.79	237.37	921.72	131.81	83.66	185.50	163.12
27	11.4	0.64	31.83	631.38	59.00	240.34	930.72	133.09	75.18	188.08	164.83
28	12.4	0.71	35.65	884.20	46.00	243.19	1173.39	167.79	64.88	190.56	166.45
29	13.2	0.70	35.01	931.25	31.80	245.95	1209.00	172.89	52.63	192.95	167.98

findings of Henson (2003) and Syahrinudin (2005) for trees aged over 26 years. The AGB as estimated in this study using the equations of Corley and Tinker (2003) and Syarinudin (2005) show an upward trend for trees from 5 to 29-years (Table 12.4). Meanwhile, the AGB obtained from Henson's (2003) calculation increased from 5 to 22, but decreased for older trees (23–29-year-old). Previously Tan et al. (2014) noted close NPP values from all the three equations for trees between 9 and 16 years old. Determining the optimal allometric equation for estimating oil palm AGB/NPP remains inconclusive. To address this, the application of the destructive method is necessary.

We associated AGB obtained through the allometry equations with the ALOS-PALSAR 2 and Sentinel 1C satellite data using MLR to derive empirical equations that can be used to scale up AGB estimation to cover the entire plantation area (Table 12.5). For the intermediate age group, AGB from all three equations were associated well with the radar data with R^2 of more than 0.92. However, for the productive age group palm trees, AGB from Syahrinudin yielded higher correlation coefficient (0.56–0.69) compared to the other two equations with R^2 ranging between 0.56 and 0.57 (Henson et al., 2012) and from 0.34 to 0.35 (Corley and Tinker, 2003). For the mature trees (26–29 years), again the equation from Corley and Tinker produced low association with radar data (R^2 0.30–0.47) compared to Henson (R^2 0.85) and Syahrinudin (R^2 0.86).

Three different equations to cover the three age classes were developed using radar properties and in situ AGB estimated using the allometry equation of Syahrinudin using the leave-one-out method (Table 12.6). This method is a special case of k-fold cross validation where k is the number of folds which is equal to the number of observations. Each sample will be used in turns, by leaving one sample out, while the remaining others are used to develop a model (Jachowski et al., 2013).

The simplest empirical equation was derived for mature palm oil trees (Table 12.6). Radar data collected using HH and HV polarizations, even without detailed texture information, accounted for approximately 88% of the variability in AGB levels. This result might be attributed to the relatively uncomplicated characteristics of mature trees in comparison to younger palms. Mature palm trees exhibit towering structures, smaller DBH, and expansive canopy coverage, albeit occasionally less dense due to pruning activities. Consequently, the effectiveness of L-band radar data is evident, particularly when HH and HV polarizations can penetrate through to the trunk of the oil palm trees. The HH returns are associated with both trunk and crown biomass, while HV returns are predominantly linked to the crown biomass.

Conversely, when considering productive aged trees, achieving a notably high and statistically significant accuracy with an R^2 value of 93% necessitated the utilization of a sophisticated model incorporating both L- and C-band SAR data, encompassing multiple polarizations and textural intricacies. Specifically, the VH and VV polarizations from Sentinel-1 satellite's C-band effectively penetrated the substantial and compact canopy of the productive-aged trees, thereby furnishing valuable biomass insights from the crown region. In a parallel context, Harrell et al. (1997) employed SIR C- and L-band multi-polarization radar data for estimating biomass in pine forests located in the southeastern United States. The L-band HH data proved highly advantageous for biomass estimation; however, the incorporation of C-band data

TABLE 12.5
Empirical Equations to Estimate Oil Palm AGB

Oil Palm Allometric Biomass Calculation	Young Oil Palm		Middle Age Oil Palm		Mature Oil Palm	
	Band Manipulation	**R^2 Value**	**Band Manipulation**	**R^2 Value**	**Band Manipulation**	**R^2 Value**
Henson (2003)	GLCM mean HH polarization	0.96	HV polarization	0.57	HH polarization	0.85
	GLCM mean HH × HH polarization	0.95	HH + HV polarization	0.56		
	HH polarization	0.94	GLCM mean HH × VH polarization	0.56		
Syahrinudin (2005)	GLCM mean HH polarization	0.96	HH – HV polarization	0.69	HH polarization	0.86
	GLCM mean HH × HH polarization	0.95	VH + VV polarization	0.58	HH × HV polarization	0.46
	HH polarization	0.94	HV/ HH polarization	0.56		
	HH × HH polarization	0.96	GLCM mean HH times VH	0.56		
			GLCM mean HH divide HV	0.57		
			GLCM mean HV divide HH	0.57		
Corley and Tinker (2003)	HH_3 × 3_64 mean	0.93	HH polarization	0.35	GLCM mean HH polarization	0.30
	GLCM mean HH × HH polarization	0.93	HH × HH polarization	0.34	GLCM mean HH × HH polarization	0.47
	HH polarization	0.92				
	HH × HH polarization	0.92				

TABLE 12.6
Three Different Equations for Three Age Classes Using Radar Properties and In Situ Aboveground Biomass (AGB) Estimated Using the Allometry Equation of Syahrinudin (2005)

Oil Palm Age	Empirical Equations
5–11 years	14091.62 + (−932.09 × HH) + (13.8 × Textured HH × HH) + (−452.75 × HH) + (−29.86 × HH × HH)
17–23 years	15717.25 + (44.23 × HH − HV) + (−0.63 × VH + VV) + (−98.97 × HV over HH) + (113.89 × Textured HH × VH) + (4839.62 × Textured HH/HV) + (11053.59 × Textured HV/HH)
>25 years	181.30 + (1.53 × HH) + (−0.06 × HH × HV)

significantly enhanced the accuracy of the estimation process. Notably, the intricate and densely covered leaves of young and productive-aged palm trees required satellite data enhancement to capture essential textural details.

The significance of the L-band becomes evident when considering intermediate aged oil palm trees, where its penetration capability reaching the trunk assumes crucial importance due to the relatively compact size of the canopy. Consequently, the L-band's penetrating capacity allows information to distinctly represent intermediate oil palm trees without intermixing signals from surrounding vegetation or soil.

In a study conducted by Toh et al. (2019), it was observed that backscattering values for both the L- and C-bands increased with advancing age of the oil palm. However, the C-band tended to saturate more rapidly, whereas the L-band's signal saturation occurred only in oil palm trees aged 20 years and above. As a result, it is apparent that a majority of the equations presented in Table 12.5 predominantly involved the L-band. This observation stresses the distinct capabilities of each band and polarization in providing information for biomass estimation. Consequently, the combination of both bands, as demonstrated in this study, yields a more robust correlation in biomass estimations, thus emphasizing the advantages of utilizing both L- and C-bands together.

The equations were employed to upscale the AGB across the entire 2,815 ha plantation area, as depicted in Figure 12.5. Concerning intermediate palm trees, AGB values spanned from 12.36 to 242 t/ha, with an average of 123.44 t/ha. Notably, the estimated value exceeded the actual average biomass by 23.81%, attributed to misclassifying intermediate oil palms as productive aged ones. In contrast, productive-aged oil palm trees exhibited AGB values ranging from 112.44 to 249.52 t/ha, averaging 160.36 t/ha, which surpasses the allometry equation-based average by approximately 11 t/ha. Regarding mature oil palm, estimated biomass ranged from 148 to 175 t/ha, compared to a recorded average of 166 t/ha. These discrepancies in biomass estimates for each age group are likely due to the inaccurate classification of oil palm age (Chapter 9). Validation results for estimated oil palm biomass are presented in Table 12.7. A high R^2 value indicates close agreement between estimated and field-calculated oil palm biomass, with the best performance observed

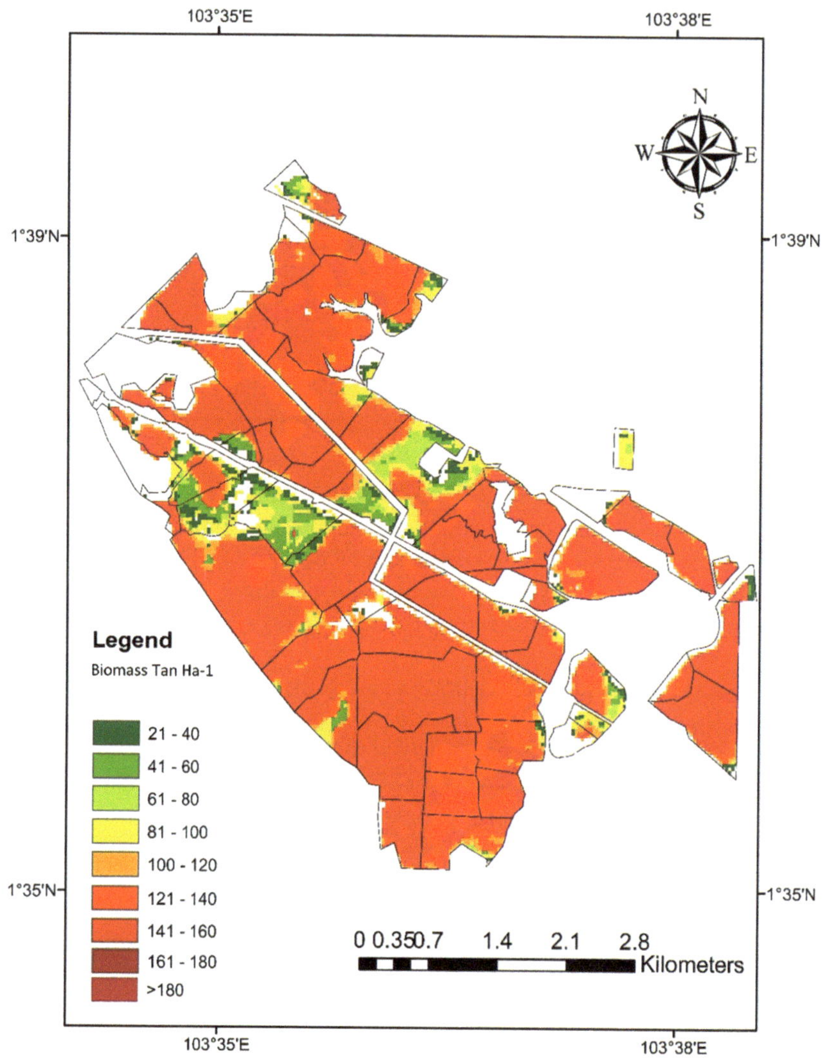

FIGURE 12.5 Estimated palm oil aboveground biomass (AGB) of Genting Plantations, Peninsular Malaysia.

TABLE 12.7
Accuracy Assessment of Estimated Aboveground Biomass (AGB)

Age Class	R^2	Mean Absolute Percentage Error (MAPE)	Percentage Root Mean Squared Error (RMSE)
Young	0.63	0.0780	13.43
Mid-age	0.79	0.0075	2.13
Matured	0.91	0.0013	0.35

in mature oil palm. Moreover, mean absolute percentage error (MAPE) values were consistently low across all age classes. Additionally, the percentage root mean squared error (RMSE) for all classes remained below 14%, signifying accurate estimations.

The average AGB estimated in this study was compared with other studies conducted in Malaysia, Indonesia, and West Africa as shown in Figure 12.6. The average AGB calculated in this study for trees between 5 and 11 years is close to the value obtained by Ng (1968) for trees with 11 years old. For trees between 17 and 23 years old, the average AGB is only slightly lower than that of the old trees. For older trees with more than 30 years old, Migolet and Goita (2020) obtained much lower values. The reason for this could be the trees are much shorter (10 m) compared to the trees considered in this study (13–17 m). Further variations in AGB across previous studies are evident (Figure 12.6). This clearly highlights the site-specific nature of oil palm plantation biomass values, heavily influenced by the unique management practices of each planting site (Asari et al., 2017). Even within the same company, such as MPOB, different plantation planning programs and varying nutrient (fertilizer) levels are implemented based on the distinct characteristics of each site. Additionally, topography exerts an effect on the biomass of oil palm plantations, resulting in differing growth rates across sites. Hence, despite the variety of planting techniques potentially impacting carbon stock in oil palm cultivation, the estimation methods for carbon also wield a significant influence.

The emergence of carbon prediction models presents opportunities to evaluate the carbon density within oil palm plantations, a crucial aspect for assessing the environmental sustainability of modified tropical landscapes. This case study illustrates the methodology and outcomes of estimating oil palm biomass, revealing the necessity for site-specific investigations due to differing biomass values resulting from distinct planting programs across various estates. We found that multiple bands from L-band

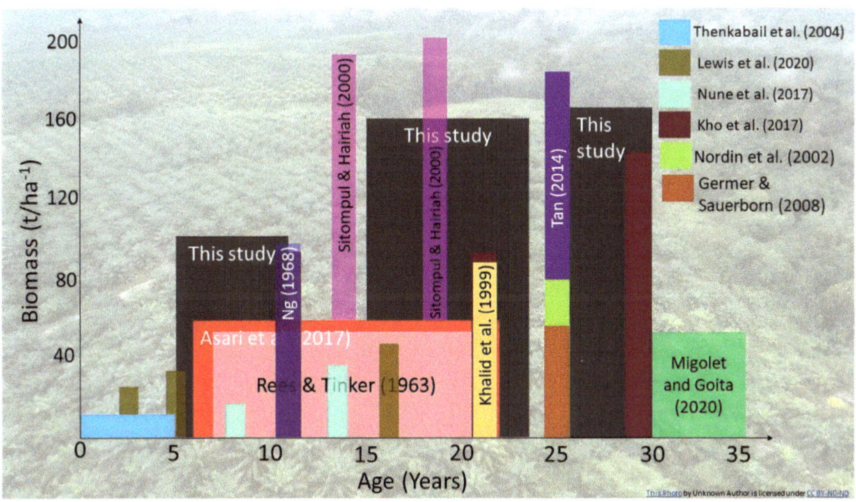

FIGURE 12.6 Comparison between palm oil aboveground biomass (AGB) estimated in this study with other studies conducted worldwide.

radar data can enhance biomass information of intermediate and mature oil palm trees, whereas, incorporation of C-band radar data improved the biomass estimation of productive aged oil palm trees. Addressing saturation challenges, this study proposes a solution by categorizing oil palm age into three classes and recommending site-specific investigations. Multiple studies have showcased the potential of remote sensing as a complementary strategy to overcome these limitations. Predictive correlations can be established between structural variables and remotely sensed data, often with local or regional relevance, though transferability in space and time can be challenging.

An array of remote sensing techniques, encompassing optical sensors, SAR, and LiDAR technologies, provide valuable tools for estimating oil palm productivity and biomass. These methodologies facilitate comprehensive insights into canopy structure, health, and carbon dynamics, thereby enhancing comprehension of oil palm ecosystems and their potential influence on carbon cycling. The Roundtable on Sustainable Palm Oil, the major accreditation body for sustainable palm oil production (Chapter 16), recognizes the importance of developing approaches for growers to monitor emissions across their estates. However, it has been recognized that emission cannot be monitored completely, or accurately, using current methodologies (RSPO, 2017). Despite existing challenges, advancements in sensor technology, data processing methods, and integration strategies hold potential for accurate and comprehensive biomass estimation on both regional and global scales. Future research endeavours should concentrate on refining these approaches, broadening their applicability, and integrating them into holistic assessments of the environmental impact of oil palm cultivation.

13 Sustainability

13.1 OIL PALM EXPANSION AND SUSTAINABILITY

The production of palm oil increased rapidly since 1970, from 2 to 84 million tons in 2022 (USDA, 2023) and forecasted to reach 111 million tons in 2026 (Malaysia Central Bank, 2021) following global demand for palm oil. Consequently, area planted with oil palm also expanded at a rate of 0.7 million ha/year between 2008 and 2017 (FAO, 2019). In 2021 the total global area covered by oil palm plantations is 28.91 million ha compared to only about 3.62 million ha in 1960 (Ritchie and Roser, 2021). Indonesia and Malaysia which produce 77.5% of global palm oil account for 69.66% of the land used for oil palm trees growth (Ritchie and Roser, 2021). With the rapid expansion of the palm oil industry and land used for its production (Figure 13.1), a range of environmental and social sustainability issues have emerged. With such continuous expansion of the industry, a lot more criticisms may be expected. Thus, it is important to discuss the sustainability of the industry in terms of all the three pillars of sustainability and propose strategies to lower the impact of the industry on the environment and social well-being.

The cultivation and milling activities of palm oil production have created extremely controversial environmental and social sustainability issues such as destruction of tropical forests and peatlands that leads to loss of natural habitats, biodiversity loss, air and water pollution, soil erosion, flooding, forest fire, greenhouse gas emission

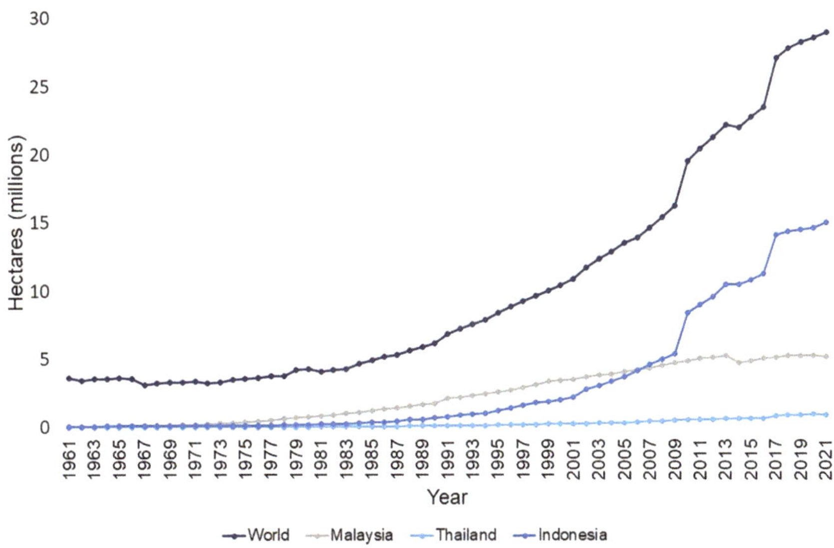

FIGURE 13.1 Areas planted with of oil palm. Data source: https://www.statista.com.

DOI: 10.1201/9780429199813-13

and climate change (Oosterveer, 2015; Jamaludin et al., 2019). A range of social issues including threats to human livelihoods, land conflicts, labour issue and other human rights violation are also debated to be caused by the expansion of the industry. Nevertheless, its contribution to the economic development and social well-being of major palm oil–producing nations, that is, Indonesia and Malaysia, is significant and cannot be over sighted (as reported in Chapter 1).

13.2 SUSTAINABLE DEVELOPMENT

Sustainability or sustainable development is a concept that emphasizes particularly the economic progress of human kind without compromising the social well-being of under privileged communities and the environmental stability. In other words, sustainable development is aimed to promote development that fulfils the current needs of human while safeguarding the requirements of future descendants. The needs of future generations can be assured only when any developments are carried out with moderate profit and without bringing severe impact to the people and the planet that they are residing in. Economy, social and environment are considered the three pillars of sustainable development, and sustainable practices are advocated as the major development tools in the industrial, agriculture and services sectors by the government institutions, private sectors and non-governmental organizations. Sustainable practices are promoted via 17 goals contained in the 2030 roadmaps for sustainable development. The 15-year strategy was deliberated by the United Nations in 2015 with the aim to attain all the 17 objectives by 2030 (Estoque, 2020). Out of the 17 goals (Table 13.1) several goals are notably relevant to the implications of palm oil industry to the economy (Goal 8), social (Goals 1, 2, 3, 5, 6, 8, 10, 15 and 16) and environment (Goals 6, 12, 13 and 15).

13.3 GEOSPATIAL TECHNOLOGY FOR SUSTAINABLE DEVELOPMENT

While implementing various strategies and actions to realize the sustainable development goals (SDGs) is essential, it is equally crucial to follow up and review the actions from time to time to track the progress, guarantee accountability of stakeholders and formulate policies for decision-making. Statistical data and indicators are essential for the purpose, and geospatial data, particularly from remote sensing, play an important role in ensuring the sustainability of the palm oil industry in data-scarce developing nations. The Inter-Agency Expert Group on the Sustainable Development Goals Indicators (IAEG-SDGs) has acknowledged the value of geospatial data in formulating several indicators of sustainability. In April 2016, IAEG-SDGs established a Working Group on Geospatial Information (WGGI) to assist nations in fully harnessing the capabilities of geospatial data and earth observation satellites within the Global Indicator Framework (https://ggim.un.org/UNGGIM-wg6/).

Remote sensing offers a constant stream of global coverage data, encompassing regions that pose accessibility challenges and supporting countries with limited database development capabilities, particularly those in the developing world. Other key attributes of remote sensing data for SDG monitoring include different spatial,

TABLE 13.1
Sustainable Development Goals

SDG Goals	Goals Description
Goal 1	No poverty
Goal 2	Zero hunger
Goal 3	Good health and well-being
Goal 4	Quality education
Goal 5	Gender equality
Goal 6	Clean water and sanitation
Goal 7	Affordable clean energy
Goal 8	Decent work and economic growth
Goal 9	Industry, innovation and infrastructure
Goal 10	Reduced inequalities
Goal 11	Sustainable cities and communities
Goal 12	Responsible consumption and production
Goal 13	Climate action
Goal 14	Life below water
Goal 15	Life on land
Goal 16	Peace, justice and string institutions
Goal 17	Partnerships for the goals

Source: https://www.undp.org/sustainable-development-goals.

temporal and spectral resolutions, access to historical data and data from various sensors are complementary to each other (Estoque, 2020). Various sensors are suited for different applications within palm oil sustainability which are explored in the next three chapters of this book. There are also processed data particularly from sensors such as moderate resolution imaging spectroradiometers (MODIS) which are readily available for quick analysis for oil palm sustainability applications.

Generally, as described in the previous chapters, geospatial technology can support palm oil industry by providing cost-effective information that supports decisions to increase yield by improving disease and pest detection, managing well resources such as water, fertilizers and pesticides, enables the restoration of inhospitable surroundings to productive land and monitoring food security. Tropical forests in the oil palm growing countries face pressure for conversion to oil palm and other associated threats such as biodiversity loss, climate change and social issues such as deprivation of livelihoods for forest-dependent communities. Geospatial technology supports sustainable production of palm oil by providing tools to map forests and monitor their changes in terms of deforestation and degradation over time, detect intrusion of oil palm into forested/protected lands and biodiversity loss and map high carbon forestry stocks. The technology is also used to monitor palm oil businesses along the supply chain to ensure their compliance with zero deforestation pledges, establish land rights, and oversee and disclose unauthorized or detrimental activities. These aspects are discussed in Chapters 14–16.

14 Environmental Sustainability

14.1 INTRODUCTION

This chapter focusses on the impact of palm oil industry including both plantation and milling processes on the environmental sustainability. The chapter is also aimed at proposing approaches to attain a sustainable palm oil production. The use of geospatial technology in addressing sustainability issues is considered in this chapter. Environmental sustainability is concerned with human's activities that can guarantee the availability of natural resources such as forest, water, coal, natural gas, stone, and soil for our next generations to live either a similar life like the current generation or a better life. Nevertheless, the current operation of the industry is blamed for causing a range of environmental issues covering forest, wetland, and peatland losses that subsequently affect forest biodiversity, air temperature, climate, water and air quality, soil erosion, flooding, and other hydrological impacts.

14.2 DEFORESTATION AND OIL PALM EXPANSION

The rapid development in palm oil industry since the 1970s has led to a significant loss of forested areas that contributed to a widespread environmental degradation and biodiversity loss in palm oil growing regions (Qaim et al., 2020). Statistics compiled by Meijaard et al. (2018) shows that 47% of deforestation in Malaysia that occurred between 1972 and 2015 was mainly for establishing oil palm plantations. However, during the same period only 16% of forested land was converted into oil palm in Indonesia because fires and timber extraction already destroyed most of the forest in the Borneo Island of Indonesia well before the introduction of palm oil cultivation (Corley and Tinker, 2016). Some industrial plantations were established on these ecologically degraded non-forests in Indonesia (Gaveau et al., 2016).

The contribution of oil palm cultivation to the loss of forest cover has been studied extensively for various time periods using a range of methods including remote sensing images (Table 14.1). Globally, the expansion of palm oil industry caused tree cover loss of about 2% (Pendrill et al., 2019), while its role in intact forest[1] loss is 0.2% (Potapov et al., 2017). Regional scale analysis shows that in Malaysia between 1990 and 2005, approximately 55%–57% of oil palm expansion occurred on forest. During this period, around 41%–45% of the expansion in Malaysia occurred on croplands especially on rubber plantations in Peninsular Malaysia (Koh and Wilcove, 2008). The high percentage of forest land that was cultivated by oil palm may include disturbed forest as Koh and Wilcove did not differentiate between undisturbed and disturbed forests, and it should be noted that their conclusion is based on data obtained from FAO and FAOSTAT. Around the same time period (1989–2008), most of the

 DOI: 10.1201/9780429199813-14

TABLE 14.1
The Impact of Oil Palm Expansion on Forest Cover in Major Palm Oil Producing Countries

Period	Oil Palm Establishment on Forested Land	Oil Palm Establishment on Other Land Over	Data and Techniques Used	References
1990–2005	55%–57% in Malaysia 56% in Indonesia	41%–45% on croplands in Malaysia	Land-use data compiled by FAO	Koh and Wilcove (2008)
1989–2008	Indonesia Intact forest—21% Secondary forest—21%	Logged forest—7% Cropland—37% Burned/cleared/bare lands—14%	Landsat TM and ETM+ Spectral unmixing model within the CLASlite programme and manual digitization from Landsat reflectance data to classify land cover. GPS for oil palm locations	Carlson et al. (2012)
2000–2010	Indonesia 11% (14.7 Ha)	-	Land cover maps produced using MODIS at 250 m—ISODATA classification and manual delineation and ALOS PALSAR (resampled to 250 m) —HH-HV-threshold method	Abood et al. (2015)
1990–2010	Indonesia, Malaysia, PNG 4.1% on undistrurbed forest (upland and swamp)	32.4% on distrubed/ degraded forest	Landsat 4, 5, and 7—on-screen digitization and visual interpretation	Gunarso et al. (2013)
2005–2015	Borneo 50% of forest loss due to oil palm plantation	-	Landsat—visual interpretations	Gaveau et al. (2016)
1973–2015	Malaysia part of Borneo 57%–60% of deforestation in for oil palm	-	Landsat 5, 7, and 8—visual interpretation, Normalized Band ration index	Gaveau et al. (2022)

oil palm expansion in West Kalimantan, Indonesia, were reported to be occurring mainly on croplands (37%), intact and secondary forests (21% each), logged forests (7%), and 14% on burned/cleared and bare lands (Carlson et al., 2012). A large-scale study covering Indonesia, Malaysia, and Papua New Guinea (PNG) stated that oil palm plantation was preferred on disturbed/degraded forest instead undisturbed forest (Gunarso et al., 2013). During 1990–2010, direct conversion of forest to oil palm was more common in Kalimantan, Sabah, and Sarawak (Gunarso et al., 2013). Similarly, Abood et al. (2015) reported that only 11% of the forests in Indonesia was used for oil palm cultivation, while as much as 25% of the forest was manipulated for timber and another 6% for mixed activities. Likewise, in a study by Gunarso et al. (2013), a land-use change analysis utilizing Landsat data revealed that during the period 1990–2010, only 4.1% of oil palm cultivation in Indonesia, Malaysia, and PNG occurred in undisturbed forests (both upland and swamp), while 32.4% took place in areas classified as disturbed or degraded forests. In the case of Indonesia, forest loss primarily resulted from unsustainable logging practices, compounded by the subsequent impact of fires (Gunarso et al., 2013).

However, later studies based on visual interpretation and manual delimitation of oil palm development from satellite imageries found that 50%–60% of forest loss in the Borneo Island was mainly occurred for expanding large-scale palm oil plantations where about 20%–21% of forest was directly converted to oil palm, while 4%–5% was used for pulpwood (Gaveau et al., 2016, 2022). In a recent investigation, Yu et al. (2023) identified the expansion of oil palm plantations across 15 Asia-Pacific nations between 2007 and 2018, encompassing countries like Indonesia, Malaysia, Thailand, and PNG. They utilized ALOS PALSAR 1 and 2 data (HH, HV, HH-HV, and HH/HV images) for mapping oil palm expansion by the means of random forest classification technique. During this period, oil palm plantations extended by approximately 12.43 ± 2.54 million ha across the Asia-Pacific region, predominantly in low-altitude and flat terrain areas. The Southeast Asian (SEA) region, particularly Borneo and Sumatra, witnessed significant oil palm distribution and growth, while Peninsular Malaysia experienced comparatively lesser expansion due to geographical factors and local regulations. Indonesia, Malaysia, and PNG were major oil palm hotspots, causing significant forest loss. Other countries like Nigeria and Latin American nations converted lesser forest area to oil palm. Brazil saw increased expansion in Pará state, mainly on cleared land (Villela et al., 2014; Lees et al., 2015).

The initial growth in oil palm areas in Indonesia and Malaysia was driven by various incentives and programmes initiated by the governments. In Malaysia, for example, the agricultural diversification programme implemented in the 1960s swiftly accelerated the cultivation of oil palm (Courtenay, 1984). Although vast areas of forests were cleared for oil palm until the end of 2000s, the trend is found to be decreasing in recent years. Indonesia, for example, had about 40% of forest clearance for planting oil palm between 2008 and 2009, but later it declined to only <15 (Austin et al., 2019). In the Borneo Island, the expansion of large-scale palm oil plantations and associated forest loss declined since 2011 (Gaveau et al., 2022) as a consequence of various efforts undertaken to curtail the intrusion of oil palm plantations into forested areas, for example, the national moratorium introduced by Indonesia in 2011 and the cancellation of permits to explore forests for oil palm in January 2022

(Yusuf et al., 2018). Consequently, the total primary forests land converted to oil palm plantation decreased from 224,100 ha in 2012 to 15,540 ha in 2021 although the crude palm oil price remained same at $ 1,000 per ton. In addition to the regulations imposed by the Indonesian government, other factors such as the market uncertainties and production cost also could be the reasons for the lower expansion of oil palm into forests (Gaveau et al., 2022).

14.3 SMALLHOLDER OIL PALM GROWERS' CONTRIBUTION TO DEFORESTATION

Smallholders are typically local farmers who cultivate oil palm up to 50 ha of land (RSPO Smallholders, 2023), but they play a vital role in global palm oil production, particularly in Indonesia and Malaysia. Estimates indicate that smallholders cultivate 30%–60% of oil palm area in Malaysia and Indonesia, and 11%–94% in Nigeria (Meijaard et al., 2018). The contribution of smallholder palm oil growers to deforestation has gained increasing attention due to the vast amount of area planted by them. Recent remote sensing studies suggest that oil palm plantations covered over 19.5 Mha globally in 2019, with about 33% being smallholder plantings (Descals et al., 2021). Smallholders, contributing nearly 50% of traded palm oil (Mutsaers, 2019), exhibit lower production yields (Euler et al., 2016), necessitating more land for equivalent palm oil volumes compared to agro-industrial plantations.

Although economic factors drive smallholders into the industry, and they are aided by government support and incentives such as credit access and land tenure arrangements (Fitzherbert et al., 2008; Koh and Wilcove, 2008), challenges like limited resources and technical knowledge lead to unsustainable land-clearing practices including forest conversion (Fitzherbert et al., 2008; Krishna et al., 2017). Studies indicate that smallholders' contributions to deforestation can be substantial. In Peru, about 30% of smallholder plantings result in deforestation (Gutiérrez-Vélez et al., 2011), and in Sumatra, 39% of smallholder expansion encroaches into forests (Lee et al., 2014). Similar to industrial actors, smallholders convert fragile ecosystems into oil palm plantations, as seen in Indonesian Borneo (Schoneveld et al., 2019).

14.4 WETLANDS DETERIORATION

The expansion of the palm oil industry has had far-reaching and often detrimental consequences for delicate ecosystems such as peatlands and wetlands. These regions, characterized by their rich biodiversity and crucial role in carbon storage, have been consistently cleared and drained for oil palm cultivation. The draining of peatlands releases substantial amounts of stored carbon, contributing to global climate change and heightening the risk of flooding in surrounding areas. The conversion of these vital ecosystems into oil palm plantations threatens unique habitats and endangers a variety of plant and animal species (Section 16.6). While efforts have been made to mitigate these impacts, it is evident that sustainable practices are essential to protect these ecosystems and maintain the overall health of the environment (refer to Chapter 16).

Two major types of wetlands, peatlands and mangroves, are often converted for oil palm plantations due to their fertile nature and abundant water supply. Peatlands, accounting for half of Earth's wetlands and covering around 3–4 million km^2 globally, offer numerous ecosystem services. These include regulating water flow, providing livelihoods for local communities, offering habitats for various flora and fauna, and acting as significant carbon sinks. Despite their critical role, peatlands are being rapidly depleted to cater for the escalating demand for palm oil. Similarly, mangroves, which provide vital coastal protection and habitat for various species, have also fallen victim to oil palm expansion, particularly in regions like Peninsular Malaysia (Kanniah et al., 2015; Xu et al., 2020).

The extent of the impact on peatlands is staggering. Research using remote sensing and satellite data has revealed alarming statistics. By the early 2000s, approximately 6% of peatland in Indonesia and Malaysia had been encroached by oil palm plantations (Koh et al., 2011). Recent studies indicate that over 20% of Asian oil palm plantations were established on peatlands by 2012, although this number has decreased since then (Yu et al., 2023). Furthermore, the global scenario reveals that 11.38% of all oil palm plantations are situated on peatlands, with significant encroachment in Indonesia and Malaysia (Yu et al., 2023).

The conversion of peatlands for oil palm brings about dire consequences. Drainage for planting oil palm and timber creates canals that lead to land subsidence, exacerbating the risk of floods. Drying out and decomposing peat for oil palm planting accelerates land subsidence, with the growth of oil palm trees near canals also contributing to sinking (Wösten et al., 2008). The compression of peat volume from the growth of oil palm trees surrounded by canals also triggers subsidence. Lowered groundwater tables resulting from drainage canals facilitate wildfires and peat degradation, as seen in Riau, Indonesia (Basuki et al., 2021).

Peatland drainage, a common practice for oil palm, leads to land sinking and releases significant amounts of CO_2, contributing to global warming (Hoyt et al., 2020; Dadap et al., 2021). The study conducted by Dadap et al. (2021) employed satellite remote sensing data to meticulously map the intricate network of drainage canals across the expansive peatlands of SEA. To undertake this endeavour, the researchers harnessed the global mosaic of RGB images captured by the PlanetScope constellation of CubeSat sensors between the months of July and September in the year 2017. These images, with a spatial resolution of 5 m, proved pivotal in discerning the narrow drainage canals that sometimes span mere metres in width. A pivotal tool in this process was a convolutional neural network (CNN), adept at identifying the elusive drainage canals within the satellite imagery. Unlike conventional pixel-by-pixel classification methods, which falter in identifying canals due to their diverse radiative spectral signatures, CNNs exhibit a crucial capability: considering the context surrounding a pixel to decipher linear shapes. In this context, the U-Net architecture combined with a recurrent CNN was implemented. This hybrid approach incorporated functional loss terms that foster global connectivity, ultimately yielding exceptional performance compared to alternative methodologies. This investigation encompassed a substantial area of approximately 157,000 km^2, encompassing the islands of Sumatra and Borneo, as well as Peninsular Malaysia.

Notably, this geographical expanse is believed to harbour approximately 53% of the total soil carbon reservoir found within tropical peatlands.

The assessment of carbon fluxes linked with subsidence data was carried out using a methodology outlined in Hoyt et al. (2020). It is noteworthy that these estimates do not serve as emissions factors; rather, they offer a rough approximation of emissions attributed to subsidence resultant from drainage. The inherent uncertainty in these estimates is partially attributed to the multi-faceted nature of subsidence rates, which stem not only from oxidation but also from compaction, shrinkage, and consolidation of peat at lower depths (Hooijer et al., 2012). Importantly, these data are representative of the period between 2007 and 2011, spanning 6–10 years prior to the drainage estimates presented in this study.

The mapping of drainage density (expressed in km/km^2) at a resolution of 1 km was also carried out within the SEA region (Dadap et al., 2021). The outcomes of the study unveiled the pervasive nature of drainage activities, extending to at least 65% of the peatlands and spanning across diverse land-use categories. Notably, substantial variations in drainage density were detected within different land-use types. Impressively, the study found that subsidence rates are 3.2 times greater in intensively drained regions compared to non-drained areas, stressing the pivotal role of drainage in influencing peat subsidence. Furthermore, the incorporation of drainage canals into a subsidence prediction model resulted in a noteworthy 30% enhancement, indicating that these canals harbour insights into subsidence patterns not entirely captured by land-use considerations alone. Consequently, the dataset established through this research stands poised to enhance predictions related to subsidence and the associated carbon emissions within peatlands, while also serving as a strategic resource for identifying areas ripe for hydrologic restoration endeavours.

Restoration of drained peatlands and implementation of responsible land-use practices are paramount to mitigating these impacts. As global climate targets loom, phasing out drainage practices in favour of sustainable solutions becomes imperative. To safeguard these vital ecosystems, a collaborative effort involving governments, industries, and local communities is required to balance the economic benefits of palm oil with the preservation of peatlands and wetlands for the well-being of the planet.

14.5 IMPACT OF DEFORESTATION AND WETLAND CONVERSION ON CARBON STORAGE

As previously mentioned, a vital ecological role of tropical forests involves the sequestration and retention of carbon. These lush forests exhibit a remarkable capacity to capture CO_2 from the atmosphere, storing it within the trunks, roots, branches, and leaves. However, the widespread deforestation on a large scale partly for oil palm establishment can trigger the release of this stored carbon. Mills et al. (2023) calculated (with data collected using eddy covariance-based flux tower measurements) an average carbon release of 1.75 ± 0.94 t C/ha/year within plots subject to moderate logging, while unsustainably logged and severely degraded plots in the tropical forest of the Malaysian part of Borneo Island exhibited a three times higher CO_2 release that persisted over 10 years after logging activities.

The carbon content held within a standing tropical rainforest range from 175 to 272 t/ha (Katayama et al., 2013; Sayer et al., 2012). In contrast, the standing carbon stocks found in oil palm plantations span a range of 2–60 t/ha (Kho and Jepsen, 2015). A comparative study by Kotowska et al. (2015), contrasting old-growth forests and oil palm plantations in Sumatra, unveiled a stark reduction of 167 t/ha in carbon pools—dropping from 195 t/ha in forests to merely 28 t/ha in oil palm plantations (Kotowska et al., 2015). In a parallel vein, the comprehensive assessment of both aboveground and belowground carbon stocks has revealed that forest conversion to oil palm and rubber results in average carbon losses of 61% and 56%, respectively (Guillaume et al., 2018). Moreover, compared to forests, monoculture plantations exhibited reduced tree biomass, slower litter decomposition, compromised root health, diminished microbial activity and biomass, along with heightened nutrient-leaching fluxes—unfavourable agricultural outcomes primarily seen in oil palm plantations (Clough et al., 2016).

Recent research conducted by Xu et al. (2022) quantified substantial forest biomass loss amounting to 50.2 ± 21.9 Tg C/year due[2] to oil palm expansion in Malaysia and Indonesia from 2001 to 2015. This investigation amalgamated various datasets extracted from satellite imagery, including annual oil palm maps (Xu et al., 2020), tree cover (Hansen et al., 2013), and annual aboveground biomass (AGB) data from Santoro et al. (2020). Their findings highlighted forest biomass losses of up to 703.4 ± 306.0 Tg C during 2001–2015 due to the encroachment of oil palm into tropical forests. Of particular note, biomass loss per unit area was notably higher in small-scale plantations (<50 ha), exhibiting a 33% increase in 2015 compared to 2022. These encroachments accounted for 23.8% of regional land-use change and forestry-related greenhouse gas (GHG) emissions (Xu et al., 2022).

Conversion from forest on mineral soil to oil palm plantations has been estimated to result in an average carbon loss of approximately 702 tonne CO_2/ha over a 30-year span, whereas conversions of forest on peatlands yield carbon losses two to seven times higher than the CO_2 emanated from mineral soil over the same period (Fargione et al., 2008). These substantial losses overshadow any potential carbon storage within an oil palm plantation as presented in Chapter 12. For instance, the conversion of peat swamp forests to oil palm resulted in the depletion of around 140 million tonnes of carbon, alongside annual emissions of roughly 4.6 million tonnes of belowground carbon due to peat oxidation. Additionally, the loss of peat swamp forests translates to the loss of the carbon sequestration service through peat accumulation, amounting to approximately 660,000 tonnes of carbon annually (Vaidyanathan, 2011).

While the planting of oil palm trees cannot serve as a full replacement for the ecological value and carbon storage potential of tropical forests, it is important to acknowledge that mature oil palm plantations do maintain a considerably higher standing carbon stock compared to other oil-producing crops such as soybeans with only 6 t/ha per 120-day crop cycle (Sayer et al., 2012). Nevertheless, it should be noted that the carbon stored by oil palm trees is transient in nature, as the accumulated carbon will be released back into the atmosphere when the mature trees are cut down for new plantings. Notably, elongated rotation cycles, such as 30 or 40 years instead of 25 years, however, can substantially decrease GHG emissions in comparison to those generated by fossil fuels (Meijide et al., 2020).

The issue of carbon debts resulted from deforestation for oil palm establishment is argued to be offset by carbon savings from producing biodiesel using palm oil instead of petroleum-based diesel (De Souza et al., 2010). Although the time required for such offsetting for a tropical forest is estimated to be 39 years, it largely depends on the previous land use such as up to 840 years when peatland is deforested and drained (Fargione et al., 2008). Consequently, relative to emissions attributed to fossil fuels, biofuel production from oil palm inadvertently can amplify net CO_2 emissions for centuries (Fargione et al., 2008).

14.6 BIODIVERSITY DECLINE

Biological diversity refers to the variety of flora and fauna species living in a particular ecosystem. Biodiversity is usually measured using species richness in terms of number and variation. The plant and animal species in SEA are highly biodiverse with a maximum of 200 tree species are found within 2 ha plot (Sodhi et al., 2010). More than 90% of species present in primary forest also found in logged forest (Pillay et al., 2022). The tropical rainforest of the Amazon houses 40,000 of plants and 5,500 vertebrate species (Pillay et al., 2022). Forest clearence aimed at planting oil palm or any other activities threatened the forest ecosystem that is rich with various flora and fauna. The impact of oil palm plantations on biodiversity is reviewed by Meijaard et al. (2018, 2020).

Observing and monitoring biodiversity has become a complex task due to the altering influence of human activities on the structure and composition of living organisms. The Convention on Biological Diversity, in place since 1992, has been advocating for measures to mitigate or diminish the rapid decline of biodiversity. However, this remains a prominent global challenge, demanding the strategic employment of modern technology for effective intervention (Reddy, 2021).

Biodiversity measurement encompasses more than just documenting species types. It encompasses a diverse array of relevant indicators, spanning changes in ecosystem structure and function, community composition, species traits, and species populations (Jetz et al., 2019). In support of biodiversity monitoring systems, the 'Group on Earth Observations—Biodiversity Observation Network' (GEO BON) has introduced the concept of 'essential biodiversity variables' (EBVs) that are amenable to regular monitoring from space. Notably, Skidmore et al. (2021) have identified nearly 120 biodiversity products that offer crucial insights into biodiversity shifts and can be derived from satellite remote sensing. As palm oil cultivation expands, it often encroaches upon ecologically sensitive areas, leading to habitat destruction, deforestation, and a decline in species richness. Remote sensing tools have emerged as indispensable instruments in documenting and quantifying the extent of biodiversity loss as a direct consequence of oil palm plantation expansion.

Globally, there exist seven distinct biodiversity prioritization frameworks,[3] namely Crisis Ecoregions, Biodiversity Hotspots, Endemic Bird Areas, Key Biodiversity Areas, Centers of Plant Diversity, Global 200 Ecoregions, and Intact Forest Landscapes (Brooks et al., 2006). To assess the implications of oil palm expansion on these conservation priority zones, Yu et al. (2023) conducted an analysis that involved overlaying the oil palm plantation areas mapped for 2018 using

ALOS PALSAR 1 and 2 satellite data in Indonesia, Malaysia, Thailand, and PNG with these biodiversity templates. Startlingly, their findings revealed that an overwhelming majority of the plantations were situated within the conservation priority zones. Specifically, approximately 97%, 72%, 65%, 8%, and 3% of oil palm plantations were found to be located within Biodiversity Hotspots, Global 200 Ecoregions, Crisis Ecoregions, Endemic Bird Areas, and Key Biodiversity Areas, respectively. In contrast, a very small proportion (less than 1%) were situated in Centers of Plant Diversity and Intact Forest Landscapes. These findings serve as compelling evidence of the potential negative impact of oil palm expansion on biodiversity in major oil palm–producing nations.

Protected areas are prohibited for agricultural production and they have very limited anthropogenic activities. Jiang and Yu (2019) studied the impact of oil palm expansion on world's protected areas that are consisted of strict nature reserve, wilderness area, and national parks. These areas are significant for biodiversity and various ecosystem services. The study found that the three major producers of oil palm (Indonesia, Malaysia, and Thailand) established oil palm plantations in the protected areas, particularly in Malaysia, lots of strict nature reserve and wilderness areas were invaded for oil palm while in Thailand more national parks were invaded.

Sadly, intact forests are also intruded by oil palm. The invasion of these areas for oil palm cultivation intensified where approximately 0.1 Mha of the areas were encroached since 1980s (Meijaard et al., 2018). Nations significantly impacted by the expansion of oil palm cultivation included Indonesia, Guatemala, Honduras, Malaysia, Nigeria, and Thailand (Meijaard et al., 2018). Among these, Indonesia and Malaysia encompass approximately 3% of the world's remaining intact forests, which have unfortunately been encroached upon by oil palm plantations. For example, the Tesso Nilo National Park Reserve in Honduras has been gradually invaded from the edge to the interior of the protected area since 2007 showing the ineffectiveness of the designation of 'protected area' to conserve biodiversity (Yu et al., 2023).

Oil palm expansion over forest ecosystems has caused loss of many species. A comprehensive review conducted by Meijaard et al. (2020) reported that species such as orangutans, gibbons, tigers, and Muscicapine (Old World) flycatchers are most affected due to oil palm expansion while some species such as the bearded pigs are becoming extinct in Peninsular Malaysia. They also stated that oil palm plantations had 99% less tree diversity than natural forests, and 65%–90% less mammal diversity. The transformation of peat swamp forests into oil palm plantations resulted in a reduction of biodiversity by 1% in Borneo, 3.4% in Sumatra, and a substantial 12.1% in Peninsular Malaysia (Meijaard et al., 2020). Qaim et al. (2020) conducted ecological survey in Jambi province, Indonesia, and detected a strong decrease in species richness of 12 different taxonomic groups of biodiversity as a consequence of intense deforestation of the lowland forest between 1990 and 2000 for smallholder oil palm establishment. Studies in Indonesia and Thailand found similarly low levels of bird diversity in oil palm and rubber monocultures (Aratrakorn et al., 2006; Prabowo et al., 2016).

Oil palm affects biodiversity mainly because oil palm is structurally simple with a single canopy; it is produced in monocultures and lacks understory vegetation. On contrary, tropical forests which is replaced by oil palm have multi-story canopy that produces woody debris and leaf litters that can also support a variety of animal species. The use of chemical substances in oil palm plantations does not provide a hospitable environment for many species to thrive. Only adaptable species such as pigs and certain types of mammalian carnivores can thrive in the inhospitable conditions found within oil palm plantations (Meijaard et al., 2020). Invasive animal species are generally introduced in oil palm cultivation to control rodents that destroy oil palm fruits and non-native cover crops and nitrogen-fixing plants to fertile the soil. Consequently, there is a potential for a rise in the population of local species like barn owls or snakes (Koh and Wilcove, 2008). It was found that only mosquitos and bees are equal or more in palm oil plantations than what is found in forest (Fitzherbert et al., 2008).

14.7 GHGs EMISSION

The expansion of the palm oil industry brings with it a series of environmental consequences, with the emission of GHGs being one of the most notable impacts. This emission, including methane, nitrous oxide, carbon dioxide, and isoprene, is particularly pronounced during forest clearance and soil disturbance for the establishment of oil palm plantations, fertilizer applications, and within palm oil mills (Carlson et al., 2012; Qaim et al., 2020). The CO_2 emission from oil palm plantations is estimated to be 56 t CO_2 eq/ha/year (MPOC, 2018).

The significance of these environmental impacts cannot be overstated. A range of studies (Murdiyarso et al., 2002; Reijnders et al., 2006; Harsono et al., 2014) highlights the role of oil palm plantations in consuming substantial amounts of mineral and nitrogen-based fertilizers. In Malaysia, for instance, the agricultural sector contributes around 5% of total GHG emissions, primarily in the form of nitrous oxides (NRE, 2015; Tang, 2019). A noteworthy experimental study conducted in the SAFE (Stability of Altered Forest Ecosystems) site in Sabah, Malaysia (Drewer et al., 2021), provides insights into the potential impact of GHG emissions arising from the conversion of logged forests to oil palm plantations. This study revealed significant soil NO_2 fluxes from both young and mature oil palm trees (2 and 12 years old, respectively), with higher fluxes compared to logged forests and riparian areas. Contrarily, oil palm plantations release a relatively modest amount of methane (-2.6 ± 17.2) compared to riparian and logged forests, with the highest emissions observed in medium-aged oil palm trees (7 years old). The variance in methane emissions could be attributed to the methane-sinking capabilities of mineral soils (Hassler et al., 2015), with logged forests displaying higher uptake rates due to their capacity to absorb methane.

The conversion of forests from mineral soil to oil palm plantations has been quantified to result in substantial carbon losses over a 30-year period (Fargione et al., 2008). Similarly, soil respiration rates are lower in oil palm plantations compared to riparian areas and logged forests due to reduced decomposing leaf litter, suggesting compromised soil health and nutrient cycling (Drewer et al., 2021; Kerdraon et al., 2020). This nutrient limitation in oil palm plantations hinders their ability to support soil microorganisms and crops, resulting in adverse ecological consequences. Furthermore, limited total carbon input in these plantations can constrain methane uptake, subsequently affecting soil nitrogen concentrations and further impeding methane uptake.

At a larger scale, across the region of Sabah, the expansion of oil palm plantations has led to an increase in measured nitrous oxide emissions and a decrease in methane fluxes since 1973, reflecting changes in land use (Drewer et al., 2021). However, there remains a need for more comprehensive research on GHG emissions in both tropical forests and oil palm plantations, considering current management practices and future sustainable alternatives to predict the impacts on ecosystem function, biodiversity, and overall reduction in GHG emissions.

Several studies have proposed options to mitigate GHG emissions and introduce sustainable palm oil production practices. These include biogas production from palm oil mill effluent (Suksong et al., 2020; Sodri and Septriana, 2022), conversion of empty fruit bunches (EFB) to biocomposts (Chang, 2014), and power generation from EFB combustion (Chiew and Shimada, 2013; Patthanaissaranukool et al., 2013).

14.8 AIR QUALITY

Oil palm plantations have a substantial influence on the local and regional air quality when the plantations are established by clearing natural forests (Pyle et al., 2011). Fires emit substantial quantities of smoke, CO_2, and noxious gases. This emission contributes to respiratory problems and heightened mortality rates as documented by Johnston et al. (2012). These issues are exacerbated during dry periods, particularly during El Niño events (Dislich et al., 2017).

Forest clearing and other slash-and-burn agricultural practices are widespread in the southern part of the SEA region, particularly in Borneo and Sumatera. These activities are most prevalent during the dry season, which extends from May to September, coinciding with the southwest monsoon. During this period, the combination of limited rainfall and prevailing south-westerly winds transport aerosols produced by biomass burning to neighbouring countries, leading to the formation of haze. This phenomenon has been extensively documented in studies conducted by Heil et al. (2007) and Pavuluri and Kawamura (2018).

To explore the origins of air masses and their transportation pathways, we conducted a 5-day backward trajectory analysis during the dry season, specifically on September 15, 2015. It's important to note that in 2015, a severe haze persisted from mid-July to October, with a peak in intensity in late September and early October due to the influence of El Niño. To illustrate the extent of biomass burning, we utilized data on hotspots (Figure 14.1), which indicate active fires as detected by the moderate resolution imaging spectrometer (MODIS) satellite sensor, across the major oil palm–producing countries, namely Indonesia and Malaysia, over 5 days period from September 14 to 18, 2015.

The HYSPLIT backward trajectories as shown in Figure 14.2 reveal that the primary source of air pollution at higher elevations was from Sumatra. The southwest monsoon transported pollutants from biomass burning in Sumatra, causing the pollution to traverse the western and central regions of Peninsular Malaysia. Consequently, both of these regions bore the brunt of the haze's significant impact.

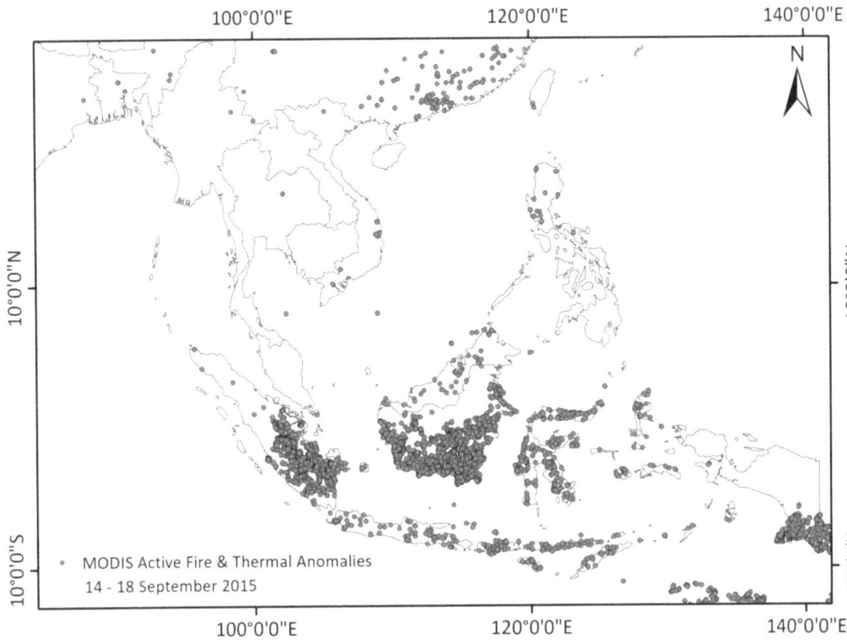

FIGURE 14.1 Active fires in the Southeast Asian region between 14 and 18 September 2015.

14.9 SOIL EROSION

Soil erosion emerges as a consequence of cultivating oil palm on newly cleared terrain featuring steep slopes exceeding 15%. The intensive establishment of palm plantations on hilly terrain can instigate erosion, presenting a pressing environmental concern. This phenomenon assumes heightened significance within oil palm plantations due to its far-reaching implications. Essential nutrients are eroded away, potentially impeding growth and oil palm yield, thereby posing a threat to food security. Remarkably, erosion rates can surge to surpass 21.26 Mg/ha/year (Satriawan et al., 2015), a loss that can outpace that observed in natural forests by a staggering 50-fold (Dislich et al., 2017).

Nonetheless, there exists a promising avenue to enhance soil health and fertility within oil palm plantations through the application of crop residue, specifically EFB (Tao et al., 2018). Furthermore, elevated levels of phosphorus and potassium originating from fertilizers have been identified in rivers and streams coursing through Borneo's oil palm plantations (Chellaiah and Yule, 2018).

A distinct study focused on the transformation of tropical forests to oil palm plantations in Sabah, Malaysia, has revealed intriguing insights into nitrogen cycling. Nitrous oxide (N_2O) emissions outweighed nitrogen (N_2) emissions following initial soil disturbance for oil palm cultivation. While gross mineralization rates remained steady in oil palm stands aged 3 months to 15 years, relative to forests, notable alterations emerged in gross nitrification and ammonium (NH_4^+) immobilization. These shifts disrupted nitrogen gas (N_2) production pathways, leading to substantial

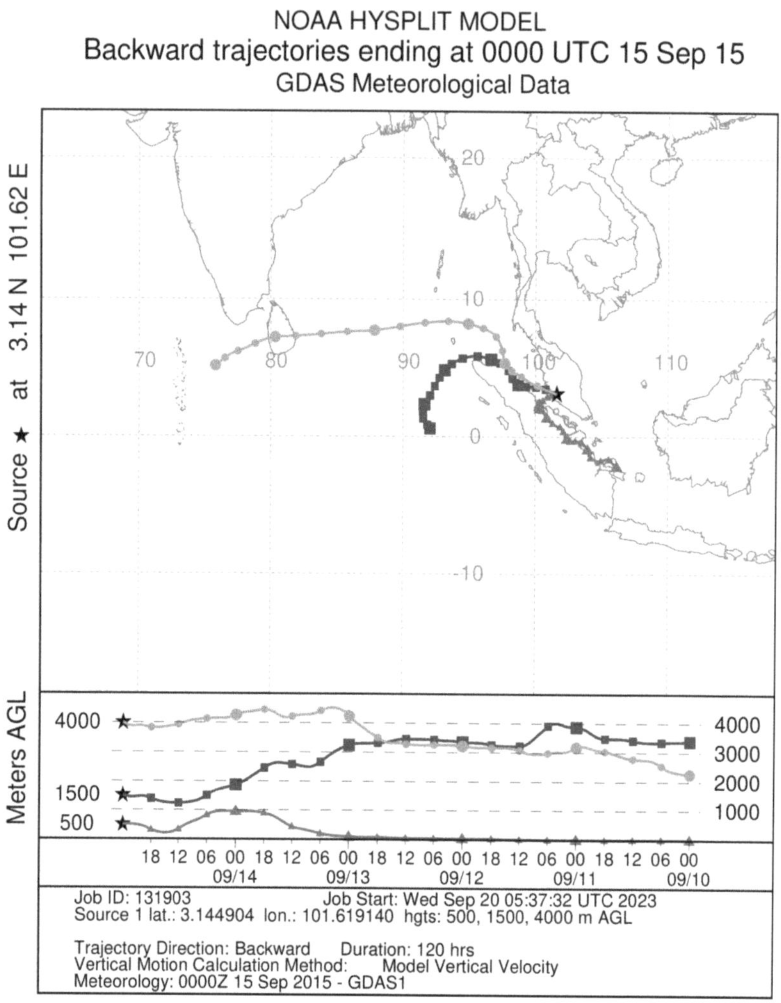

FIGURE 14.2 A 5-day backward trajectory using the HYSPLIT model.

reductions (exceeding 90%) in denitrification and 'anammox' rates in newly planted oil palm areas compared to primary forests. These findings provide valuable groundwork for further research aimed at refining estimates of regional nitrogen turnover and loss in SEA, with potential global implications for nitrogen biogeochemical cycling (Hamilton et al., 2016).

14.10 WATER QUALITY

Water pollution arising from the oil palm industry primarily emanates from the utilization of fertilizers within plantations and the discharge of palm oil mill effluent. This pervasive issue has been highlighted by Qaim et al. (2020). Moreover, the expansive cultivation of oil palm has regional implications for water quality.

Notably, excessive fertilizer application can lead to nitrate pollution (Comte et al., 2012), while alterations in water flow patterns may induce intermittent water scarcity in communities adjacent to oil palm estates (Merten et al., 2016). A significant ecological concern is the unprocessed release of millions of tons of palm oil mill effluent, which constitutes a contaminated amalgamation of crushed shells, water, and fat residues, into watercourses (Comte et al., 2012). Cultivated land adjacent to water bodies can wield significant impact on water quality and quantity due to siltation (Silalertruksa et al., 2017). Cordoba et al. (2019) found that oil palm's influence on ecosystem services in the Brazilian Amazon was most acutely perceived in terms of water availability, as well as water quality.

Conversion of natural forests into oil palm plantations amplifies the runoff and sediment export, predominantly due to the loss of riparian buffers, decreased ground cover, and the establishment of dense road networks. The distinctive features of streams coursing through plantations include warmer temperatures, shallower depths, increased sand content, and elevated turbidity. These alterations are associated with diminished aquatic species abundances, such as dragonflies (Anisoptera), compared to forested areas. Chemical substances, including fertilizers and pesticides, used within plantations further compromise water quality and aquatic habitats. While modern mills minimize effluent release, its discharge into local water bodies has engendered detrimental effects on communities and aquatic ecosystems. It is noteworthy that certain hydrological interventions, such as constructing flood-control channels and sedimentation ponds for palm oil effluent, can offer benefits for specific water birds.

The Igan River in Sarawak, a site featuring peatland planted with oil palm, serves as a poignant example of water quality degradation. Measurements of parameters like dissolved oxygen, chemical oxygen demand, biochemical oxygen demand, total suspended solids, and ammonia-nitrogen have revealed the deterioration of water quality along this river (Gandaseca et al., 2011).

14.11 HYDROLOGY AND OTHER IMPACTS

The drainage of peatlands and wetlands for the establishment of oil palm plantations disrupts natural hydrological cycles, potentially causing repercussions for surrounding forests and various habitats. Consequently, safeguarding water quality hinges on the crucial preservation and restoration of riparian buffers and reserves within oil palm plantations.

The expansion of oil palm across watershed areas can trigger floods. Merten et al. (2020) linked the establishment of oil palm plantation from forest within the Tembesi River catchment area in Sumatera, Indonesia, to deteriorating floods (increased frequency and intensity). When the forests were cleared and soil was compacted by heavy vehicles or topsoil material lost through erosion, the infiltration capacity of the soil reduced and resulted in excessive surface runoff. In the Jambi province, where the Tembesi River flows, 2 million acres of forests were cleared from 1990 to 2013 to make way for oil palm and rubber cultivation (Merten et al., 2020). The groundwater levels within oil palm and rubber plantations exhibited diminished capacity for water storage compared to the original forests. This condition can lead to reduced yield and increased vulnerability to flooding (Merten et al., 2020).

In our recent study conducted in several indigenous peoples' villages in Peninsular Malaysia (unpublished), we discovered that deforestation that occurred approximately 20 years ago for oil palm cultivation has resulted in soil erosion and flooding. Through focus group discussions with the local residents, we learned that the previously deforested lands were not fully replanted. Only areas at lower elevations were reforested, leaving the hilly regions devoid of trees. These uncovered hilly areas have become prone to landslides, particularly during the rainy season, especially in December, which coincides with the northeast monsoon season.

Landslides have caused significant soil erosion, with the eroded soil eventually finding its way into the rivers that flow towards the villages. The increased deposition of sediments in these rivers has led to a shallowing of the riverbeds, making them less conducive for fish populations. Consequently, the villagers have experienced a reduction in fish, both as a food source and as a means of income. Moreover, these shallow rivers, have become a primary cause of flash floods in the villages, as depicted in Figure 14.3. For example, in 2006, more than 70 families were affected by a flash flood in one of the villages (Segamat Kecil). During the focus group discussions, the village head, Mr. Ariffin, emphasized the urgent need for responses from local authorities, particularly during the monsoon season. The diminished forest coverage also poses a significant threat to local wildlife, including elephants as they have lost their natural habitat.

14.12 CLIMATE

The microclimate within oil palm plantations undergoes alteration due to their monoculture nature, characterized by a singular canopy. This absence of a dense, multi-layered canopy contributes to elevated air temperatures and enhanced evapotranspiration (Fan et al., 2019). Notably, the air within these plantations registers an average temperature increase of 2°C and a 12% decrease in humidity compared to forested areas (Meijide et al., 2018). Across Borneo Island, a pattern of increased temperatures and diminished rainfall has been observed since the mid-1970s, potentially linked to forest clearance for oil palm expansion (McAlpine et al., 2018).

Climate change can have significant impacts on palm oil productivity, and these impacts are a growing concern for the palm oil industry. Some of the key impacts are summarized in Table 14.2.

14.13 INTERCROPPING IN OIL PALM PLANTATIONS

The environmental consequences of oil palm plantations as discussed in the previous sections can be addressed through intercropping practice which is a sustainable agriculture strategy to enhance productivity and ecological resilience of oil palm plantations. Intercropping in oil palm plantations involves the farming of other crops in close proximity with oil palm trees simultaneously in the same land (ETA and TBI, 2021). The integration of livestock, such as cattle, within oil palm plantations also can bring benefits of diminished reliance on fertilizers and herbicides. Intercropping in oil palm plantations offers various environmental benefits besides increasing farmers' income without compromising the yield of palm oil

FIGURE 14.3 Flooding and waterlogging in indigenous villages located in the northern region of Johor State, Peninsular Malaysia.

(Namanji et al., 2021; Slingerland et al., 2019). For example, interplanting legume crops such as groundnuts, beans, and peas with oil palm can enrich the soil and provide nutrients through nitrogen fixation. Sapalina et al. (2022) discovered how intercropping system (Mucuna legume) within oil palm plantation in north Sumatera enhanced bacteria in the soil which is an indication of improved soil health and soil organic carbon.

TABLE 14.2
The Impact of Climate Change on Palm Oil Industry

Climate Change	Impact	References
Rainfall- timing and amount	Water availability (prolonged drought or excessive rainfall) can affect palm fruit production	Akram et al. (2022)
Increased temperature	Elevated evaporation rates and increased water demand can reduce yield	Okoro et al. (2017)
Drier and warmer climate	Amplifies the susceptibility to fires, particularly in peat ecosystems, culminating in smoke pollution	Crippa et al. (2016)
Drier and warmer climate	Proliferation of pests and diseases that affect oil palm, such as the oil palm rhinoceros beetle and Ganoderma fungus	Murphy et al. (2021)
Sea level rise	Sea level rise of 2 m, for example, can reduce oil palm area 8% and productivity by almost 8%	Sarkar et al. (2020)

Intercropping with other cover crops like Mucuna particularly in the high terrain slopes can serve to mitigate soil erosion. Cultivating crops like bananas, vegetables, and rice within oil palm plantations dominated by young palm trees as it is practiced in Malaysia and Indonesia (Pasmans, 2017) reduces the need for extensive weeding, as attention is primarily required around individual trees. On the other hand, shade-tolerant trees like cacao and vanilla are intercropped within mature oil palm plantations (ETA and TBI, 2021). Moreover, planting fruit and timber trees offers pivotal services such as microclimate regulation, nutrient enhancement, and fuel provision on top of supplementing the farmers with extra income. Successful examples of oil palm intercropped with timber tree such as *Shorea leprosula* are such as in Sumatra, Indonesia (ETA and TBI, 2021).

Blending different plant species within oil palm is a nature-based solution to alleviate the adverse impacts of deforestation on biodiversity and other ecosystem functioning such as climate regulation. It is well known that the conversion of structurally complex forests into monoculture oil palm plantations alters the air temperature and humidity (Meijide et al., 2018). The impact of intercropping on the microclimate was studied by Donfack et al. (2021) with the help of geospatial technology, that is, thermal images recorded by a drone in Sumatera. Their study demonstrated that the size and structural complexity of various native tree species planted within the plantation influenced the microclimate in the oil palm plantation; however, it may take longer period to significantly reduce the temperature and/or increase humidity. Other benefits of intercropping include enhancing soil infiltration by adding legume crops (Agusta et al., 2020).

These studies highlight the potential of intermixing different plant species within oil palm plantations to enhance their environmental performance besides improving the farmers' income and food security. This strategy, however, is currently implemented by small plantation holders but not preferred by the large corporations due to various reasons (ETA and TBI, 2021).

The utilization of palm oil is also debated due to societal considerations, as it affects the rights of workers and local communities residing in plantation landscapes which is explored in the next chapter.

NOTES

1 Intact forests are subset of primary forest that are extremely rich in biodiversity because they are little disturbed by anthropogenic activities, thus they are protected and prioritized for conservation (Meijaard et al., 2018).
2 1 Tg = 1 million tonnes.
3 Biodiversity prioritization templates are structured methodologies incorporating ecological, spatial, and socioeconomic data to assess the relative importance of different areas or species for conservation efforts. Various information sourced from biodiversity databases, scientific literature, GIS data, species distribution maps, habitat quality assessments, and socio-economic data related to human interactions with ecosystems are used (Brooks et al., 2006).

15 Oil Palm and Social Sustainability

15.1 SOCIAL SUSTAINABILITY

Palm oil industry particularly plantation and milling has been accused for creating social sustainability violation although the industry has been showing to improve people's financial situation and their well-being to some extent. Social sustainability is one of the three pillars of overall sustainability, and it emphasizes on the welfare of individuals, groups and communities as a whole. It covers the design, establishment and development of societal infrastructure that fulfils the physical, social, emotional, health and equity needs that reinforce individual prosperity and community attachment (Dempsey et al., 2011; Ngan et al., 2022). McKenzie (2004) characterizes social sustainability as the progression and revolution focusing on human values and quality of life and the processes enabling them. This pursuit includes safeguarding and developing the life standards for the present as well as future generations (Chiu, 2003). Social sustainability, therefore, encompasses economic development that is aligned with harmonious progress of the society that creates a conducive environment for the coexistence of culturally diverse groups while promoting social integration and enhancing life standard for all (Caulfield et al., 2001). Essential aspects for a sustainable society cover fair employment and labour practices, human rights, and societal and product responsibility, as emphasized by the Global Reporting Initiative in 2020.

Unlike the environmental sustainability as reported in Chapter 14, measuring social sustainability is challenging as it is susceptible to cultural and societal values (Arcagni et al., 2021). Social sustainability is commonly measured using indicators such as human health and well-being, housing and environmental health, education, job opportunities, human rights, social justice, demographic shift and social integration (Pasaribu and Vanclay, 2021; Santika et al., 2021). The palm oil industry frequently faces accusations of creating various issues related to land grabbing, human rights, poor education, work place conditions and welfare of the indigenous community, estate workers and rural communities (Alam and Begum, 2015; Begum et al., 2019; Pasaribu et al., 2021; Santika et al., 2021). These issues are addressed under various goals of the sustainable development framework (SDGs) (Chiriacò et al., 2022); therefore, the palm oil industry must formulate strategies to solve the issues and achieve the goals of the 2030 Agenda (refer to Chapter 16). This chapter provides a review of the impact of palm oil industry on social sustainability including a case study in Malaysia in the following sub-sections. Although it was originally developed for environmental and natural resource studies, the geospatial technology has profound applications, including the examination of social sustainability. Nevertheless, its usage to study social sustainability within the context of palm oil

DOI: 10.1201/9780429199813-15

industry is limited. Thus, in this chapter, the technology's potential for investigating social sustainability of palm oil industry is outlined.

15.2 EMPLOYMENT AND INCOME GENERATION

Undoubtedly, palm oil industry makes a substantial contribution to the economic development of palm oil–producing countries and plays a significant role in increasing income and alleviating poverty of people particularly farmers and labourers. The industry plays a crucial role in boosting the economy of the countries by increasing the gross domestic product (GDP) and foreign exchange earnings. In 2021, the global trade in palm oil and its by-products reached USD 49 billion, and in the same year, palm oil contributed to 2.5% of Malaysia's GDP (Statistica, 2023), while Indonesia produces over 30 million tons of palm oil annually, making 4.5% of its GDP and offering job opportunities to 3 million people (UNDP, 2019). These two countries dominate the world export of palm oil with 88% (Table 15.1). Similarly, the contribution of palm oil to the national economic development and social well-being in other smaller nations like Guatemala, Papua New Guinea, Columbia, Honduras and Cote d'Ivoire is significant with the export of at least 300,000 million tons of palm oil annually (Table 15.1). Palm oil is a primary industry and a major agricultural export of these countries globally.

Palm oil export has brought revenues to the producing countries which in turn significantly contribute to local rural development by creating jobs (stable and predictable income) not only in the farms but also in palm oil processing mills, increasing rural household income and developing infrastructure (Santika et al., 2019a). In Indonesia, palm oil industry contributed more than 63% of small holders' household income (Susila, 2004), and in Ghana and Uganda, people working in the palm oil sector are generally better paid compared to other agricultural sectors (Ahmed et al., 2019). Similar trends are also found in many other oil palms–growing countries (Qaim et al., 2020; Sibhatu, 2023). Notably, palm oil industry has empowered economically underprivileged women particularly in Africa to generate income. In Ghana and Cameroon, for example, women involve in managing small-scale palm oil processing mills (Awusabo-Asare and Tanle, 2008; Nkongho et al., 2014).

TABLE 15.1
World Exports of Palm Oil in 2022

Country	World Export (%)	Exports (000 MT)
Indonesia	56	28,200
Malaysia	32	15,800
Thailand	2	900
Guatemala	2	860
Papua New Guinea	2	780
Colombia	1	700
Honduras	1	430
Cote d'Ivoire	1	300

Source: USDA Foreign Agricultural Service (2023).

15.3 LIVELIHOOD IMPROVEMENT

The industry, either large corporate companies or small holder farmers, not only improved the workers financial situations but also significantly improved the workers livelihood. In Indonesia, for example, Chrisendo et al. (2022) and Santika et al. (2021) reported that palm oil industry improved infrastructure in rural areas such as electricity, schools, market places and healthcare centres. There is a difference between villages that grow oil palm and that do not involve in palm oil industry in terms of improved housing conditions, children health/nutrition and establishment of small industries (Santika et al., 2019b; Qaim et al., 2020). Similar situation also occurred in Latin America, Ghana and Columbia, where municipalities involved in oil palm cultivation experienced better access to food and other daily necessitates including healthcare, reduced infant mortality, etc. compared to municipalities without oil palm cultivation (Castiblanco et al., 2015; Abrams et al., 2019; Ahmed et al. 2019).

All these advantages of economic development, poverty eradication, improved livelihood and women empowerment resulted from the palm oil industry are counterbalanced by its extensive impacts to the environment (as discussed in Chapter 14) and significant social challenges as provided below.

15.4 LAND DISPUTES AND COMMUNITY CONFLICTS

The expansion of oil palm areas usually drives conflict between large oil palm plantations and local community and indigenous people. Conflicts and disputes usually occur between large plantation companies with local people or between the government and the local people. The main reasons are commonly unfulfilled promises and compensations by the government or plantation companies to provide substitute lands and essential infrastructure to the local people. Other reasons include forest boundary encroachment and illegal operations by small holders (Abram et al., 2017; Qaim et al., 2020). There have also been instances where lands have been seized from indigenous communities, bypassing customary land rights and, in certain cases, resorting to the use of force (Castellanos-Navarrete et al., 2021).

The establishment of oil palm plantations in Indonesia was aimed to convert underutilized land, previously held under customary tenure by villagers, into more productive terrain that would generate increased revenue for them (Li, 2018). However, a concerning trend has emerged, with over 600 land disputes surfacing in Indonesia involving clashes between local communities and palm oil companies (Colchester, 2010). About one-third of the disputes is connected to large palm oil companies that violating land tenure rights and unfulfilled commitments to provide vital infrastructure to the local communities (Rutheen et al., 2017).

15.4.1 Case Study

A case study encompassing seven indigenous villages in the northern region of Johor state in Peninsular Malaysia (Figure 15.1) has spotlighted the land predicaments encountered by these communities. A couple of decades ago, forest clearance was carried out for the establishment of oil palm plantations, which, over time, left these villagers with an inadequate land base to support their lives. Insights were gathered

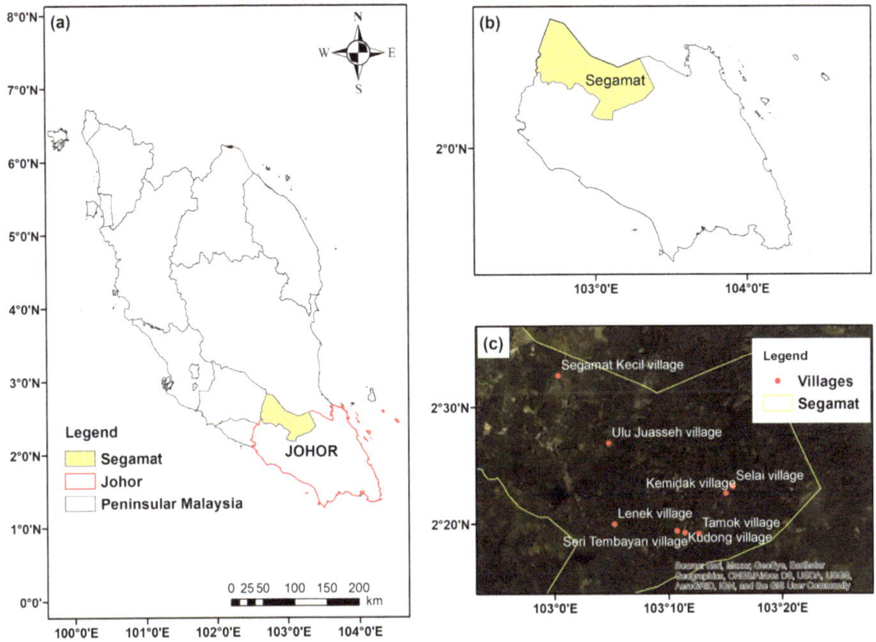

FIGURE 15.1 Map of Peninsular Malaysia (a) and the location of the district (Segamat) (b) where the study was conducted. Seven indigenous people villages (c) that were investigated to understand the impact of deforestation (for oil palm plantation) on their social well-being.

through surveys, focus group discussions (FGDs), informal conversations with forest rangers and consultations with the forestry officer (Figure 15.2), revealing that villagers undertook deforestation to make space for oil palm trees, rubber trees, other agricultural crops like fruits and residential construction (Figure 15.3).

The historical deforestation, spanning around two decades, resulted in a reduced forest area, consequently giving rise to land-related challenges for the villagers. Following a government resettlement initiative launched over two decades ago, each family was allocated approximately 4–8 acres of land as part of the compensation package. However, these land allocations proved insufficient as family sizes expanded. For instance, in one of the villages (Segamat Kecil), the population surged from 92 to 700 individuals, leading to numerous families being left landless.

Adding to the predicament is the absence of designated land reserves for indigenous people. The villagers from Segamat Kecil noted that during the 2011 election, promises were made by the Malaysian government to allocate them 1,000 acres of land, but the allocated land was situated far from their village. In the case of another village (Lenek village), a request for 500 acres of land was submitted, yet it remains unfulfilled. Consequently, the villagers have resorted to encroaching upon forested areas to expand their land holdings, primarily for cultivating oil palm trees and boosting agricultural productivity, as well as constructing houses. Currently, the primary focus of these indigenous communities is on oil palm cultivation, given its faster growth and higher-income potential compared to other crops.

FIGURE 15.2 The author conducting survey to understand the impact of deforestation (for oil palm plantation) on indigenous people's social well-being.

15.5 INEQUALITIES IN WEALTH

Although palm oil industry contributes to the economic welfare of people, there is an unequal benefit distribution among the rural communities. Using village-level socio-economic data collected in Kalimantan, Santika et al. (2019b) discovered that not all communities are economically benefitted from oil palm cultivation. Villages that had knowledge and experience with market economy even before starting oil

FIGURE 15.3 Forest clearance by the villagers at their backyard.

palm cultivation benefitted more compared to villages centred on subsistence agriculture economy. In such cases, any changes in their livelihood standards are only temporary, and notably, villages located in remote locations do not gain any profits from the cultivation of oil palm. They get higher benefits by utilizing products and services delivered by forests than oil palm. Inequalities also occur due to some communities that get better access to capital to establish oil palm plantations than other communities in Indonesia and Malaysia (Colchester, 2010). In Jambi, Sumatera particularly, differences in wealth occur because migrants received land titles and therefore credits to establish and/or expand their farms compared to the local people (Gatto et al., 2017). The resulted low income within the marginalized communities further affects their living conditions, leading poor housing conditions and housing shortages mainly in Indonesia and Malaysia (Ayompe et al., 2021).

15.6 LABOUR AND CHILDREN EXPLOITATION AND POOR WORKING CONDITIONS

Labour exploitation covering migrants and children by large plantation companies is reported to be a significant social sustainability issue in the palm oil industry. Labour exploitation usually occurs when the plantation companies set to produce unrealistically high yield that coerce workers including children. In Indonesia, labour manipulation occurs towards people migrating from other parts of Indonesia to palm oil–producing regions of Kalimantan and Sumatera. The workers are abused with extremely low wages, remote working locations, poor housing, restricted movement and without contractual agreements (Fair Labor Association, 2018).

An investigation conducted across 12 oil palm plantation companies in Indonesia by Pasaribu et al. (2021) revealed that children were involved as a workforce to assist their parents to collect fallen palm fruits. This happened due to limited availability of play areas or after school care for the children. This situation was worsened when children working in the plantations are not protected with safety gears (Li, 2018). Most of the plantation companies did not have clear policies or protocols on children's or human rights (Pasaribu et al., 2021). In Malaysia, 80% labours working in oil palm plantations are consisting of foreigners from Indonesia, Bangladesh, Nepal, India and Myanmar (Fair Labor Association, 2018). They are reported to be highly vulnerable to human rights violation issues such as confiscation of travelling documents, dangerous transit routes, poor living conditions and other facilities for workers, and insufficient implementation of legal procedures (Fair Labor Association, 2018).

15.7 CLEAN WATER AVAILABILITY

One vital feature of social sustainability is the availability and accessibility to clean water. Clean water is absolutely necessary for drinking and cooking, and its availability, therefore, determines human health and well-being. Other fundamental services of clean water for human include reducing poverty by irrigating agricultural areas, providing food sources from rivers and lakes, reducing waterborne diseases and thereby lessening infant mortality rates, and improving overall life expectancy. Despite these livelihood enabling roles, clean water is hardly available to certain communities in the developing countries as a consequence of palm oil industry.

Diverse actions, including the expansion of oil palm plantations into forested regions and areas adjacent to rivers and water bodies, the redirection of pesticides and harmful chemicals from processing facilities into streams, and the sedimentation resulting from land disruption during plantation establishment, collectively impact both the volume and purity of water resources, influencing their accessibility for communities. Research conducted in Indonesia and Uganda revealed adverse impacts on water quality and local fishing sector (Carlson et al., 2014). For example, Carlson et al. (2014) discovered oil palm–dominated catchments in Kalimantan exhibited 4°C warmer temperature and increased suspended sediments (550 times) compared to watersheds in intact forests. Sadly, in some cases, oil palm plantations do not just change the size and direction of rivers/stream to accelerate water movement (Kang and Kanniah, 2022) but bury them for establishing new plantations.

These actions affected the availability of clean water and fish stock to people in adjacent areas especially the indigenous communities who depend on rivers for clean water, food and cultural rituals (World Rainforest Movement, 2022). In the Riau province for example, the World Rainforest Movement (2022) reported that fish catches decreased substantially due to oil palm cultivation in the riparian zones. Clean water availability is also a challenge in other oil palm–growing nations like Cameroon where people there consume contaminated water that is harmful to their health (World Rainforest Movement, 2022).

15.8 GEOSPATIAL TECHNOLOGIES FOR SOCIAL SUSTAINABILITY

Geospatial technology has been used widely in recent years for socio-economic development and to operationalize SDG targets through high-resolution and low-cost data, although their direct use in the context of palm oil industry is not readily available. Nevertheless, the technologies have potential for detecting water sources and their pollution (Avtar et al., 2020), monitoring household poverty/slum development (Jean et al., 2016; Kuffer et al., 2016; Duque et al., 2017; Mahabir et al., 2018; Weide et al., 2022), assessing slavery (Boyd et al., 2018), estimating and allocating population and assessing people's well-being and other infrastructure are relevant to the industry.

15.8.1 Mapping Water Bodies

In the sphere of the palm oil industry, the utilization of remote sensing methodologies and data holds promise in appraising flood (discussed in Section 3.3, Chapter 3), the dispersion and magnitude of water across diverse water bodies, as well as assessing their quality. This potential stems from the industry's demonstrated impact on both the volume and attributes of water, as elaborated in Section 15.9. A range of satellite sensors providing data in the high–medium–coarse spatial resolution and water extraction methods based on threshold, spectral indices and machine learning can be manipulated for the purpose (Chen et al., 2022). Generally, the near-infrared (NIR) and shortwave infrared (SWIR) bands record significantly lower reflectance for water compared to other features.

Various algorithms devised using these and other bands are used to identify water bodies. Noteworthy examples include the normalized difference water index (NDWI) computed using the green and NIR bands for water extraction. The modified NDWI index calculated using the red- and medium-infrared bands could identify and separate water bodies with different turbidity levels (Xu, 2006). Further, any confounding effects of terrain, that is, shadows on water bodies extraction can be overcome when various information extracted from satellite data are combined. For instance, Guerschman et al. (2011) integrated NDWI with NDVI, SWIR and relief data to improve the extraction of water from hilly areas.

The automated water extraction index developed by Feyisa et al. (2014) combined blue, green, NIR, and SWIR bands from Landsat 5 TM sensor and bathymetric datasets to estimate alterations in the volume and surface areas of water bodies volume and surface area at the global scale. Similarly, Pekel et al. (2016) harnessed the extensive Landsat archive to quantify shifts in global surface water over a 32-year span. They analysed millions of images using a combination of expert systems, visual interpretations and evidential reasoning based on spectral information covering NDVI and colour intensity analysis using SWIR, NIR, red and NIR/green/blue bands. With these combination of methods, they could identify and contrast the permanently disappeared with emerged water bodies between 1984 and 2015.

Radar remote sensing offers notable capability in penetrating through clouds, dense canopies and tree trunks of forests and oil palm plantations. This capacity for penetration is essential to map and monitor rivers flowing through the forest and oil palm plantations and to distinguish smoother terrains like water bodies and soil from rougher surfaces like shrubs and trees. In our previous study, we proved the influence of L-band radar to separate water bodies from other land uses including oil palm plantations in the region of Peninsular Malaysia (Najib and Kanniah, 2020). Compared to optical remote sensing, radar remote sensing holds superiority due to its interactions with diverse surface roughness, dielectric properties and moisture content. In ALOS PALSAR 2 HH polarization images, water manifests as a black appearance due to its smooth surface that reflects signals away from the receiver. Using a simple thresholding method, we demonstrated almost twice low backscattering value for water than other land uses, rendering it discernible from oil palm plantations.

These types of analyses using remote sensing data could provide valuable support in identifying and monitoring changes in the area occupied by water bodies, including their potential reduction or complete disappearance due to the establishment of oil palm plantations. The findings could prove instrumental in aiding sustainability accreditation organizations such as RSPO in matters related to water management and the development of policies for conservation efforts.

15.8.2 Water Quality Monitoring

Detecting the quality of water serves a paramount importance to ensure clean water is available for various consumption by people living within and nearby oil palm plantations. Remote sensing aids to map water quality parameters such as the concentration and distribution of suspended sediments, nutrients, chlorophyll/alga, dissolved organic matters, oil spills, agrochemical, extraction of byproducts (palm oil mill effluents) and temperature at broad spatial and temporal scales. These parameters notably optically active constituents such as sediments, chlorophyll and organic matters are retrieved by establishing a connection between a subset of their concentration (samples) in water and the satellite/aircraft signal that is reflected back from the parameters, known as water leaving radiance (Figure 15.4). Other pollutant constituents such as nitrogen, phosphorus, chemical oxygen demand (COD), and dissolved oxygen (DO) do not hold a direct relationship with the radiation (Yang et al., 2022). Instead, their retrieval depends on their association with optically active constituents. Thus, using remote sensing for water pollution assessment is largely dependent on accurately computing reflectance from the constituents. A range of techniques covering empirical, semi-empirical, physical and artificial intelligence are used to retrieve the relationship between water leaving radiance and the quantity of pollutant constituents (Adjovu et al., 2023; Yang et al., 2022).

Suspended sediments generated during forest clearance and the establishment of oil palm plantations and construction of infrastructure like roads and harvesting paths to facilitate crop management constitute one of the major pollutants in the rivers (Carlson et al., 2014). Remote sensing sensors operating in the spectral range spanning 580–680 nm and 700–900 nm are used to monitor sediment concentration due to their sensitivity to differences in sediment properties. Suspended sediments typically absorb more strongly in the blue and green regions (around 480–560 nm) and

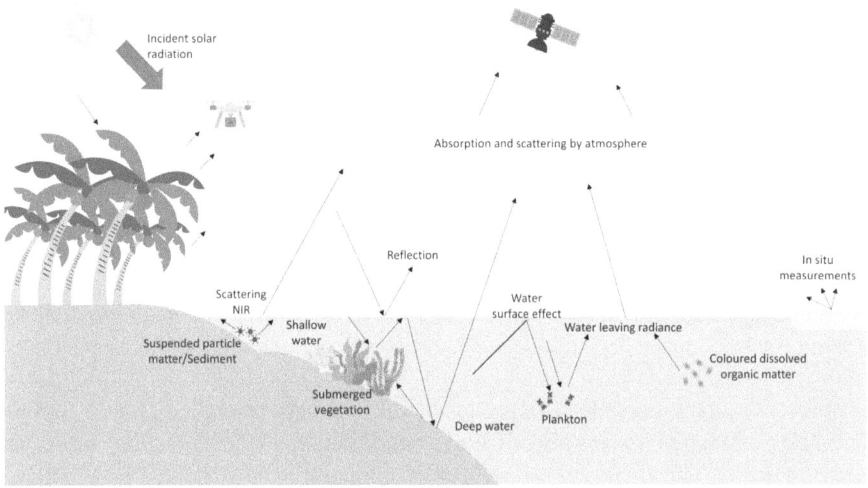

FIGURE 15.4 Remote sensing of water pollution detection.

scatter light in the NIR region (~700–900 nm). Larger particles specifically lead to increased scattering of light within the 580–680 nm range. These spectral bands are also sensitive to other water constituents that are linked to sedimentation processes, such as chlorophyll-a and coloured dissolved organic matter (CDOM). Monitoring changes in these constituents can provide insights into sediment dynamics. Light in these spectral ranges can penetrate the water column to varying depths, allowing for the assessment of suspended particles throughout the water column. So, sensors such as Landsat Thematic Mapper (TM) and Operational Land Imager (OLI), Satellite pour l'Observation de la Terre (SPOT), Thermal Emission and Reflection Radiometer (ASTER), Moderate Resolution Imaging Spectroradiometer (MODIS) and Sentinel-3A Ocean and Land Colour Instrument (OLCI) have been used to estimate sediment concentration in rivers and highly turbid estuaries (Yang et al., 2022; Yu et al., 2022).

The concentration of chlorophyll-a (Chl–a) in water serves as an indicator of eutrophication, a phenomenon driven by the influx of nutrients (nitrogen and phosphorus) originating from oil palm plantations. The amount of Chl-a is estimated using the band ratio method (NIR to Red) and other advanced artificial intelligence methods such as neural network and support vector regression (empirical models). Please refer to Adjovu et al. (2023) and Yang et al. (2022) for a latest review on the use of remote sensing for water quality analysis.

15.8.3 Poverty Mapping within Oil Palm Plantations

Poverty and slum development are critical social sustainability issues faced by many workers in the palm oil industry around the world. The communities affected by palm oil industry live in poverty and slum conditions, lacking access to basic amenities such as proper housing, schools, cultural and religious centres, and road networks.

Thus, it is imperative to track, map and monitor poverty of these people for urging corporate companies to take initiatives to uplift their living conditions and guiding mitigation efforts against deforestation that is responsible for the poverty of indigenous people in particular. Mapping the exact locations of poverty can inform proper land-use planning and support equitable compensation, enhance transparency and enable the involvement of local communities in decision-making processes. Most importantly, it can help to identify accountable businesses causing such situations and help policymakers to formulate and implement appropriate local and global policies such as the SDGs that are aimed at allocating resources equally and improve the living conditions and well-being of people living in poor conditions or at slum areas.

Detecting poverty/hotspots within or adjacent to oil palm plantations using remote sensing involves analysing various socio-economic and environmental indicators to assess living conditions and economic well-being. Mapping slum areas (indication of poverty) became possible with the availability of high-resolution remote sensing data. High-resolution images provide fine-grained details of the plantation landscape, enabling the identification of small structures, temporary shelters and informal housing arrangements that are common in slum areas. There has been a growing availability of multi-spectral and panchromatic very-high-resolution sensors operating in the visible, NIR and radar wavebands such as IKONOS, Worldview, Quickbird, Sentinel, PALSAR and Terra SAR-X that has enabled enhanced object-level analysis (Kuffer et al., 2016). Alongside commercially available very-high-resolution imagery, starting from 2005, universal web-based access to such imagery has been provided by Google Earth Engine. However, this access does not encompass the original spectral bands, thereby limiting the potential for thorough analysis. Medium and coarse resolution satellite sensors also pave ways in the slum mapping studies at large spatial extent using Sentinel (Agyemang et al., 2023), MODIS (Puttanapong et al., 2022), The Operational Linescan System (OLS) flown on the Defense Meteorological Satellite Program (DMSP) (Jean et al., 2016; Puttanapong et al., 2022) and The Suomi National Polar-orbiting Partnership (SNPP) Visible Infrared Imaging Radiometer Suite (VIIRS) Day-Night Band (DNB) sensor (VIIRS/DNB) (Puttanapong et al., 2022).

Most of the studies conducted using remote sensing focussed on mapping slum in the urban areas (Kuffer et al., 2016; Mahabir et al., 2018; Duque et al., 2017). Some studies have highlighted the feasibility of remote sensing for slum detection at small-scale rural areas (Weide et al., 2022) and large-scale rural poverty in African continent (Jean et al., 2016), Pakistan (Agyemang et al., 2023) and Thailand (Puttanapong et al., 2022). These studies used several geospatial indicators such as land cover, land surface temperature, built up areas, night-time light intensity, digital surface models (DSMs) and distance to main roads that show strong correlation to poverty.

For oil palm–related poverty detection, remotely sensed images can be classified using a range of techniques to delineate and map buildings or dwellings in the plantation area. Multi-scale strategies utilize methodologies and approaches that can distinguish features by considering characteristics that arise across various spatial scales. Frequently employed multi-scale techniques in slum analysis encompass measures such as fractal and lacunarity assessments (Mahabir et al., 2018). Textural information based on intensity and colour of pixels, which vividly depict the structure of objects and their spatial organization within an image, is also widely used to identify slum areas in satellite images.

One of the most commonly used methods has been based on the grey level co-occurrence matrix for extracting slums at the settlement level since these measures are better able to detect the high density of dwellings typical to slums. Within object-based image analysis (OBIA), an image is perceived as a composition of entities. Attributes like dimensions, form, texture, interactions with adjacent entities and diverse amalgamations of these attributes are harnessed to facilitate the process of extracting these entities. The OBIA technique demonstrated effective performance in extracting features such as roofs and roads at the settlement level. When applied to local slum mapping, OBIA techniques displayed commendable proficiency in extracting information based on both areas and individual objects (Kuffer et al., 2016).

Other data sources employed to differentiate and demarcate slum areas from their surroundings based on height of buildings/dwellings are the DSMs. Light detection and ranging (LiDAR) data can be utilized for the purpose as LiDAR systems can capture a large number of laser points per unit area, resulting in high-spatial resolution data. This density of points allows for detailed and accurate representation of object heights and shapes and enables slum area representation in three dimensions.

Drones possess the capability to manoeuvre beneath cloud cover to capture intricate settlement features, yet achieving comprehensive coverage across entire plantation could pose significant computational and financial challenges. Nighttime light intensity data retrieved from VIIRS DNB nighttime sensor mounted on the SNPP satellite began to provide nighttime lights data with higher spatial resolution (~750 m) and better data quality can reveal locations of slum areas within and adjacent to oil palm plantations (https://www.ncei.noaa.gov/maps/VIIRS_DNB_nighttime_imagery/).

Pixel- and object-based digital image classification techniques are not efficient in the complex environment of slum areas that has high spectral diversity, small and clustered objects, and diverse morphological characteristics. Thus, machine learning algorithms are increasingly being employed for slum area detection (Jean et al., 2016; Puttanapong et al., 2022; Agyemang et al., 2023). These techniques enable iterative learning from extensive and diverse training datasets. Machine learning techniques such as the random forest and support vector machines are reported to have highest predictive accuracy of slum area even when working with moderately sized datasets (Kuffer et al., 2016; Duque et al., 2017).

Although studies have highlighted the potential of geospatial technology in detecting and mapping poverty-stricken regions, limitations exist in terms of generalizing methods across diverse geographical areas due to the intricate and varied nature of slums. Each slum area varies in size, shape, orientation and construction materials for dwellings. Consequently, textural patterns extracted from one image might not be transferable to another. Hence, rules devised for extracting slums in one image are often specific to that scene, significantly constraining their application to other locations. Challenges also encompass mixed pixels (features) in sub-meter resolution images; however, recent commercial sensor technologies provide data with centimetre-level spatial resolution, thus enhancing object recognition in slums.

Advancements in machine learning and artificial intelligence hold promise for automating the analysis of very-high-resolution remote sensing data, facilitating swifter and more precise identification of poverty-stricken and slum areas. Striking a balance between spatial resolution requirements and the cost and accessibility of high-quality satellite imagery presents a challenge. Additionally, the establishment of open-access databases like Google Earth Engine, the emergence of geospatial data sources like volunteer geographic information and technological progress in geo-sensor networks providing current data and analysis tools can empower policy-makers, researchers and NGOs to effectively monitor these areas.

15.8.4 Mapping Indigenous Lands

Geospatial technology plays a vital role in helping indigenous communities resolve land conflicts with plantation owners and governments. It accurately maps and documents traditional land boundaries, clarifying land tenure rights (Moore, 2022). For instance, the Amazon Geo-Referenced Socio-Environmental Information Network (RAISG) collaborates with indigenous communities in the Amazon basin, creating maps that depict land boundaries, cultural features and natural landmarks (https://www3.socioambiental.org/geo/RAISGMapaOnline/). These maps include data on protected areas, indigenous territories, mining sites, roads, fire hotspots, deforestation and more. Participatory mapping is common, allowing communities to contribute their traditional knowledge to these spatial datasets (Figure 15.5).

Drones equipped with high-resolution cameras and remote sensing technology capture detailed images of indigenous lands, aiding in land surveying, monitoring changes and gathering evidence against illegal activities. For example, the Shipibo-Conibo people in Peru have successfully used drones to document illegal logging and mining, resulting in reduced deforestation rates (González and Kröger, 2023). In regions with limited internet access, The Rain Forest Foundation, USA, has established data hubs in Lima, Peru, to receive satellite information, which indigenous monitors can access via smartphones. These data, with a 30-m resolution, help track deforestation and can be shared with governmental authorities. The technology is also employed for documenting healthcare issues (González and Kröger, 2023).

Additionally, geospatial technology assists in creating formal land tenure documentation, crucial for legal recognition and protection of indigenous land rights. Initiatives like the Tenure Facility in Indonesia (https://thetenurefacility.org/timeline/indonesia/) combine geospatial technology with legal advocacy to support indigenous communities in mapping and securing their land rights.

To achieve a genuine social sustainability, it is crucial to support workers' rights, protect their well-being, guarantee their representation in decision-making processes and harness the potential of geospatial technology for mapping and analysis. The next chapter presents strategies/actions undertaken globally by the governments, corporations, local communities and civil society that pave the way for a future where the cultivation of oil palm aligns harmoniously with the well-being of people and the planet.

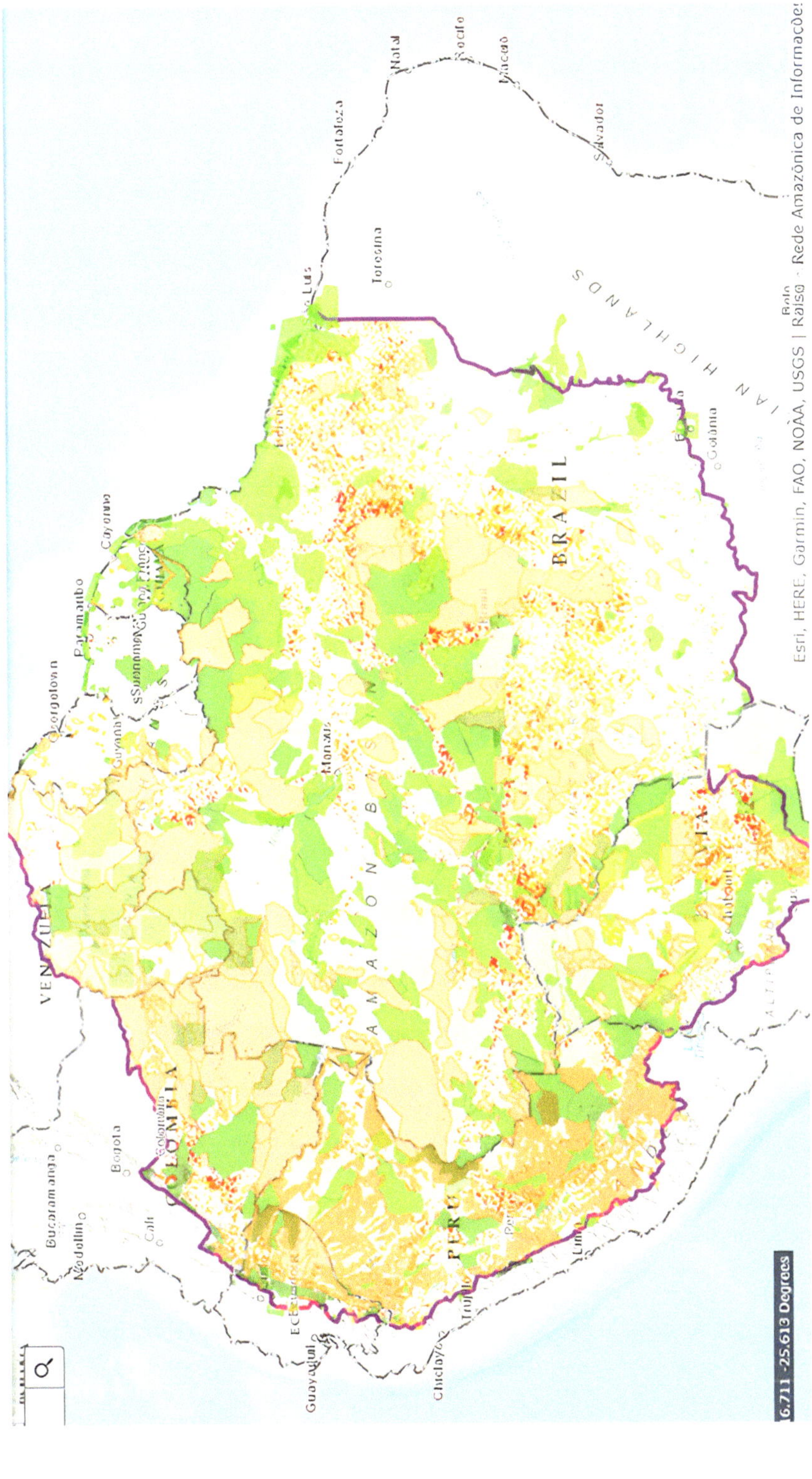

FIGURE 15.5 Geospatial data of indigenous region of the Amazonia: Green represents natural areas, brown delineates boundaries, purple signifies base locations, and red highlights deforested areas (https://www3.socioambiental.org/geo/RAISGMapaOnline/).

16 Sustainable Palm Oil

16.1 INTRODUCTION

The previous two chapters highlighted a range of environmental and social implications of the palm oil industry that urge adoption of sustainable practices. With the forecasted annual increase of 2.2% in palm oil production between 2022 and 2026 (ReportLinker, 2023), the main producers of Indonesia and Malaysia are pressured to adopt sustainable practices along the supply chain due to the European Union's concern and ban on unsustainable palm oil production (Pacheco et al., 2017; Taheripour et al., 2019). Many researchers and oil palm organizations believe that restricting the expansion of oil palm tree planting and disallowing the import of products and biofuels derived from palm oil are unwise actions as palm oil is rather a cheap vegetable oil that will have a growing demand. Conversely palm oil production can be made sustainable both environmentally and socially through policy options that require oil palm growers (large plantations companies and smallholders) to halt deforestation, stop forest burning, not to grow on peatlands, and not to exploit workers. Policy instruments are also applicable across the palm oil sector's supply chains; producers/growers, refiners, processors, traders, distributors, and manufacturers. Thus, this chapter is aimed at discussing various policy options and promotion of geospatial technology to attain sustainability (both environment and social) in the palm oil sector.

16.2 SUSTAINABILITY CERTIFICATION

The process of ensuring a product, procedure, or service is complied with specific standards is referred to as "certification" (Dankers, 2003). Sustainability certification could serve as a valuable tool for supporting oil palm plantations, milling, and refinery companies, in the responsible production of palm oil without causing harm to the environment and communities.

In the certification process, an independent party assesses the stakeholders throughout the supply chain to verify their compliance with the standards established by certification bodies. The stakeholders should also demonstrate that they have instigated measures to mitigate any such adverse effects to the environment and communities. The motivation for these certifications emanates from civil society organizations, which have coerced oil palm supply chains to produce palm oil without deforestation. Notably, the European Union has established stringent sustainability criteria for consumer goods and biofuels, focusing on the use of deforestation-free palm oil. Consequently, as many as 400 companies from various agricultural sectors including oil palm had adopted sustainable certifications by 2017 (Donofrio et al., 2017).

DOI: 10.1201/9780429199813-16

The predominant international body responsible for accrediting sustainable palm oil production is the Roundtable on Sustainable Palm Oil (RSPO). In addition to RSPO, there is the European Sustainable Palm Oil (ESPO) initiative, established by the Dutch Oils and Fats industry in 2015, with the aim of achieving 100% sustainable palm oil-based product consumption in Europe by 2020 (IDH, 2023). Another international certification relevant to sustainable biofuel and biomass production from palm oil is the International Sustainability and Carbon Certification (ISCC). On the national level, Indonesia and Malaysia have instituted their own sustainable certifications for palm oil production, namely the Indonesian Sustainable Palm Oil (ISPO) and the Malaysian Sustainable Palm Oil (MSPO) certifications. The following subsections provide concise descriptions of these certification systems and compare and contrast the key principles to ensure a sustainable palm oil industry.

16.2.1 The Roundtable on Sustainable Palm Oil

RSPO is an international non-profit organization formed by multistakeholders consisting of representatives from oil palm producers, palm oil processors, traders, manufacturers, retailers, banks and investors, and non-governmental organizations that advocate environmental sustainability, nature conservation, and social issues (https://rspo.org/who-we-are/). RSPO is responsible for implementing international standards that it developed to produce palm oil without harming the environment and social well-being (Apriani et al., 2020). It was established in 2004 according to the Swiss Civil Code and is headquartered in Switzerland, with its secretariat located in Kuala Lumpur and a satellite office in Jakarta (RSPO, 2016).

RSPO's standards contain 7 operating principles (covering 3 impact goals), 42 criteria, and 172 indicators as shown in Table 16.1 (RSPO, 2020). The first three principles are focused on prosperity, while principles 4, 5, and 6 emphasize on people and principle 7 underlines planet (Table 16.1). RSPO encompasses three certification systems: (i) Producer/grower certification (Principles & Criteria certification) which is applied to plantations and associated mills, addressing the primary environmental and social consequences of palm oil production, (ii) supply chain certification; and (iii) RSPO independent smallholder (ISH) standards certification (https://rspo.org/as-an-organisation/certification/). The supply chain certification is intended to verify that fresh fruit bunches traded in the supply chain, from plantation to processing mills to refineries to manufacturers, are sourced in a sustainable way (RSPO, 2020). This certification also requires individuals who are not palm oil producers, but claim more than 500 RSPO credits[1] for a period of 12 months to show compliance to the supply chain certification (https://rspo.org/as-an-organisation/rspo-credits/i-am-a-buyer/checks-and-audits/).

Smallholders include farmers cultivating oil palm, with/without other crops for basic needs with land size below 50 ha (RSPO, 2020). The smallholder's certification is aimed to introduce sustainable practices in their operations and also to assist independent smallholders to improve their livelihoods by increasing their palm oil yield and by getting them exposed to international markets, and control land conversion and associated risks. The certification processes guarantee oil palm businesses demonstrate transparency and traceability of sustainably sourced palm

TABLE 16.1
Principles of Four Major Sustainable Certification Systems for Operating a Sustainable Palm Oil Industry

Certifications/Category	RSPO	ISCC	ISPO	MSPO
Customary rights- traditional rights and land tenure systems of indigenous communities and local populations	Principle 4 emphasizes the requirement for obtaining Free, Prior, and Informed Consent (FPIC) from communities impacted by land acquisition or development activities	No explicit mention about customary rights but, it focuses on social aspects, such as labour rights and fair treatment (ISCC, 2022).	Requires companies to conduct social mapping and consultation processes to secure FPIC of affected communities prior to land conversion	Principle 3 requires palm oil companies to respect and recognize the rights of indigenous population and local communities in compliance with pertinent Malaysian laws and regulations (MPOCC, 2022).
Social safeguards- Important to ensure the protection of workers' rights, communities, and indigenous peoples within the palm oil industry	RSPO has robust social safeguard mechanisms in place to address human rights, labour conditions, and community engagement (Principles 4–6). It requires its members to abide by the RSPO Principles and Criteria, which cover social aspects such as fair treatment, health and safety, and community involvement (RSPO, 2020). Example: The RSPO New Planting Procedure (NPP) promotes social safeguards by mandating social assessments, community consultations, and gaining FPIC before new plantation development.	No specific provisions/ guidelines on social safeguards, but emphasizes broader social sustainability and fair treatment of workers. Example: ISCC-certified company Musim Mas collaborates with NGOs and local communities in Indonesia to implement social programmes focusing on education, healthcare, and women empowerment (Musim Mas, n.d.).	Includes specific requirements for social safeguards in its certification framework. It emphasizes the rights and well-being of labourers, local populations, and indigenous communities. For example, PT Darmex Agro, an ISPO-certified company, has implemented various social programmes, including building schools, providing healthcare facilities, and supporting community development in the regions where it operates (PT Darmex Agro).	MSPO has provisions for social safeguards, promoting responsible labour practices and community engagement. It requires certified companies to comply with social standards, including fair treatment, health and safety, and respect for the rights of workers and communities. Example: Sime Darby Plantation, an MSPO-certified company, has implemented a comprehensive sustainability policy that includes social commitments such as decent working conditions, community development, and respect for indigenous rights (Sime Darby Plantation, n.d.).

(Continued)

TABLE 16.1 (*Continued*)
Principles of Four Major Sustainable Certification Systems for Operating a Sustainable Palm Oil Industry

Certifications/Category	RSPO	ISCC	ISPO	MSPO
Gender and discrimination- Addressing gender equality and eliminating discrimination is essential for promoting inclusive and equitable practices in the industry.	RSPO emphasizes on gender equality and inclusivity in its Principles and Criteria. It encourages members to promote fair treatment, equal opportunities, and non-discrimination in the workplace. Example: RSPO-certified company Wilmar International has implemented initiatives to empower women through capacity building, training, and leadership programmes, promoting gender equality within its operations (Wilmar International, n.d.).	ISCC integrates social aspects, including gender equality, within its sustainability certification. However, it does not have specific provisions or guidelines exclusively focused on gender-related issues. Example: ISCC-certified company New Britain Palm Oil in Papua New Guinea has implemented programmes to support female employees, including education and healthcare initiatives (New Britain Palm Oil, n.d.).	ISPO recognizes gender equality and inclusion within its certification framework. It promotes fair treatment and equal opportunities for all workers, irrespective of gender. Example: PT. Dharma Satya Nusantara Tbk, an ISPO-certified company, has established gender-inclusive policies and practices, providing training and capacity building opportunities for women within its workforce (PT. Dharma Satya Nusantara Tbk, n.d.).	MSPO acknowledges the importance of gender equality and non-discrimination in its certification standards. It promotes fair treatment, equal opportunities, and women's empowerment. Example: MSPO-certified company Cargill Palm Products Sdn Bhd has implemented initiatives to empower women, including financial literacy programmes and vocational training, to enhance their participation in the workforce (Cargill, n.d.).

(*Continued*)

TABLE 16.1 (*Continued*)
Principles of Four Major Sustainable Certification Systems for Operating a Sustainable Palm Oil Industry

Certifications/Category	RSPO	ISCC	ISPO	MSPO
Smallholders- land <50Ha	The RSPO Smallholder Standard is designed to address the unique challenges faced by smallholders, encompassing matters concerning land ownership, financial access, and technical expertise. The standard promotes capacity building, training, and inclusion of smallholders in sustainable supply chains (Principle 5 of RSPO).	No specific provisions or standards tailored exclusively for smallholders. While it emphasizes the importance of social sustainability and fair treatment, it does not address the unique challenges faced by smallholder palm oil producers. The lack of specific guidelines may limit the assistance and capacity building necessary to enable smallholders to meet sustainability criteria.	Recognizes the significance of smallholders and includes provisions in its certification framework to support their involvement in sustainable palm oil production. It encourages smallholders to adopt sustainable practices, provides technical assistance, and facilitates access to financing and market opportunities. However, the effectiveness of these measures depends on proper implementation and support from relevant stakeholders.	Recognizes the role of smallholders and encourages their participation in the certification process (MPOC, 2021). It provides guidance and support to smallholders through training programmes, capacity building initiatives, and financial assistance schemes. MSPO empowers smallholders and improve their livelihoods by promoting sustainable practices and enhancing their market access. The United Plantations Group in Malaysia actively engages smallholders, providing training, access to credit, and technical assistance, resulting in improved productivity, income, and social development (United Plantations, n.d.).

(*Continued*)

TABLE 16.1 (*Continued*)
Principles of Four Major Sustainable Certification Systems for Operating a Sustainable Palm Oil Industry

Certifications/Category	RSPO	ISCC	ISPO	MSPO
Access to remedy- Access to remedy is crucial for ensuring that affected parties have avenues for addressing grievances and seeking justice within the realm of palm oil production	RSPO has formed a comprehensive grievance mechanism addressing complaints and facilitate access to remedy. It includes a formalized process for submitting grievances, independent review, and mediation. Example: RSPO's complaints panel has resolved numerous cases, including disputes related to land tenure conflicts and human rights violations, ensuring affected parties have access to justice (RSPO, 2020).	ISCC acknowledges the importance of access to remedy and encourages certification bodies to have grievance procedures in place. However, it does not have a centralized mechanism specifically dedicated to addressing grievances within the palm oil sector. Example: ISCC-certified companies are expected to have their own grievance resolution mechanisms tailored to their operations and supply chains (ISCC, 2022).	ISPO has established a grievance mechanism to provide access to remedy for affected parties. It includes channels for submitting complaints, investigation procedures, and sanctions for non-compliance. Example: ISPO's Complaints Handling Unit has resolved cases related to land disputes, environmental issues, and labour rights violations, ensuring affected parties can seek redress (https://www.indonesiapalmoilfacts.com/ispo/)	MSPO incorporates a grievance mechanism within its certification framework to address complaints and provide access to remedy. It includes procedures for complaint submission, investigation, and resolution. Example: MSPO-certified companies have resolved grievances related to labour rights, land conflicts, and environmental concerns through their established grievance mechanisms (Malaysia Palm Oil Certification Council, 2019).

(*Continued*)

TABLE 16.1 (*Continued*)
Principles of Four Major Sustainable Certification Systems for Operating a Sustainable Palm Oil Industry

Certifications/Category	RSPO	ISCC	ISPO	MSPO
Environmental safeguards- crucial for ensuring the sustainability and conservation of natural resources within the palm oil industry.	RSPO places a strong emphasis on environmental safeguards through its Principles and Criteria. It addresses deforestation, biodiversity conservation, and responsible land use practices. Example: The RSPO GreenPalm Certification programme allows palm oil producers to trade palm oil with sustainable certification, promoting transparency and traceability throughout the supply chain.	ISCC includes environmental aspects within its sustainability certification, focusing on greenhouse gas emissions reduction and resource efficiency. Example: ISCC-certified company IOI Group in Malaysia has implemented various environmental initiatives, such as waste management and biodiversity conservation programmes (IOI Group, n.d.).	ISPO has specific requirements for environmental safeguards in its certification framework. It promotes responsible land use, protection of high-conservation-value areas, and biodiversity conservation. Example: ISPO-certified company PT. SMART Tbk has implemented a zero-burning policy, water management practices, and conservation programmes in Indonesia (PT. SMART Tbk, n.d.).	MSPO integrates environmental safeguards within its certification standards. It addresses issues like deforestation, biodiversity, and water management. Example: United Plantations Berhad, an MSPO-certified company, has implemented various sustainable practices, including conservation of riparian reserves and wildlife corridors (United Plantations, n.d.).

(*Continued*)

TABLE 16.1 (*Continued*)
Principles of Four Major Sustainable Certification Systems for Operating a Sustainable Palm Oil Industry

Certifications/Category	RSPO	ISCC	ISPO	MSPO
Quality assurance- Ensuring quality assurance is crucial for maintaining the integrity and reliability of sustainable palm oil certification initiatives.	RSPO has a rigorous quality assurance system to maintain the credibility of its certification. It includes independent audits, stakeholder engagement, and a complaints and disputes procedure. Example: RSPO's Quality Assurance Program (QAP) ensures that accredited certification bodies adhere to rigorous assessment criteria, resulting in consistent and reliable certification outcomes (RSPO, 2021).	ISCC incorporates quality assurance measures within its certification framework to ensure reliable and consistent outcomes. It includes accreditation of certification bodies and regular monitoring of their performance. Example: ISCC certification is granted only after thorough evaluation of the certification body's competence, independence, and performance against ISCC's requirements (ISCC, 2022).	ISPO emphasizes quality assurance through its certification process, ensuring compliance with national standards and regulations. It includes independent verification and assessment of certification bodies. Example: ISPO's certification bodies undergo evaluation and supervision to ensure they meet the requirements for competency, independence, and quality of assessment (Ministry of Agriculture, Republic of Indonesia, 2019).	MSPO incorporates quality assurance measures to uphold its certification standards. It includes certification body accreditation, regular monitoring, and evaluation of their performance. Example: MSPO's certification bodies are subject to stringent requirements and undergo regular assessment and monitoring to ensure compliance with MSPO standards (Malaysia Palm Oil Certification Council, 2019).

(*Continued*)

TABLE 16.1 (*Continued*)
Principles of Four Major Sustainable Certification Systems for Operating a Sustainable Palm Oil Industry

Certifications/Category	RSPO	ISCC	ISPO	MSPO
Transparency and traceability – are essential for promoting accountability and ensuring the integrity of sustainable palm oil certification initiatives.	RSPO emphasizes transparency and traceability as fundamental principles of its certification. It requires members to disclose information regarding their supply networks, including the source of palm oil and compliance with RSPO standards. Example: RSPO-certified company Golden Agri-Resources has implemented the High Carbon Stock (HCS) Approach, enhanced traceability and ensuring transparency in its supply chain (Golden Agri-Resources, n.d.).	ISCC incorporates transparency and traceability within its certification framework, focusing on the origin and sustainability of biomass-based products. It requires documentation and verification of supply chain information. Example: ISCC-certified company Sime Darby Plantation tracks its supply chain through advanced digital platforms, promoting transparency and traceability (Sime Darby Plantation, n.d.).	ISPO emphasizes transparency and traceability through its certification standards. It requires palm oil companies to disclose information about their operations, including land use and compliance with regulations. Example: ISPO-certified company PT SMART Tbk uses satellite technology to monitor and track its plantations, enhancing transparency and traceability (PT SMART Tbk, n.d.).	MSPO emphasizes transparency and traceability as part of its certification standards. It requires companies to maintain records and documentation of their operations, ensuring traceability throughout the supply chain. Example: MSPO-certified company IOI Corporation Berhad implements a comprehensive traceability system, including the use of blockchain technology, to enhance transparency and traceability (IOI Corporation Berhad, n.d.).

Source: RSPO (2020), ISCC (2022), ISPO (2021), and MPOC (2021).

oil/ingredients within the industry (Laurance et al., 2010), from the location of production through their supply chains (Watts et al., 2021). Stakeholders are assessed for sustainability using the standards every five years. Once certified, they are subject to annual assessments to ensure ongoing compliance. The producer seeking RSPO certification bears the cost of audits. Following the initial five-year assessment, the primary evaluation will be repeated.

To date RSPO had certified 4.5 million ha of land and of this 6.7% are conserved and protected, while the certified palm oil production (CSPO) increases annually at 6.3% (RSPO Impact Report, 2022). In terms of the number of companies, a total of 534 companies/growers (https://rspo.secure.force.com/membership/RSPOCertSearch) and 6,872 supply chain (https://rspo.org/search-members/supply-chain-certificate-holders/) had been certified by April 2023 (RSPO, 2022). The members of RSPO (stakeholders of the industry who joins RSPO on a volunteer basis) must uphold to the principles of transparency, commit to conserve natural resources, and protect their employees', smallholders', and other communities' welfare. They are also committed to the responsible development of new plantings (https://rspo.org/as-an-organisation/membership/). Violating the Code of Conduct may lead to the potential cancellation of membership.

16.2.2 International Sustainability and Carbon Certification

The ISCC is another accreditation system that plays a crucial role in the European Union's Renewable Energy Directive. Its primary goal is to guarantee that all crops, such as palm oil used in biofuel production, originate from lands free from deforestation and are not sourced from peatlands or regions with high carbon content and rich biodiversity (https://www.iscc-system.org/). The German Federal Ministry of Food and Agriculture and 250 global stakeholders consisting of corporations, international associations, NGOs, and research institutions around the world supported the establishment of the foundation in 2010 in Berlin (https://www.iscc-system.org/about/who-we-are/history/). ISCC encompasses certification processes for agricultural and forestry biomass, biogenic waste and leftover materials, circular resources, and renewable sources, all aimed at ensuring the integrity of the biofuel industry and promoting environmental cleanliness (https://www.iscc-system.org/). The ISCC PLUS is a certification system established in 2012 for other sectors covering packaging, chemicals, food, and feed. Promoting sustainable agriculture is also part of ISCC's vision on top of its other visions to create a carbon neutral world and circular economy (https://www.iscc-system.org/). Agricultural and forestry feedstocks that are certified under ISCC are corn, camelina, canola/rapeseed, cereal, cotton, palm, shea nuts, soy, sugar beet, sugarcane, sunflower, wood, waste, and residues. The ISCC has developed a guidance document for the certifications of waste and residues from palm oil mills and empty fruit bunches (ISCC, 2022). Similar to RSPO, ISCC also emphasizes the substantial role of smallholders in the renewable energy sector by setting up an independent smallholder (ISH) certification programme and helps the farmers to increase their productivity and income (https://www.iscc-system.org/certification/independent-smallholder-certification/). As of April 2023, >45,000 stakeholders from 130 nations have been certified by the ISCC sustainability certification (https://www.iscc-system.org/). ISCC is advantageous because a single ISCC audit is sufficient to cover most feedstock markets as it is aligned with crucial platforms and industry standards. The certification is based on six principles that cover efficient production of palm oil, that protect the environment and conserve land with high conservation and carbon values, respect human rights and social development, legal operation, and safe working environment (Table 16.1). ISCC employs geospatial technology to verify land use change via global risk assessment services (GRAS) tool which is discussed in Section 16.5.2. The implementation of ISCC resulted in improved waste management, increased use of renewable energy, and reduced deforestation and land use change (ISCC, 2022). In terms of social sustainability, occupational safety and living conditions for employees improved. Additionally, their professional knowledge and skills saw enhancements, and more smallholders became involved in the certification process (ISCC, 2022).

16.2.3 The Indonesia Sustainable Palm Oil Certification

The ISPO certification was developed by the Indonesian Ministry of Agriculture in 2011. It was further enhanced through the Presidential Regulation Number 44 of 2020 with the aim of promoting sustainable palm oil production in alignment with Indonesian environmental, legal, and social regulations and for export to the global market (Purwanto, 2020). The enhanced ISPO certification introduced a new principle (Principle 6) that strongly emphasizes on transparency. This enhancement is expected to facilitate traceability within the supply chain, allowing for the identification of the source of crude palm oil (CPO) or fresh fruit bunches (Purwanto, 2020). A new indicator termed "Free, Prior and Informed Consent" was incorporated into the certification to resolve any land disputes that may arise due to the expansion of oil palm plantations (Purwanto, 2020). It also aims to expedite efforts to reduce Indonesia's greenhouse gas emission compared to ISPO. ISPO is obligatory for all oil palm growers, including those involved in energy production, as well as smallholders. Large plantations should have obtained certification by 2014, while smallholders are required to do so by the end of 2025 (Winarni et al., 2014).

The Presidential Regulation comprises 7 principles, 28 criteria, and 15 subcriteria (Table 16.1) covering various aspects related to the palm oil industry such as (i) licensing system and plantation management; (ii) technical guidelines for palm oil cultivation and processing; (iii) environmental management and monitoring; (iv) responsibilities for workers; (v) social and community responsibility; (vi) strengthening community economic activities; and (vii) sustainable business development (Purwanto, 2020). Smallholder businesses who fail to obtain ISPO certification by 2025 will have to face temporary business suspension, whereas, large plantations could incur penalties and face the possibility of losing their operating license. With these improvements, ISPO is expected to be in line with the international standard of RSPO or ISCC. By March 2020, 32% of Indonesia's total oil palm plantations (16.38 million ha) have been certified to be operating sustainably. However, only 2.75% of the total 763 ISPO-certified growers are smallholders as of early 2021 (Dharmawan et al., 2021).

16.2.4 The Malaysia Sustainable Palm Oil Certification

Malaysia, the second largest producer of palm oil, also established a sustainable certification system for operating the palm oil industry in an environmentally and socially sustainable manner. The MSPO certification was introduced by the government in 2013. MSPO is operated by the Malaysian Palm Oil Certification Council, an independent non-profit organization (Malaysian Palm Oil Certification Council, 2019). Although the implementation of MSPO was on a voluntary basis in the beginning, it was later enforced to all categories (companies with/without RSPO certification, independent, and organized smallholders) to implement MSPO by 31 December 2019. The MSPO contains 7 principles, 31 criteria, and 101 indicators (Table 16.1) for oil palm plantations, organized and independent[2] smallholders, and palm oil mills to conduct their operation sustainability (Malaysian Palm Oil Certification Council, 2019).

The MSPO certification is applicable to the administration of oil palm plantations, smallholdings, and palm oil processing facilities. Supply chain certification is pertinent to companies engaged in processing, trading, or manufacturing palm oil sourced from certified oil palm management units. This certification traces the palm oil product's journey from the mill to the final product, maintaining the segregation of certified palm oil products from uncertified ones. This ensures the continuity of certification value throughout the entire value chain, up to the end customer. The MSPO standards are also reviewed every 5 years once with annual monitoring audits, similar to the RSPO standards. As of May 2023, a total of 6.4 million ha including certified planted area, infrastructure, buildings, roads, and conservation areas have been certified by MSPO (Malaysian Palm Oil Certification Council, 2023). From these 1.3 million ha cover smallholders land, both independent and organized smallholders.

16.3 PRINCIPLES OF CERTIFICATIONS

There are several main principles (first column in Table 16.1) that most of the certification systems consider for producing a sustainable palm oil sector. These principles are covering the three sustainability pillars namely the environment, economy, and social. The principles are listed in Table 16.1 along with an analysis of the strengths and limitations of each of the sustainability certifications.

In summary, with respect to the environmental safeguards, all the four certifications set strong and clear criteria to address deforestation, and responsible land use practices. ISCC places a strong emphasis on climate change mitigation by reducing GHG emissions. RSPO, ISPO, and MSPO demonstrate a greater social safeguard to address labour rights, community engagement, indigenous peoples' rights, and smallholder inclusion. ISCC has less explicit provisions, while MSPO needs more scrutinized enforcement to protect customary rights. ISCC lacks specific provisions, potentially hindering the ability of smallholders to meet sustainability criteria effectively. ISCC also must provide stronger guidelines on community consent and land acquisition.

In terms of fair labour standards, the RSPO standard sets forth explicit and obligatory targets for companies pursuing certification. In contrast, ISCC offers comprehensive guidance on employment conditions, although not all of its criteria are deemed mandatory. ISCC categorizes its criteria into major and minor requirements, stipulating that compliance with all major criteria and a minimum of 60% of minor criteria is necessary to attain certification. Conversely, the remaining three certifications demand full adherence to all criteria for certification or the submission of an approved time-bound plan to rectify minor non-compliances.

In the matter of gender equality and discrimination, RSPO, ISPO, and MSPO demonstrate commitment by promoting fair treatment, equal opportunities, and women's empowerment. ISCC includes social aspects, but gender-related provisions are not explicitly highlighted. RSPO, ISPO, and MSPO demonstrate their commitment to providing access to remedy by establishing grievance mechanisms within their certification frameworks. They enable affected parties to seek justice and redress grievances related to social, environmental, and labour issues. While ISCC acknowledges the importance of access to remedy, it relies on certified companies to develop their own grievance resolution mechanisms.

All the certification systems except for ISCC have comprehensive quality assurance systems in place to uphold the integrity and reliability of their sustainable palm oil certifications. ISCC incorporates quality assurance through accreditation and performance evaluation. Similarly, RSPO, ISPO, and MSPO show dedication to transparency and traceability within their certification initiatives, requiring disclosure of supply chain information and promoting accountability. They implement measures such as traceability systems and advanced technologies to enhance transparency. ISCC, while focusing on transparency and traceability, may have fewer specific requirements for palm oil traceability within its broader sustainability framework. In a comparative desk study carried out by the Forest People Programme (2017), RSPO scored higher points in terms of providing a strong and clear requirement for palm oil companies to achieve a sustainable palm oil production, followed by ISCC, MSPO, and ISPO. ISPO generally provides fewer, less detailed criteria. It could be argued that the lack of clear guidance makes ISPO considerably weaker than its counterparts.

16.4 EFFECTS OF CERTIFICATION SYSTEMS TO PROMOTING SUSTAINABILITY

The certification systems as discussed in the previous section have resulted in some notable positive effects on the environment and communities. Through the RSPO certification, an estimated 4.5 million ha oil palm plantations have been accredited and 6.7% of them has been conserved and protected from conversion into oil palm (RSPO Impact Report, 2022). According to the RSPO Impact Report of 2022, the yearly growth rate of certified sustainable palm oil (CSPO) production was 6.3%, and in 2021, around 25% of CPO received certification as being devoid of activities detrimental to the environment or local communities, such as deforestation (https://rspo.org/updated-cspo-and-cspko-production-and-sales-data/). Other positive impacts include reduction of greenhouse gas emission (GHG) and assistance to independent smallholders through the certified scheme (RSPO Impact Report, 2022). A regional scale study conducted by Carlson et al. (2018) revealed a 33% reduction in deforestation in Indonesia between 2001 and 2015 as a result of the implementation of RSPO certification. Similarly, Saswattecha et al. (2015) noted that RSPO-certified producers in the Tapi Basin of Thailand exhibited a notable reduction (23%–34%) in environmental impact when compared to their non-certified counterparts (58%–75%). The RSPO-certified smallholders especially from Colombia and Indonesia are found to be adopting sustainable practices such as using organic fertilisers, reduced usage of herbicides and allocation of conservation sites within their farms (Hutabarat et al., 2018; Furumo et al., 2020).

In terms of social impacts, it is extremely important for the certifications to improve smallholders' social conditions by respecting their land rights via free, prior, and informed consent, introducing best management approaches to increase palm oil productivity, income, and their working conditions (Ibrahim et al., 2020). Smallholders often operate on smaller landholdings and are dispersed across regions, which makes their collective impact significant in terms of land use and conservation. In Indonesia, smallholders provide employment opportunities which reduced poverty among the local communities and contributed to rural development (Sayer et al., 2012). Their adoption of sustainable practices, such as practicing agroforestry, using integrated pest management and organic fertilizers can contribute to reducing deforestation, promoting biodiversity conservation, and mitigating the environmental impacts associated with the industry (Fitzherbert et al., 2008; Koh and Wilcove, 2008).

The inclusion of smallholders in sustainable palm oil production is vital for promoting social inclusivity and improving livelihoods (Ibrahim et al., 2020). Smallholders often comprise of marginalized communities and rural population who depend on palm oil cultivation for their income and livelihood. By engaging smallholders in sustainability initiatives, such as through certification schemes like RSPO, it provides them with access to training, technical assistance, and market opportunities (Dewi et al., 2017). Economically, sustainability certifications can offer smallholders improved market access and higher prices for their products. Certification schemes often create opportunities for smallholders to sell their sustainably produced palm oil at premium prices, as there is a growing demand for certified sustainable products in the international markets (Meijaard et al., 2018). For example, RSPO-certified smallholders in Malaysia have reported receiving higher prices for their palm oil compared to their non-certified counterparts (Bernama, 2023). Access to these markets can enhance the economic viability of smallholder operations and increase their incomes.

RSPO-certified smallholders in Indonesia have implemented best management practices to reduce GHG from peatland areas (Basiron et al., 2016). Among the practices include avoiding new planting on deep peat soil, using controlled drainage techniques, and employing mechanical land-clearing approaches instead of fire. Such practices can help to protect the natural ecosystem and alleviate the impact of climate change. Nevertheless, independent smallholders face difficulties to embrace the certification due to financial constraints, lack of access to market, limited technical knowledge on sustainable practices, complex documentation, and so on (Brandi et al., 2015).

Despite the existence of sustainability certifications since 2004, they have yet to fully safeguard the environment. Although plantations have obtained certification, there is no substantive evidence to validate their commitment to prevent forest loss, conserve biodiversity, and reduce the incidences of fire (Carlson et al., 2018; Morgans et al., 2018). A long-term (1984–2020) remotely sensed time-series data revealed that the majority of recently certified oil palm plantations in Sumatra and Borneo were hotspots for biodiversity in the 1990s (Gatti and Velichevskaya, 2020). Gatti et al. (2019) reported approximately 40% of the RSPO concession area had been deforested and affected by fire between 2001 and 2016 in the major palm oil–producing countries. The deadline for achieving the target for zero deforestation policy was in 2020, but many of the main palm oil–producing companies and their supply chain are still yet to achieve the target. According to the international conservation charity, the Zoological Society of London (ZCL), only 42% of the companies have monitored deforestation in their operations (https://www.spott.org/news/zsl-report-finds-many-palm-oil-companies-failing-to-meet-2020-zero-deforestation-targets/).

Although smallholders have gained some benefits following the certification schemes, there are challenges and limitations that can hinder their full impact on smallholders. The expenses and intricacies associated with certification procedures can present formidable obstacles for smallholders who have limited resources and technical capabilities (Dewi et al., 2017). The management costs and fees for oil palm plantations have risen, further increasing the financial burden on smallholders. To ensure the economic viability of certification, a substantial increase of at least 14.3% in fresh fruit bunches yield and a 9.4% rise in prices are necessary, as demonstrated in a field study conducted on the Amanah Independent Oil Palm Smallholders Association in the Ukui District, Indonesia (Hutabarat et al., 2018). It is noteworthy that firms with sustainability certification achieved nearly 2% higher profitability compared to those without certification (Hafizuddin-Syah et al., 2018). It is crucial to address these challenges by providing adequate training, technical assistance, and financial support tailored to the needs of smallholders.

The certifications did not reduce poverty and fulfil other basic needs of oil palm plantation workers. Morgans et al. (2018) performed a comparative analysis on 535 oil palm plantations including those certified and uncertified by RSPO in Indonesian Borneo between 1999 (pre-certification) and 2014 (post certification). They highlighted that certification only improved the yield of palm oil but did not significantly reduce poverty or increase health care facilities. In another study Santika et al. (2021) evaluated poverty across 36,311 villages in Sumatera and Kalimantan, Indonesia, between 2000 and 2018, periods covering prior to establishment of oil palm plantations and several years after they were certified by RSPO. Their study found RSPO certification did not reduce poverty in villages that were depending on subsistence livelihood before oil palm plantations were established. These studies suggest that additional resources must be provided by the certification schemes to uplift workers livelihood.

In a collaborative report published by NGOs in 2021 (Griffiths and Jiwan, 2021), it was highlighted that persistent social and environmental issues persist within the Indonesian palm oil industry and the global palm oil supply chains. These issues encompass land conflicts, deforestation, and governance weaknesses. The report, which investigated the connections between companies and financiers, emphasizes the insufficient attention given to identifying and addressing human rights impacts within business operations and investments. It strongly advocates for the enactment of new legislation in both importing and producing nations to mandate due diligence by these stakeholders and hold them accountable (Griffiths and Jiwan, 2021). Furthermore, the NGO Earthworm, in their efforts across Indonesian provinces like North Sumatra, East Kalimantan, and Aceh, has uncovered risks related to child labour, low wages, excessive working hours, and unfair employment conditions within the palm oil sector (Earthworm, 2020). In response, they have developed guidelines aligned with RSPO and ISPO standards to assist companies in enhancing workers' rights and working conditions (Earthworm, 2020).

16.5 GEOSPATIAL TECHNOLOGY FOR CERTIFICATIONS

Among all the sustainability principles/criteria (Table 16.1), principles that emphasize on deforestation and conversion of natural ecosystems that have high conservation values (HCV) have a global impact due to GHG emissions, global warning, and climate change. Thus, these principles accentuate the importance of conducting comprehensive environmental assessments and monitoring. The transparency and traceability of the companies' sourcing of palm oil are also heavily considered in the principles to ensure they do not purchase palm oil that was produced from deforested lands or from areas that had HCV such as forest (primary and secondary), peat land, water bodies, riparian zones, and steep slopes.

Geospatial technology has emerged as a valuable tool in enhancing transparency, accountability, and environmental responsibility. By utilizing satellite imageries and Geographic Information System (GIS) technology, stakeholders can monitor land-use changes (conversion of HCV into oil palm), identify deforestation, separate deforestation from replanting and maintenance programmes, monitor expansion of oil palm into designated peat and other protected areas, and identify suitable areas for oil palm expansion.

Transparency and traceability are used as tools in many of the sustainability certification systems to identify and track down where the source (location) of palm oil that companies use to refine and manufacturers use to produce consumer goods comes from (Garcia-Torres et al., 2019). It is a requirement for palm oil producers and consumer goods companies (producing products from palm oil) to prove to the EU importers as well as consumers that they do not contribute to environmental degradation (Gunawan et al., 2021).

The supply chain (Figure 16.1) of palm oil is consisting of large plantations and smallholders located usually within 50 km from the mills because raw fruits must be treated within 2 days from the time of collecting. The raw fruits are transported via roads to oil mill and crush plants. The CPO is then transported to refineries around the world by ship and on its way, it may collect more CPO from thousands of plantations/smallholders covering different countries before reaching the refineries. The refined oil is sold to consumer product companies to produce various products for consumers' use. In order to track down the origin (location and distribution) of

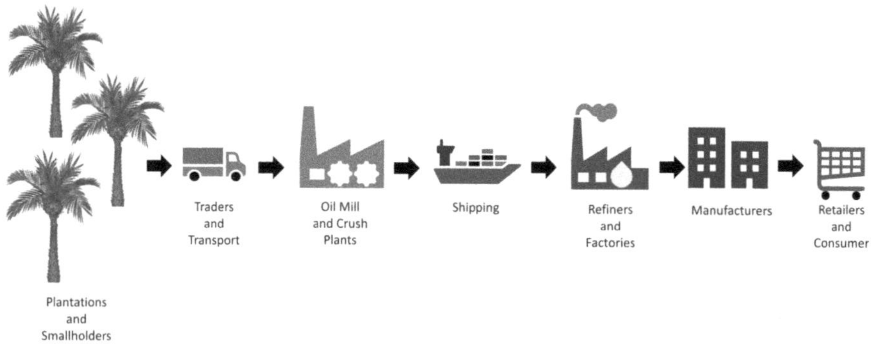

FIGURE 16.1 Palm oil supply network.

palm oil along the supply chain, relevant information such as the list of mills that are connected in the supply network of companies must be made available without any bias, delay, and discrimination for common understanding and decision-making (Parris et al., 2016). These data are integrated with satellite monitoring data to analyse deforestation in the vicinity of the mills.

Satellite remote sensing offers a cost-effective approach to monitor historical land-use changes covering large plantations. The high-medium resolution imageries are able to detect and track deforestation activities, conversion of forests to plantations, and encroachment into protected areas. By analysing historical and real-time satellite data, certifications can identify areas of concern, evaluate the level of changes in land use/land cover, and initiate appropriate actions for sustainable palm oil production.

16.5.1 Remote Sensing for Land-Use Change Analysis

Principle 7 of RSPO (Protect, Conserve, and Enhance Ecosystems and the Environment) specifically necessitates the application of geospatial analysis to investigate changes in land use. This ensures the avoidance of HCV ecosystems for oil palm cultivation (https://rspo.org/as-an-organisation/our-standards/). RSPO mandates its grower, processor, and trader members to submit all their concession maps in a GIS (.shp file) format via the GeoRSPO interactive mapping platform (source: https://rspo.org/as-an-organisation/tools/georspo/). This platform was developed through a partnership between RSPO and the World Resources Institute (WRI) (https://www.wri.org/) and plays a critical role in identifying and monitoring environmental issues, such as land clearing and forest fires.

These GIS data, in conjunction with satellite images and additional geospatial information like topography and water bodies, are utilized to assess potential environmental breaches among RSPO members (https://rspo.org/how-gis-technology-helps-keep-rspo-members-accountable/). Apart from concession boundaries, GeoRSPO also encompasses other data elements, such as company and plantation names, countries, mill data, certification status, deforestation, tree cover, climate data, and high-resolution satellite imagery which are used for sustainability certification purposes.

Although RSPO and other sustainability certifications are in place, non-compliant land clearance continues to occur perhaps due to dearth of a knowledge with regard to the RSPO Principles and Criteria particularly HCV assessment requirements and procedures. Consequently, in 2015 RSPO developed Remediation and Compensation Procedure (RaCP). This procedure allows RSPO members to conduct HCV assessments of their concessions before embarking on new land development for planting oil palm trees or any infrastructure developments. The HCV assessment is mandatory to growers (smallholders and plantation companies) who report instances of non-compliant land clearance since November 2005 before they becoming eligible for RSPO certification. RaCP facilitates restoration of cleared HCV areas and, when applicable, compensate for the conservation values that were lost.

The oil palm growers are required to calculate their environmental and social liability and based on the liability, the companies should propose and later must implement any appropriate compensation and remediation plans. RSPO will monitor the companies after two years from the date of their certification to ensure the plans are implemented. The RSPO members are required to conduct a land-use change analysis (LUCA) in order to calculate the final conservation liability (FCL). The LUCA report submitted by the growers to RSPO is assessed by independent reviewers to ensure the FCL is correctly calculated. In the following section, the author shares her experience as a consultant for RSPO in preparing LUCA reports and as an independent reviewer assessing LUCA reports submitted by oil palm growers.

According to the RSPO compensation scheme guidelines (https://rspo.org/as-an-organisation/tools/remediation-and-compensation/), palm oil plantation companies should disclose LUCA within their concession for four periods: November 2005 to 30 November 2007, November 2007 to 31 December 2009, January 2010 to May 2014, and after May 2014. Areas that have undergone clearance without a prior HCV assessment are categorized into four vegetation coefficient values, determined by the complexity of the vegetation structure and their ecological roles (Table 16.2). The total land cleared for oil palm plantation is multiplied by the coefficient values to obtain the FCL. The quantity of FCL depends on the type of vegetation, which is determined by its conservation value.

Land use during various time frames, as mentioned earlier, is typically determined through the use of satellite imagery. Some satellite data, such as that captured by Landsat satellites, can cover early RSPO reporting periods (as early as 2005), while other satellites, such as Sentinel, Advanced Land Observing Satellite (ALOS), and Satellite pour l'Observation de la Terre (SPOT), can cover more recent periods. These satellites provide imagery at a spatial resolution ranging between 10 and 30 m, which is sufficient for monitoring land use at the plantation scale. Notably, images from the Landsat series of satellites are commonly utilized by RSPO members due to their extensive temporal coverage and free availability.

Most growers conduct land-use classification on satellite imagery through visual interpretation and assess the accuracy of the classification results using data collected in the field with Global Positioning System (GPS). Occasionally, a few companies have utilized biomass data estimated from Synthetic Aperture Radar data. They assign biomass values to vegetation coefficients, where areas with biomass <70 t/ha receive a vegetation coefficient value of 0.0, areas with biomass between 70 and

TABLE 16.2
Vegetation Coefficient Values of Different Vegetation Types

Vegetation Coefficient Value	Description
1.0	Forests with a structurally complex nature, which may include primary forests, regenerating areas, or selectively logged forests with prominent high canopy elements.
0.7	Natural forests that are structurally deteriorated but still functional ecologically
0.4	Agroforestry with various plants species
0	Single-species tree and non-tree plantations, as well as other degraded areas that are permanently cultivated, developed, or open land without any cultivation.

Source: https://rspo.org/as-an-organisation/tools/remediation-and-compensation/.

140 t/ha are assigned a vegetation coefficient value of 0.7, and locations with biomass >140 t/ha receive a high vegetation coefficient value of 1.0.

LUCA for RSPO periods starting from November 2005 to after May 2014 involves several key steps (Figure 16.2). Initially, satellite imagery is downloaded

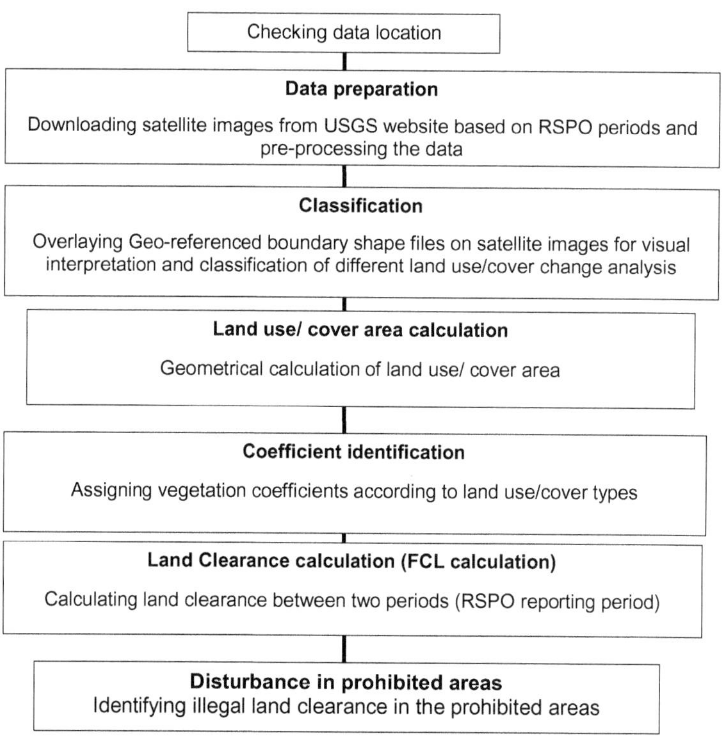

FIGURE 16.2 The review process flow within the Remediation and Compensation Procedure (RaCP).

and processed, which includes geometric correction and radiometric calibration to convert raw digital data into reflectance values. The primary method for classifying land use during this period is a combination of visual interpretation, spectral recognition techniques, and digital image classification.

During visual interpretation, satellite images are examined using various elements such as object shape, colour, texture, and reflectance values across spectral bands. This process includes not only true colour (RGB) interpretation but also false colour composite bands to distinguish between soil and vegetation. Google Earth imagery corresponding to RSPO evaluation periods is utilized to assess interpretation results, especially when reviewers are situated in different countries and lack direct access to the areas.

Historical land-use information provided by companies prior to 2005 serves as a guide for analysing land-use changes within company concessions. Once different land uses are identified in the images, each is assigned one of four vegetation coefficient values (Table 16.2). The FCL calculation considers the company's ownership history and the date of HCV assessment as the cut-off date.

A close examination for any disturbances in the oil palm plantation within prohibited areas, such as riparian zones, water bodies, deep peat, and steep slopes (>25°), is also conducted. If growers are found to have violated these regulations by expanding their palm areas into the prohibited zones, they are required to remediate the entire land that has been planted with oil palm. Remediation and compensation for non-compliant land clearance are mandatory steps for growers to attain RSPO certification.

16.5.2 Remote Sensing-Based Palm Oil Traceability Systems

There are several geospatial technology systems operating worldwide to trace the source of palm oil for sustainable production. These systems are described in the following sub-sections.

16.5.2.1 Starling Platform

The Starling system, developed in 2016, is jointly operated by the Earthworm Foundation (a Swiss-based non-profit organization) and Airbus (a European multinational aerospace corporation specializing in the design, manufacture, and sale of commercial aircraft and helicopters). It is used not only to trace the origin of palm oil but also to track various other commodities, including cocoa, forestry, and coffee (European Space Agency, 2021). The Starling system integrates satellite imageries and in-situ data to track changes in land cover and disruptions to forest cover in near real time for the purpose of verifying whether palm oil producers are genuinely procuring sustainable palm oil. The platform now covers 22 countries around the world, representing 7 million km^2 (European Space Agency, 2021).

The Starling platform utilizes satellite imagery primarily from SPOT, Landsat, and Sentinel 2 satellites. Landsat data, available since the 1970s, and Sentinel data, available since 2015, enable Airbus to analyse land cover changes and monitor whether palm oil plantations are encroaching upon forests or high-risk areas. These satellite datasets offer temporal resolutions of 26 days (SPOT), 16 days (Landsat), and 5 days (Sentinel), and frequent updates in line with their temporal resolutions through Starling.

This allows for the timely detection of forest cover changes by comparing images from two different time periods, prompting users to take necessary actions, such as discontinuing purchases from companies responsible for deforestation. Sentinel 2 satellites are particularly noteworthy for their twin satellites with 180-degree separation, enabling a broad swath width of 580 km. This is advantageous compared to SPOT (60–80 km) and Landsat (185 km) satellites. These satellites complement each other, offering long-term data (Landsat), high-temporal resolution suitable for cloud-free data in tropical regions (Sentinel), and high-spatial resolution for detailed mapping, even at small-scale forest cover loss (SPOT). Furthermore, the extensive number of spectral bands, especially from Landsat (8 bands) and Sentinel (13 bands), aids in mapping various types of vegetation and distinguishing forests from other vegetation types.

Prominent private companies, including Nestle, Ferrero, Hershey, and PZ Cussons, have employed the Starling service to assess their supply chains and support their zero-deforestation commitments (Earthworm, 2021). Notably, Ferrero was an early adopter of satellite technology for monitoring deforestation in the palm oil industry, successfully tracing 100% of its palm oil sources using the Starling system from 2013 to 2016 (Earthworm, 2021). Nestle, another major player, utilized satellite technology to trace 91% of its supply chain by 2021, sourcing palm oil from deforestation-free lands. The remaining 9% could n't be verified as deforestation-free due to limited transparency from farmers (Earthworm, 2021). Companies like Nestle also rely on independent auditors to confirm deforestation data obtained from satellite imagery.

Colombian palm oil supplier Alianza Team, which serves Nestlé, used remote sensing and the Starling programme to confirm a 100% deforestation-free supply chain (https://www.starling-verification.com/). Remote sensing technology played a vital role in helping multinational corporations make informed decisions and implement mitigation strategies to reduce deforestation, protect forests, and promote sustainability in the oil palm industry. Despite the Starling programme's monitoring of over 5.8 million km^2 of oil palm plantations since 2016, using 20 years of historical data to detect potential expansion over forested areas (https://www.starling-verification.com/), a collective commitment from more companies is essential to establish a truly sustainable palm oil industry.

16.5.2.2 Global Risk Assessment Services (GRAS)

Similar to Starling, GRAS is another initiative backed by the German Federal Ministry of Food and Agriculture through the Agency for Renewable Resources (https://www.gras-system.org/). Its purpose is to evaluate the sustainability of biofuel companies. The European Union mandates all companies that sell biofuels to them to receive sustainability certification to prove that their product is not acquired at the expense of deforestation, biodiversity loss, loss of high carbon stock, and social ethics. The collaborating partners of the platform among others include the ISCC, Germany's National Aeronautics and Space Research Centre, the Nature Conservancy, Department of Agriculture and Agri-Food, Canada, and the Energy Resources Centre of the University of Illinois and Casa da Floresta, a private environmental consulting in Brazil.

GRAS also uses remote sensing technology to evaluate alterations to land use and detect deforestation and land deterioration to verify deforestation-free supply chains. Biodiversity datasets, including protected areas such as strict nature reserves, wilderness areas, national parks, and natural monuments or features, HCV areas, and high-risk areas like habitat or species management areas, protected landscapes or seascapes, and protected areas with sustainable natural resource use, are utilized by GRAS to overlay on satellite images to support conservation. Protected areas are classified as no go areas where no commodity cultivation/expansion is allowed in high priority protected areas (e.g. protected areas with IUCN category I–III[3]) and if these areas were found to be cultivated with oil palm, for example, the companies will not be certified for ISCC.

The main image processing technique used by GRAS is the enhanced vegetation index (EVI) calculated from high-temporal resolution (twice daily) satellite images such as the moderate resolution spectroradiometer (MODIS). Interpretation of the fluctuating EVI values can assist in assessing land-use change since 2000. Land-use change can also be assessed using heat maps produced from Landsat TM/ETM and MODIS sensors. Other data from active sensors such as ALOS PALSAR and light detection and ranging (LiDAR) are also used for identifying deforested areas for situations where the optical data are obscured by clouds. The frequent revisit time of MODIS data enables the determination of the exact time of changes occurred in the land use. Meanwhile, Sentinel data, with its high-spatial resolution of 10 m, enables the detection of various land-use transformation categories, along with their precise areal changes. This capability is crucial for demonstrating compliance with sustainability requirements.

In addition to land-use change, GRAS also computes carbon stock of biomass to identify if HCV areas are disturbed for plantation purposes. GRAS also integrates social data such as child and forced labour and slavery information to identify if the biofuel is produced by manipulating human rights. Combining alteration in land use, carbon inventory, ecological diversity, and social data, a detailed risk assessment called GRAS index is computed from 0 to 1 with 0 indicating lowest sustainability and 1 representing highest sustainability (https://www.gras-system.org/). The project covers regions in the world that have large coverage of forest such as the Central and South America, Southeast Asia, the EU, the United States, and Canada. GRASS monitoring tool is used to monitor Principle 1 of ISCC which emphasizes on zero deforestation (Table 16.1). GRAS data are updated on a quarterly basis in a GIS database and maps visualization tool on a Google maps Platform (https://www.gras-system.org/gras-platforms/gras-tool/) which can be accessed by the public, businesses, investors, governments, and NGOs for assessing if biofuels are produced sustainability.

16.5.2.3 Global Forest Watch (GFW)

The Global Forest Watch (GFW) programme (https://www.globalforestwatch.org/) significantly contributes to palm oil certification efforts by offering essential data and tools for monitoring and evaluating deforestation and forest degradation linked to palm oil production. GFW is an online platform created by the WRI. It combines satellite imagery from all available Landsat sensors, including up to Landsat 8, along with other data sources, advanced image processing algorithms, and cloud computing (Hansen et al., 2013), enabling real-time tracking of changes in forest cover and land use (Shea, 2022). The radar for detecting deforestation (RADD) alerts embedded on the GFW online platform are designed to utilize Sentinel 1 radar data for continuous monitoring of forest cover loss, without the cloud interruption that is pertinent in Landsat data (Reiche et al., 2021). The high-temporal frequency (every 6–12 days) and spatial detail can detect forest cover changes in the tropics within 10 m. The disappearance of forest cover is identified through the analysis of the VV and VH backscatter data from the Sentinel C-band sensor, employing the probabilistic algorithm (Reiche et al., 2018). This enables the detection of forest loss even within the smallholder farms. However, it is important to recognize that erroneous notifications, particularly within swamp forests, may arise because C-band radar is sensitive to fluctuations in moisture levels (Reiche et al., 2018). GFW project is funded by large palm oil–producing companies in Indonesia and Malaysia like Bunge, Cargill, Golden Agri-Resources (GAR), Mondelēz International, Musim Mas, Nestlé, PepsiCo, Sime Darby Plantation, Unilever, and Wilmar (https://data.globalforestwatch.org/).

GFW provides up-to-date and reliable information on deforestation, allowing stakeholders, including palm oil companies, certification bodies, and auditors, to monitor and verify compliance with "no deforestation" commitments. By using the platform, companies can identify areas of forest loss and take necessary actions to mitigate the threat of forest loss in their supply networks. The platform promotes transparency in the palm oil sector by making forest-related data publicly accessible. This allows various stakeholders, including civil society organizations and consumers, to access information on deforestation hotspots, protected areas, and land-use changes. It enhances accountability and encourages companies to adopt sustainable practices in order to safeguard their reputation and maintain access to markets. This contributes to the due diligence process mandated by palm oil certification schemes.

16.5.2.4 CrossCheck

Another intriguing platform, based on satellite data, for tracing and tracking a deforestation-free palm oil supply chain is CrossCheck. This system was developed by Sime Darby, one of the major palm oil–producing companies based in Malaysia, in 2019. (https://smart.simedarbyplantation.com/gisportal/apps/webappviewer/index.html?id=36e336808bb24fe6981847e4ddfecd94). This platform utilizes satellite data to provide evidence of deforestation by combining satellite data with GPS locations and ownership details of palm oil mills that supply CPO to refineries, along with biodiversity data, to identify high-risk areas in terms of forests and biodiversity. Remote sensing data are instrumental in advancing the company's objectives, including the realization of its No-Deforestation, No-Peat, and No-Exploitation (NDPE) policy, as well as the implementation of its Responsible Agriculture Charter and Human Rights Charter (https://www.cspo-watch.com/sime_darby_sustainable_palm_oil.html).

16.5.2.5 Satelligence

Satelligence is a geospatial sustainability company that monitors and assesses the sustainability of agricultural commodities, including palm oil (https://satelligence.com/). The company offers a variety of services and products relevant to palm oil certification and sustainable sourcing efforts for some of the world's largest palm oil corporations such as Musim Mas, Mondelez International, Sime Darby, Bunge, and R.E.A Holdings. Satelligence integrates freely available optical and radar satellite data from 1984 to the present with supply chain data, which includes among others geolocated farms, mills, factories, refineries, and ownership details. They also incorporate auxiliary data such as local knowledge and field trips to monitor land cover changes on the ground. By analysing changes in land cover over time, they can identify areas of concern and provide accurate and timely information on deforestation risks. Advanced artificial intelligence techniques are used for monitor deforestation across sourcing landscapes with real-time notifications. For example, Satelligence used 7–10 m high-resolution optical (Landsat and Sentinel-2) and radar (Sentinel-1) and machine learning techniques to map the distribution of palm oil (smallholder and plantations), intrusion of palm oil into protected areas, and palm oil age in areas owned by Musim Mas in Acheh, Indonesia (https://satelligence.com/musim-mas).

16.5.3 Challenges in Using Geospatial Techniques for Palm Oil Certifications

Remote sensing technology has demonstrated its worth as an effective tool for evaluating and tracking forest depletion resulting from palm oil production (Payne, 2021). Nonetheless, it is crucial to acknowledge the constraints of remote sensing technology in the context of RaCP for accurate impact assessments and effective compensation and measures. One of the primary limitations of remote sensing technology is its spatial resolution. Currently, Landsat images are widely used due to their long-term availability and being free of charge. However, these data may not accurately identify and delineate small-scale forest loss, which can pose challenges in precisely assessing the extent of forest loss. This, in turn, may lead to inadequate compensation or remediation measures for affected areas.

Remote sensing imagery can also miss forest loss in areas covered by clouds or due to striping effects, particularly in Landsat 7 images. This can impact the assessment process by potentially excluding certain deforested areas from compensation and remediation calculations. Incomplete coverage may compromise the fairness and effectiveness of the RaCP.

The temporal resolution of Landsat data, with a 16-day interval and frequent cloud cover in maritime regions such as Indonesia and Malaysia, may not provide adequate monitoring frequency. Consequently, data from outside the RSPO monitoring period might be necessary to represent the monitoring periods accurately. This can impact the accuracy of impact assessments and the effectiveness of ongoing remediation efforts. Interpreting remote sensing data demands expertise and may involve subjective judgment. Distinguishing specific land-use changes attributed to palm oil production from other factors or natural variations can be challenging, potentially leading to errors in impact assessments and compensation calculations. Land covers that share similar spectral reflectance, such as old and young shrubs or other vegetation types, are difficult to discriminate. Misclassification can influence the assignment of vegetation coefficient values, such as assigning a coefficient value of 1 for old shrubs and 0.4 for young shrubs. This, in turn, can impact the FCL calculation.

Remote sensing data need validation and should be complemented with ground truthing, involving field surveys and on-site data collection. The logistical challenges of accessing oil palm plantations, especially in remote or densely vegetated areas, can limit the extent of ground truthing, potentially affecting the accuracy of impact assessments and remediation planning. Oil palm plantations often exist within complex land cover mosaics, including mixed cropping systems and intercropping with other crops. Remote sensing may struggle to accurately distinguish oil palm plantations from surrounding land cover types, making it challenging to attribute damages and determine compensation accurately.

The costs associated with remote sensing technologies and high-resolution imagery can be prohibitive for some organizations or regions. Additionally, access to up-to-date and high-quality imagery may be limited in certain areas, hindering the consistent and timely monitoring required for effective RaCP. It is important to acknowledge these limitations and integrate remote sensing technology with other data sources and field assessments to ensure a comprehensive and accurate evaluation of damages and an effective RaCP.

The primary challenge in using satellite imagery to establish deforestation-free supply chains lies in acquiring traceability information from various levels, including farms, mills, refineries, and manufacturers, and effectively integrating it with forest loss data. Earthworm, an organization actively involved in this field, places a strong emphasis on establishing direct commercial relationships with Tier 1[4] suppliers as an initial step (source: www.airbus.com/newsroom/stories/starling.html). This approach allows them to directly engage with mills that supply CPO to manufacturing companies like Nestle, underscoring the importance of traceability and building trust with these suppliers.

Another method for obtaining traceability data involves sharing satellite-monitored forest loss data with Tier 1 suppliers and encouraging them to conduct on-site investigations to verify the occurrence and source of deforestation. By collaborating with local stakeholders, these suppliers can then take measures to address and prevent deforestation.

One of the substantial challenges in establishing deforestation-free supply chains is effectively monitoring forest loss attributed to smallholders. In Malaysia and Indonesia, smallholder oil palm plantations constitute over 40% of the total oil palm plantations (source: https://rspo.org/as-a-smallholder/). However, monitoring these smallholder areas is seldom carried out due to factors such as their diminutive size, scattered and fragmented distribution, remote locations, irregular shapes, susceptibility to rapid changes, and frequent intermingling with other agricultural commodities (Pribadi et al., 2023).

To tackle this challenge, several techniques have been employed to map smallholder oil palm plantations. These methods encompass a combination of field surveys integrated with unmanned aerial vehicles (UAVs), digital classification of high-resolution satellite data sourced from platforms like SPOT, Sentinel, and Google Earth, the utilization of machine learning and deep learning algorithms, as well as participatory mapping (Descals et al., 2019, 2021; Okarda et al., 2018; Pribadi et al., 2023). Nevertheless, mapping smallholder farmlands continues to be an intricate task, as numerous farmers cultivate expanses of land that are widely distributed, and not all of them have obtained legal authorization for their activities (Qaim et al., 2020). The absence of surveillance in smallholder plantations presents a notable impediment to attaining sustainable palm oil cultivation.

16.5.3.1 Future Outlook

Sustainable palm oil certifications, supported by remote sensing and GIS, offer a pathway towards a more sustainable future for the industry. Notably, SKYwatch, a new satellite imaging platform from Canada, has attracted substantial investment, including from the global palm oil industry. The collective efforts of stakeholders and technological advancements are instrumental in overcoming challenges and achieving long-term sustainability goals in palm oil production.

NOTES

1 Organizations that involve directly in the supply chain, banks/investors and NGOs that buy, use, and sell palm oil worth of ≥ 500 metric tonnes.
2 Independent smallholders are oil palm land owners who manage their farm by themselves, while organized smallholders' lands are managed by government agencies.
3 Category Ia: Strict Nature Reserve, Category Ib: Wilderness Area, Category II: National Park and Category III: Natural Monument or Feature (https://www.iucn.org/resources#resources).
4 Tier 1 suppliers are companies or organizations that are directly involved in the cultivation and processing of CPO.

17 Future Outlook

17.1 CURRENT AND FUTURE CHALLENGES

The palm oil industry faces and will continue to confront numerous challenges as it endeavours to enhance sustainable yields. Currently, smallholders are experiencing a plateau in yields at approximately 3–4 metric t/ha, while commercial plantations are encountering a similar bottleneck at around 4–5 t/ha (Murphy, 2014), although there is a potential to increase productivity up to 18 t/ha (Woittiez et al., 2017). These challenges include vulnerability to pests and diseases, the threat of climate change, and limited access to advanced agricultural technologies, especially for small-scale farmers (Murphy et al., 2021). The industry also struggles to produce palm oil in a sustainable manner without causing any detrimental effects to the environment and its workers. Overcoming these hurdles is essential for the industry to ensure long-term sustainability and minimize its environmental and social impacts.

One promising solution to address the multifaceted challenges encompassing the economic, environmental, and social sustainability aspects of the palm oil industry lies in harnessing spatial data. This valuable data is acquired through cutting-edge geospatial technologies such as remote sensing, GIS, and GPS. When geospatial data is coupled with other data collected by IOT devices, Artificial Intelligent (AI)-driven analysis, and distributed ledger technology such as block chain, innovative solutions can be created for comprehensive understanding, monitoring, and management of plantations to enhance palm oil yield and concurrently realize environmental and social sustainability goals.

17.2 WAY FORWARD TO INCREASE PALM OIL YIELD

Satellite imagery, coupled with AI-driven analytics, empowers the assessment of crop health, enabling timely responses to issues such as disease outbreaks and nutrients deficiencies as detailed out in Chapters 11 and 12. In the future, the identification and treatment of individual oil palm trees as production units may become possible, particularly with high spatial resolution data, as demonstrated in Chapter 8. Productivity of individual oil palm trees is important for maximizing yield as higher productivity per individual tree means each tree produces more oil, which can lead to increased overall oil yield.

Improving the productivity of individual trees allows for the optimal use of resources such as water, nutrients, and land. It means that resources are efficiently allocated to trees that can generate higher yields, and reducing waste and environmental impact. When individual trees are more productive, there is less need to expand plantations into new areas, reducing deforestation and habitat destruction. This contributes to the conservation of biodiversity and mitigates the environmental impact of palm oil production.

DOI: 10.1201/9780429199813-17

In the coming years, palm trees will be outfitted with compact sensors, each bearing an individual identification tag and capable of storing a wealth of data. This information will encompass essential details like planting dates, precise GPS locations, production statistics, fertilizer utilization, irrigation records, and a host of other critical data points. This groundbreaking approach mirrors a cloud-based healthcare system designed to proactively monitor and address the well-being of these trees (Selvaraj and Sundaravaradhan, 2020; He et al., 2020). Sensors on palm trees can track growth, nutrient levels, and pest infestations. Data from these sensors can help plantation managers make informed decisions about irrigation, fertilizer application, and disease management, providing both preventive and corrective measures as needed.

Robots and automation technologies are increasingly finding applications in oil palm plantations, ushering in improvements in efficiency while mitigating reliance on manual labour and the associated costs. This transition towards automation aims to boost overall productivity, a need exacerbated by the challenges arising from the Covid-19 pandemic, which stressed the vulnerabilities of Malaysia's heavy dependence on labour in both plantations and mills.

A notable innovation in this domain is the deployment of robots with telescopic arms, as demonstrated by Arikapudi and Vougioukas (2021). These robots are equipped to reach higher into the oil palm trees to harvest ripe fruit bunches (FFBs) often located at significant heights. This advancement results in a substantial increase in productivity by reducing the necessity for manual harvesting, particularly in the case of taller trees.

The telescopic-arm robots are programmable to selectively harvest only ripe FFBs while leaving unripe ones on the tree. This precision not only ensures higher fruit quality but also minimizes wastage, a pivotal consideration in the palm oil industry. Moreover, these robots are furnished with a suite of advanced features, including continuous tracks for mobility, integrated GPS systems pre-loaded with optimized harvest routes, and built-in weight scales. These attributes collectively enhance their efficiency in the harvesting process.

Crucially, the harvested data can be seamlessly uploaded in real-time to a cloud-native platform, where it can be centralized and integrated with data from other sources, such as weather stations, palm tree sensors, and remote sensing imagery. This comprehensive data ecosystem proves invaluable for agronomists and state controllers, offering a multifaceted tool to optimize plantation management, monitor environmental conditions, and ensure compliance with industry standards.

The outlook for drones in the palm oil industry holds the promise of transformation, with the potential to deliver a multitude of benefits in plantation management and monitoring. Drones equipped with advanced sensors and imaging technology will continue to capture high-resolution imagery, allowing farmers to monitor tree health, detect diseases, and optimize fertilizer and irrigation use. Drones equipped with robotic arms can assist in harvesting ripe fruit bunches from tall oil palm trees. They can reach heights that are challenging for human harvesters and improve efficiency. Drones can be used to monitor and assess the environmental impact of oil palm plantations. They can track deforestation, land-use changes, and compliance with sustainable practices, helping the industry adhere to environmental regulations and certifications. Drones will continue to play a crucial role in data collection and integration. Data collected by drones, including imagery, thermal data, and

GPS coordinates, can be integrated into cloud-based platforms alongside data from other sources, such as sensors and weather stations, for comprehensive plantation management.

17.3 ADVANCED TECHNOLOGIES FOR BOOSTING SUSTAINABILITY EFFORTS

In response to the growing environmental and social concerns associated with palm oil production, advanced technologies offer sustainable solutions. Blockchain technology, renowned for its tamper-proof and decentralized ledger, plays a pivotal role in enhancing traceability and transparency within palm oil supply chains. Blockchain records transactions across multiple computers, operating on the fundamental principles of transparency, security, and immutability (Rahmani et al., 2022). This innovation ensures full transparency while safeguarding privacy, benefiting both workers and consumers within the palm oil supply chain.

The integration of blockchain technology into the palm oil supply chain, as depicted in Figure 17.1, offers numerous advantages. It empowers plantation owners to securely record various data points within a blockchain ledger, encompassing employee details such as passports, work contracts, and insurance, cultivation data (including planting dates, fertilizer usage, real-time harvest data, and land management techniques), and working conditions. This approach draws parallels with practices in the corporate sector involving migrant workers, where blockchain technology has been employed to combat human slavery (Christ and Helliar, 2021).

Within the palm oil industry, smart contracts running on blockchain platforms utilized by plantation companies can automate payments to workers upon completing harvests. This payment data can also be securely stored in the blockchain. Additionally, crucial harvest data pertaining to fruit bunch sources and processing methods is meticulously recorded on the blockchain (Figure 17.1). Sustainable certification details, such as RSPO and ISCC compliance, are transparently added to the blockchain, ensuring accountability.

The blockchain further facilitates transparency by meticulously tracking every step in transportation, from source to destination (Figure 17.1). QR codes affixed to product packaging seamlessly link consumers to blockchain data, offering insights into origin and sustainability certification. Anyone within the supply chain, including consumers, can verify blockchain data, thereby promoting responsible sourcing. This blockchain-based data additionally supports ongoing assessments of sustainability compliance.

By recording each stage of the supply chain, blockchain technology ensures comprehensive traceability and substantially reduces the risk of fraud. Smart contracts automatically validate compliance with sustainable production criteria, enabling the labelling of compliant products. Furthermore, blockchain provides real-time visibility into the supply chain, mitigating illegal activities while maintaining the cryptographic security necessary to safeguard sensitive data.

Through the automation of processes such as payments, inspections, and customs clearance, blockchain minimizes the need for intermediaries, streamlining operations. Most importantly, it fosters trust by offering verifiable information regarding palm oil sourcing. Regulators and auditors gain real-time access to monitor

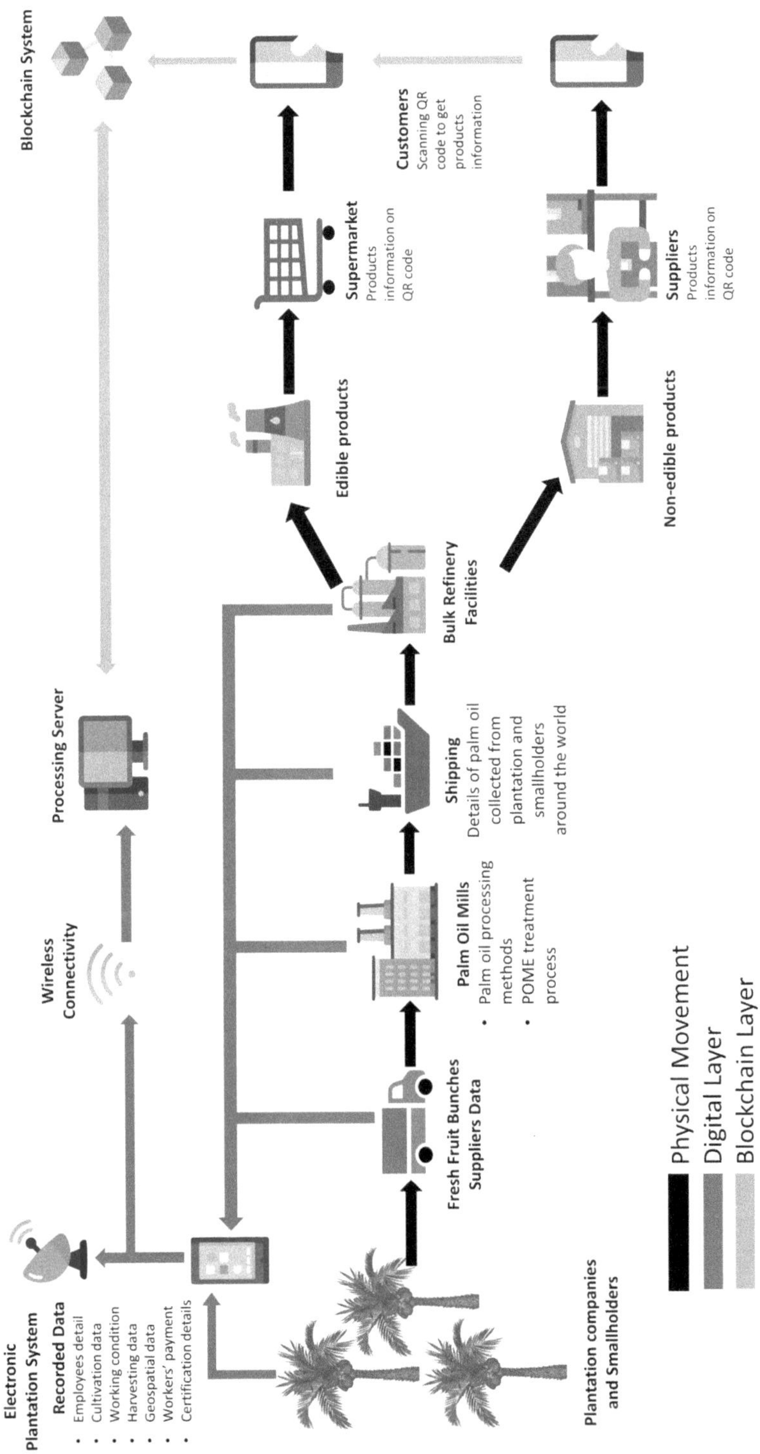

FIGURE 17.1 An illustrative blockchain system for the palm oil industry.

compliance, simplifying audit processes and ensuring adherence to sustainability standards. Secure data sharing among stakeholders promotes efficient decision-making and collaboration.

While the adoption of blockchain technology in the palm oil industry is still in its early stages, it holds significant promise for enhancing sustainability, accountability, and transparency throughout the supply chain. As the industry continues to grapple with environmental and social concerns, blockchain can play a vital role in promoting responsible palm oil production and trade.

17.4 DATA COLLECTION AND SHARING

Data availability plays a pivotal role across various aspects of the palm oil industry. In-situ data pertaining to oil palm, encompassing GPS locations, areal extent based on age, biometric information like palm tree height, trunk size, canopy size, infection rates, and yield, among others, are indispensable for calibrating and validating models constructed using remotely sensed data.

Predicting palm oil yield, for instance, advanced AI modelling techniques, such as deep learning, depend on the utilization of extensive data inputs collected from a wide array of sources. These sources encompass satellite indices, GIS data providing insights into geographical conditions, publicly accessible weather forecasts, and plantation-specific data related to crop management. These diverse data sources serve as the fundamental building blocks for training the model, empowering it to generate recommendations aimed at enhancing planting conditions. Such recommendations may include optimizing fertilizer and water usage for improved outcomes. Utilizing diverse data sources, Digital Twin technology enhances sustainability in the palm oil industry by offering real-time insights, optimizing processes, and enabling better decision-making. Digital twins enable precision agriculture, predictive maintenance, supply chain optimization, ecosystem monitoring, and stakeholder collaboration, reducing environmental footprint through simulated plantation conditions, machinery failure prediction, optimized supply chains, and collaborative sustainable practices.

Currently, there are no initiatives or platforms available for gathering comprehensive palm oil–related data and curating them with a standardized protocol, as seen with initiatives like GEOGLAM for crops such as soy and maize (https://earthobservations.org/geoglam.php). GEOGLAM has implemented an in-situ data coordination strategy to leverage partner investments and ensure data are curated with a standard protocol.

Similar platforms can be established within the palm oil industry, facilitating collaboration among palm oil companies, NGOs, and government agencies to create a shared platform for accessible geospatial data pertaining to palm oil production. Collaboration is indispensable for optimizing data collection and harnessing information effectively within the palm oil industry. Open data sharing fosters trust among consumers and stakeholders, bolstering the credibility of sustainability claims and promoting responsible practices.

Potential strategies to address the challenges of in-situ oil palm–related data collection and improve data reliability include crowdsourcing. Crowdsourcing presents an innovative and efficient solution for the acquisition of field-based data, particularly in the context of oil palm production. Mobile phones equipped with various

TABLE 17.1
Crowdsourcing Methods to Collect Oil Palm Data

Method of Data Collection	Person in Charge	Types of Data
Tailored mobile apps	Farmers/plantation owners	Real time updates on plantation status: GPS coordinates, crop health, pest infestations, yield data.
Surveys and questionnaires	Local communities residing near oil palm plantations	To get insights about land use, social impacts, and environmental concerns tied to oil palm cultivation
Satellite image tagging	Volunteers	Mark areas of land-use changes, deforestation, and plantation expansion.
OpenStreetMap	Volunteers	Update geospatial data concerning oil palm locations, road networks, mills, and other pertinent features

sensors, such as cameras and GPS, serve as powerful platforms for crowdsourced data collection. Mobile applications, exemplified by the GPS-video-GIS app, can collect a wealth of information, including crop types, planting dates, irrigation practices, and anticipated yields, all enriched with precise geolocation data. Embracing crowdsourcing methods can significantly enhance the breadth and efficiency of data collection processes related to oil palm cultivation. Table 17.1 lists some possible crowdsourcing methods to gather oil palm–related data. Incorporating crowdsourcing methodologies into the domain of oil palm production represents a dynamic approach to data collection. It empowers stakeholders at all levels, fosters community engagement, and leverages modern technology to accumulate a comprehensive and nuanced understanding of the multifaceted aspects associated with oil palm cultivation. However, quality checks in crowdsourced data are important for ensuring accuracy, reliability, and credibility.

Data contributed by the stakeholders on a shared platform can undergo advanced processing, yielding increasingly refined and precise insights to tackle ever-evolving challenges. The integration of AI and machine learning algorithms with geospatial technology can significantly enhance the analysis and interpretation of extensive spatial data. For example, when growers' data is interconnected across all palm oil–producing regions, market price predictions gain a significant boost in accuracy and accessibility. Consequently, this increased accuracy empowers informed executive decision-making processes.

In conclusion, the future outlook of the palm oil industry is marked by significant strides in increasing oil palm yields, a steadfast commitment to boosting sustainability efforts, and an innovative approach to data collection and sharing. With the integration of technologies such as drones and combined intelligence, the industry is self-assured to achieve greater efficiency, transparency, and environmental responsibility. As stakeholders collaborate on a global scale to address the complex challenges facing palm oil production, there is a strong sense of optimism that the industry will continue to evolve, adapt, and ultimately thrive in a manner that balances economic growth with ecological stewardship, ensuring a sustainable and responsible future for this vital commodity.

References

Abdullah, A. F., & Muhadi, N. A. (2015). GIS data collection for oil palm (DaCOP) mobile application for smart phone. *ISPRS Annals of the Photogrammetry, Remote Sensing and Spatial Information Sciences*, 2, 165–168.

Abood, S. A., Lee, J. S. H., Burivalova, Z., Garcia-Ulloa, J., & Koh, L. P. (2015). Relative contributions of the logging, fiber, oil palm, and mining industries to forest loss in Indonesia. *Conservation Letters*, 8(1), 58–67.

Abram, N. K., MacMillan, D. C., Xofis, P., Ancrenaz, M., Tzanopoulos, J., Ong, R., … Knight, A. T. (2016). Identifying where REDD+ financially out-competes oil palm in floodplain landscapes using a fine-scale approach. *PLoS One*, 11(6), e0156481.

Abram, N. K., Xofis, P., Tzanopoulos, J., MacMillan, D. C., Ancrenaz, M., Chung, R., … Knight, A. T. (2014). Synergies for improving oil palm production and forest conservation in floodplain landscapes. *PLoS One*, 9(6), e95388.

Abrams, J., Pischke, E. C., Mesa-Jurado, M. A., Eastmond, A., Silva, C. A., & Moseley, C. (2019). Between environmental change and neoliberalism: The effects of oil palm production on livelihood resilience. *Society & Natural Resources*, 32(5), 548–565.

Abrams, M., Tsu, H., Hulley, G., Iwao, K., Pieri, D., Cudahy, T., & Kargel, J. (2015). The advanced spaceborne thermal emission and reflection radiometer (ASTER) after fifteen years: Review of global products. *International Journal of Applied Earth Observation and Geoinformation*, 38, 292–301.

Adão, T., Hruška, J., Pádua, L., Bessa, J., Peres, E., Morais, R., & Sousa, J. J. (2017). Hyperspectral imaging: A review on UAV-based sensors, data processing and applications for agriculture and forestry. *Remote Sensing*, 9(11), 1110.

Aderungboye, F. O. (1977). Diseases of the oil palm. *Pans*, 23(3), 305–326.

Adjovu, G. E., Stephen, H., James, D., & Ahmad, S. (2023). Overview of the application of remote sensing in effective monitoring of water quality parameters. *Remote Sensing*, 15(7), 1938.

Aghighi, H., Azadbakht, M., Ashourloo, D., Shahrabi, H. S., & Radiom, S. (2018). Machine learning regression techniques for the silage maize yield prediction using time-series images of Landsat 8 OLI. *IEEE Journal of Selected Topics in Applied Earth Observations and Remote Sensing*, 11(12), 4563–4577.

Agusta, H., Handoyo, G. C., & Sudaryanto, M. T. (2020). Cover crops and frond piles for improving soil water infiltration in oil palm plantations. In *IOP Conference Series: Earth and Environmental Science* (Vol. 460, No. 1, p. 012045). IOP Publishing. https://doi.org/10.1088/1755-1315/460/1/012045

Agyemang, F. S., Memon, R., Wolf, L. J., & Fox, S. (2023). High-resolution rural poverty mapping in Pakistan with ensemble deep learning. *PLoS One*, 18(4), e0283938.

Ahmad Safwan, A. B., & Zareen, Z. (2019). Challenges of smart farming in oil palm plantation in Malaysia: An overview. *Konvensyen Kebangsaan Kejuruteraan Pertanian Dan Makanan*, 280–282.

Ahmadi, P., Mansor, S., Farjad, B., & Ghaderpour, E. (2022). Unmanned Aerial Vehicle (UAV)-based remote sensing for early-stage detection of Ganoderma. *Remote Sensing*, 14(5), 1239.

Ahmed, A., Dompreh, E., & Gasparatos, A. (2019). Human wellbeing outcomes of involvement in industrial crop production: Evidence from sugarcane, oil palm and jatropha sites in Ghana. *PLoS One*, 14(4), e0215433.

Ahmed, A., Ishak, M. Y., Uddin, M. K., Abd Samad, M. Y., Mukhtar, S., & Danhassan, S. S. (2021). Effects of some weather parameters on oil palm production in the Peninsular Malaysia. *Preprints*, 2021, 2021060456. https://doi.org/10.20944/preprints202106.0456.v1

Ahmed, A., Mohd, Y. B. I., & Abdullah, A. M. (2021). Oil palm in the face of climate change: A review of recommendations. *In IOP Conference Series: Earth and Environmental Science* (Vol. 646, No. 1, p. 012065). IOP Publishing. https://doi.org/10.1088/1755-1315/646/1/012065

Aholoukpè, H., Dubos, B., Flori, A., Deleporte, P., Amadji, G., Chotte, J. L., & Blavet, D. (2013). Estimating aboveground biomass of oil palm: Allometric equations for estimating frond biomass. *Forest Ecology and Management*, 292, 122–129.

Akram, H., Levia, D. F., Herrick, J. E., Lydiasari, H., & Schütze, N. (2022). Water requirements for oil palm grown on marginal lands: A simulation approach. *Agricultural Water Management*, 260, 107292.

Alam, A. F., Er, A. C., & Begum, H. (2015). Malaysian oil palm industry: Prospect and problem. *Journal of Food, Agriculture & Environment*, 13(2), 143–148.

Alometri, P. T. M., Korom, A., Phua, M., & Matsuura, T. (2016). Relationships between crown size and aboveground biomass of oil palms: An evaluation of allometric models. *Sains Malaysiana*, 45(4), 523–533.

ALOS, (2023). ALOS global digital surface model "ALOS World 3D -30m (AW3D30)". *ALOS*. https://www.eorc.jaxa.jp/ALOS/en/dataset/aw3d30/aw3d30_e.htm.

Altunel, A. O. (2019). Evaluation of TanDEM-X 90 m digital elevation model. *International Journal of Remote Sensing*, 40(7), 2841–2854.

Amani, M., Ghorbanian, A., Ahmadi, S. A., Kakooei, M., Moghimi, A., Mirmazloumi, S. M., … Brisco, B. (2020). Google earth engine cloud computing platform for remote sensing big data applications: A comprehensive review. *IEEE Journal of Selected Topics in Applied Earth Observations and Remote Sensing*, 13, 5326–5350.

Ampatzidis, Y., De Bellis, L., & Luvisi, A. (2017). iPathology: Robotic applications and management of plants and plant diseases. *Sustainability*, 9(6), 1010.

Anderegg, J., Tschurr, F., Kirchgessner, N., Treier, S., Schmucki, M., Streit, B., & Walter, A. (2023). On-farm evaluation of UAV-based aerial imagery for season-long weed monitoring under contrasting management and pedoclimatic conditions in wheat. *Computers and Electronics in Agriculture*, 204, 107558.

Ang, Y., Shafri, H. Z. M., Lee, Y. P., Bakar, S. A., Abidin, H., Mohd Junaidi, M. U. U., … Samad, M. N. A. (2022). Oil palm yield prediction across blocks from multi-source data using machine learning and deep learning. *Earth Science Informatics*, 15(4), 2349–2367.

Anzum, R., & Naeem, J. (2021). Leveraging LoRaWAN technology for smart agricultural monitoring of Malaysian palm oil plantation. In *IOP Conference Series: Earth and Environmental Science* (Vol. 756, No. 1, p. 012052). IOP Publishing. https://doi.org/10.1088/1755-1315/756/1/012052

Apichatmeta, K., Sudsiri, C. J., & Ritchie, R. J. (2017). Photosynthesis of oil palm (*Elaeis guineensis*). *Scientia Horticulturae*, 214, 34–40.

Apriani, E., Kim, Y. S., Fisher, L. A., & Baral, H. (2020). Non-state certification of smallholders for sustainable palm oil in Sumatra, Indonesia. *Land Use Policy*, 99, 105112.

Arab, S. T., Noguchi, R., Matsushita, S., & Ahamed, T. (2021). Prediction of grape yields from time-series vegetation indices using satellite remote sensing and a machine-learning approach. *Remote Sensing Applications: Society and Environment*, 22, 100485.

Aratrakorn, S., Thunhikorn, S., & Donald, P. F. (2006). Changes in bird communities following conversion of lowland forest to oil palm and rubber plantations in southern Thailand. *Bird Conservation International*, 16(1), 71–82.

Arcagni, A., Fattore, M., Maggino, F., & Vittadini, G. (2021). Some critical reflections on the measurement of social sustainability and well-being in complex societies. *Sustainability*, 13(22), 12679.

Ardila, J. P., Bijker, W., Tolpekin, V. A., & Stein, A. (2012). Multitemporal change detection of urban trees using localized region-based active contours in VHR images. *Remote Sensing of Environment*, 124, 413–426.

Arikapudi, R., & Vougioukas, S. G. (2021). Robotic tree-fruit harvesting with telescoping arms: A study of linear fruit reachability under geometric constraints. *IEEE Access*, 9, 17114–17126.

Asari, N., Suratman, M. N., & Jaafar, J. (2017). Modelling and mapping of above ground biomass (AGB) of oil palm plantations in Malaysia using remotely-sensed data. *International Journal of Remote Sensing*, 38(16), 4741–4764.

Asian Agri. (2023). Asian Agri's Topaz seeds result in higher incomes for smallholders. Retrieved from Asian Agri: https://www.asianagri.com/en/media-publications/articles/topaz-quality-seeds-of-asian-agri/.

Austin, K. G., Mosnier, A., Pirker, J., McCallum, I., Fritz, S., & Kasibhatla, P. S. (2017). Shifting patterns of oil palm driven deforestation in Indonesia and implications for zero-deforestation commitments. *Land Use Policy*, 69, 41–48.

Austin, K. G., Schwantes, A., Gu, Y., & Kasibhatla, P. S. (2019). What causes deforestation in Indonesia?. *Environmental Research Letters*, 14(2), 024007.

Avila-Diazgranados, R. A., Daza, E. S., Navia, E., & Romero, H. M. (2016). Response of various oil palm materials (Elaeis guineensis and Elaeis oleifera× Elaeis guineensis interspecific hybrids) to bud rot disease in the southwestern oil palm-growing area of Colombia. *Agronomía Colombiana*, 34(1), 74–81.

Avitabile, V., Herold, M., Heuvelink, G. B., Lewis, S. L., Phillips, O. L., Asner, G. P., … Willcock, S. (2016). An integrated pan-tropical biomass map using multiple reference datasets. *Global Change Biology*, 22(4), 1406–1420.

Avtar, R., Komolafe, A. A., Kouser, A., Singh, D., Yunus, A. P., Dou, J., … Kurniawan, T. A. (2020). Assessing sustainable development prospects through remote sensing: A review. *Remote Sensing Applications: Society and Environment*, 20, 100402.

Avtar, R., Suab, S. A., Syukur, M. S., Korom, A., Umarhadi, D. A., & Yunus, A. P. (2020). Assessing the influence of UAV altitude on extracted biophysical parameters of young oil palm. *Remote Sensing*, 12(18), 3030.

Awusabo-Asare, K., & Tanle, A. (2008). Eking a living: Women entrepreneurship and poverty reduction strategies: The case of palm kernel oil processing in the Central Region of Ghana. *Norsk Geografisk Tidsskrift-Norwegian Journal of Geography*, 62(3), 149–160.

Ayompe, L. M., Schaafsma, M., & Egoh, B. N. (2021). Towards sustainable palm oil production: The positive and negative impacts on ecosystem services and human wellbeing. *Journal of Cleaner Production*, 278, 123914.

Azam, S. H. M. N., Tahir, W., & Abdullah, J. (2021). Assessment of SRTM, ASTER and IFSAR digital elevation model (DEM) in oil palm plantation river derivation and basin delineation. In *ICCOEE2020: Proceedings of the 6th International Conference on Civil, Offshore and Environmental Engineering (ICCOEE2020)* (pp. 56–65). Singapore: Springer. https://doi.org/10.1007/978-981-33-6311-3_7

Aziz, N., Omar, W. A. H. I. D., Kassim, R., & Kamarudin, N. (2012). Remote sensing measurement for detection of bagworm infestation in oil palm plantation. *MPOB Information Series*, 589, 597.

Baharim, M. S. A., Adnan, N. A., Mohd, F. A., Seman, I. A., Izzuddin, M. A., & Abd Aziz, N. (2022, July). A review: Progression of remote sensing (RS) and geographical information system (GIS) applications in oil palm management and sustainability. *In IOP Conference Series: Earth and Environmental Science* (Vol. 1051, No. 1, p. 012027). IOP Publishing.

Balasundram, S. K., Memarian, H., & Khosla, R. (2013). Estimating oil palm yields using vegetation indices derived from Quickbird. *Life Science Journal*, 10(4), 851–860.

Balasundram, S. K., Robert, P. C., Mulla, D. J., & Allan, D. L. (2006). Relationship between oil palm yield and soil fertility as affected by topography in an Indonesian plantation. *Communications in Soil Science and Plant Analysis*, 37(9–10), 1321–1337.

Baldocchi, D. D. (2003). Assessing the eddy covariance technique for evaluating carbon dioxide exchange rates of ecosystems: Past, present and future. *Global Change Biology*, 9(4), 479–492.

Barbouchi, M., Lhissou, R., Abdelfattah, R., El Alem, A., Chokmani, K., Ben Aissa, N., … Bahri, H. (2022). The potential of using Radarsat-2 satellite image for modeling and mapping wheat yield in a semiarid environment. *Agriculture*, 12(3), 315.

Basiron, Y., Shamsudin, I., & Mielke, T. (2016). Oil palm and the sustainability of peatland in Southeast Asia: an Ulu Masen case study. In M. Osaki & R. Tsuji (Eds.), *Sustainability Assessment Tools in Higher Education Institutions* (pp. 207–226). Springer International Publishing.

Baskauf, C. J., & Eickmeier, W. G. (1994). Comparative ecophysiology of a rare and a widespread species of Echinacea (Asteraceae). *American Journal of Botany*, 81(8), 958–964.

Baskett, J.P. (2016). *Myanmar Oil Palm Plantations. A Productivity & Sustainability Review. Report no. 28 of the Tanintharyi Conservation Programme, a Joint Initiative of Fauna & Flora International and the Myanmar Forest Department*. Fauna & Flora International, Cambridge.

Basri, M. W. (1993). Life history, ecology and economic impact of the bagworm, Metisa plana Walker (Lepidoptera: Psychidae) on the oil palm, Elaeis guineensis Jacquin (Palmae) in Malaysia. PhD thesis, University of Guelph.

Basuki, I., Budiman, A., Netzer, M., Safitri, R., Maulana, R., Nusirhan, T. S. E., & Bernal, B. (2021, February). Dynamic of groundwater table, peat subsidence and carbon emission impacted from deforestation in tropical peatland, Riau, Indonesia. *In IOP Conference Series: Earth and Environmental Science* (Vol. 648, No. 1, p. 012029). IOP Publishing. https://doi.org/10.1088/1755-1315/648/1/012029

Beese, L., Dalponte, M., Asner, G. P., Coomes, D. A., & Jucker, T. (2022). Using repeat airborne LiDAR to map the growth of individual oil palms in Malaysian Borneo during the 2015–16 El Niño. *International Journal of Applied Earth Observation and Geoinformation*, 115, 103117.

Begum, H., Alam, A. F., & Awang, A. H. (2019). Sustainability of Malaysian oil palm: A critical review. *International Journal of Environment and Sustainable Development*, 18(4), 409–429.

Benjamin, M., & Chee, K. H. (1995). Basal stem rot of oil palm-a serious problem on inland soils. *MAPPS Newsletter*, 19(1), 6.

Bernama (2023). RSPO helps smallholders market sustainable Palm Oil to the global market, https://www.nst.com.my/news/nation/2023/08/947105/rspo-helps-smallholders-market-sustainable-palm-oil-global-market, Accessed 3 March 2024.

Bhardwaj, A. (2019). Assessment of vertical accuracy for TanDEM-X 90 m DEMs in plain, moderate, and rugged terrain. *Multidisciplinary Digital Publishing Institute Proceedings*, 24(1), 8.

Bloomberg, (2023a). Floods and old trees in Malaysia set to tighten palm oil market. *Bloomberg*, March 11, 2023. https://www.bloomberg.com/news/articles/2023-03-11/floods-and-old-trees-in-malaysia-set-to-tighten-palm-oil-market#xj4y7vzkg.

Bloomberg, (2023b). Palm oil hits three-month low on weaker soy oil and demand worry. https://www.bloomberg.com/news/articles/2023-09-21/palm-oil-extends-drop-on-losses-in-soyoil-and-demand-concerns#xj4y7vzkg (Accessed 14 Oct 2023)

Boursianis, A. D., Papadopoulou, M. S., Diamantoulakis, P., Liopa-Tsakalidi, A., Barouchas, P., Salahas, G., … Goudos, S. K. (2022). Internet of things (IoT) and agricultural unmanned aerial vehicles (UAVs) in smart farming: A comprehensive review. *Internet of Things*, 18, 100187.

Boyd, D. S., Jackson, B., Wardlaw, J., Foody, G. M., Marsh, S., & Bales, K. (2018). Slavery from space: Demonstrating the role for satellite remote sensing to inform evidence-based action related to UN SDG number 8. *ISPRS Journal of Photogrammetry and Remote Sensing*, 142, 380–388.

Brandi, C., Cabani, T., Hosang, C., Schirmbeck, S., Westermann, L., & Wiese, H. (2015). Sustainability standards for palm oil: Challenges for smallholder certification under the RSPO. *The Journal of Environment & Development*, 24(3), 292–314.

Breure, C. J. (2010). Rate of leaf expansion: A criterion for identifying oil palm (Elaeis guineensis Jacq.) types suitable for planting at high densities. *NJAS: Wageningen Journal of Life Sciences*, 57(2), 141–147.

Brooks, T. M., Mittermeier, R. A., Da Fonseca, G. A., Gerlach, J., Hoffmann, M., Lamoreux, J. F., … Rodrigues, A. S. (2006). Global biodiversity conservation priorities. *Science*, 313(5783), 58–61.

Cai, Z., & Vasconcelos, N. (2018). Cascade r-cnn: Delving into high quality object detection. In *Proceedings of the IEEE Conference on Computer Vision and Pattern Recognition* (pp. 6154–6162).

Carlson, K. M., & Garrett, R. D. (2018). Environmental impacts of tropical soybean and palm oil crops. In *Oxford Research Encyclopedia of Environmental Science*. https://doi.org/10.1093/acrefore/9780199389414.013.234

Carlson, K. M., Curran, L. M., Asner, G. P., Pittman, A. M., Trigg, S. N., & Marion Adeney, J. (2013). Carbon emissions from forest conversion by Kalimantan oil palm plantations. *Nature Climate Change*, 3(3), 283–287.

Carlson, K. M., Curran, L. M., Ponette-González, A. G., Ratnasari, D., Ruspita, Lisnawati, N., … Raymond, P. A. (2014). Influence of watershed-climate interactions on stream temperature, sediment yield, and metabolism along a land use intensity gradient in Indonesian Borneo. *Journal of Geophysical Research: Biogeosciences*, 119(6), 1110–1128.

Carlson, K. M., Curran, L. M., Ratnasari, D., Pittman, A. M., Soares-Filho, B. S., Asner, G. P., … Rodrigues, H. O. (2012). Committed carbon emissions, deforestation, and community land conversion from oil palm plantation expansion in West Kalimantan, Indonesia. *Proceedings of the National Academy of Sciences*, 109(19), 7559–7564.

Carlson, K. M., Heilmayr, R., Gibbs, H. K., Noojipady, P., Burns, D. N., Morton, D. C., … Kremen, C. (2018). Effect of oil palm sustainability certification on deforestation and fire in Indonesia. *Proceedings of the National Academy of Sciences*, 115(1), 121–126.

Carolita, I., Sitorus, J., & Rizqi, I. S. (2015). Classification of vegetation and oil palm age using SPOT6: The case of tana Laut South Kalimantan, Indonesia. In *Proceedings of the 36th Asian Association on Remote Sensing* (Vol. 47, pp. 399–405). IOP Publishing Ltd.

Carr, M. K. V. (2011). The water relations and irrigation requirements of oil palm (Elaeis guineensis): A review. *Experimental Agriculture*, 47(4), 629–652.

Carter, C., Finley, W., Fry, J., Jackson, D., & Willis, L. (2007). Palm oil markets and future supply. *European Journal of Lipid Science and Technology*, 109(4), 307–314.

Castellanos-Navarrete, A., de Castro, F., & Pacheco, P. (2021). The impact of oil palm on rural livelihoods and tropical forest landscapes in Latin America. *Journal of Rural Studies*, 81, 294–304.

Castiblanco, C., Etter, A., & Ramirez, A. (2015). Impacts of oil palm expansion in Colombia: What do socioeconomic indicators show?. *Land Use Policy*, 44, 31–43.

Caulfield, J., Polèse, M., Stren, R., and Polese, M. (2001). The social sustainability of cities: Diversity and the management of change. *Canadian Public Policy*, 27, 23–58. doi: 10.2307/3552480.

Chang, S. H. (2014). An overview of empty fruit bunch from oil palm as feedstock for bio-oil production. *Biomass and Bioenergy*, 62, 174–181.

Chapman, R., Cook, S., Donough, C., Lim, Y. L., Ho, P. V. V., Lo, K. W., & Oberthür, T. (2018). Using Bayesian networks to predict future yield functions with data from commercial oil palm plantations: A proof of concept analysis. *Computers and Electronics in Agriculture*, 151, 338–348.

Chellaiah, D., & Yule, C. M. (2018). Riparian buffers mitigate impacts of oil palm plantations on aquatic macroinvertebrate community structure in tropical streams of Borneo. *Ecological Indicators*, 95, 53–62.

Chemura, A., van Duren, I., & van Leeuwen, L. M. (2015). Determination of the age of oil palm from crown projection area detected from WorldView-2 multispectral remote sensing data: The case of Ejisu-Juaben district, Ghana. *ISPRS Journal of Photogrammetry and Remote Sensing*, 100, 118–127.

Chen, J., & Ran, X. (2019). Deep learning with edge computing: A review. *Proceedings of the IEEE*, 107(8), 1655–1674.

Chen, J., Chen, S., Fu, R., Li, D., Jiang, H., Wang, C., … Hicks, B. J. (2022). Remote sensing big data for water environment monitoring: Current status, challenges, and future prospects. *Earth's Future*, 10(2), e2021EF002289.

Chen, K., Wang, J., Pang, J., Cao, Y., Xiong, Y., Li, X., … Lin, D. (2019). MMDetection: Open mmlab detection toolbox and benchmark. *arXiv Preprint arXiv:1906.07155*. https://doi.org/10.48550/arXiv.1906.07155

Cheng, G., & Han, J. (2016). A survey on object detection in optical remote sensing images. *ISPRS Journal of Photogrammetry and Remote Sensing*, 117, 11–28.

Cheng, L., Weiguang, G., Bo, S., Jun, L., Wei, W., Ying, C., … Qun, D. (2021). Development of BeiDou satellite-based augmentation system. *Navigation*, 68(2), 405–417.

Cheng, Y., Yu, L., Cracknell, A. P., & Gong, P. (2016). Oil palm mapping using Landsat and PALSAR: A case study in Malaysia. *International Journal of Remote Sensing*, 37(22), 5431–5442.

Cheng, Y., Yu, L., Xu, Y., Liu, X., Lu, H., Cracknell, A. P., … Gong, P. (2018). Towards global oil palm plantation mapping using remote-sensing data. *International Journal of Remote Sensing*, 39(18), 5891–5906.

Cheng, Y., Yu, L., Xu, Y., Lu, H., Cracknell, A. P., Kanniah, K., & Gong, P. (2019). Mapping oil palm plantation expansion in Malaysia over the past decade (2007–2016) using ALOS-1/2 PALSAR-1/2 data. *International Journal of Remote Sensing*, 40(19), 7389–7408.

Cheng, Y., Yu, L., Zhao, Y., Xu, Y., Hackman, K., Cracknell, A. P., & Gong, P. (2017). Towards a global oil palm sample database: Design and implications. *International Journal of Remote Sensing*, 38(14), 4022–4032.

Chiew, Y. L., & Shimada, S. (2013). Current state and environmental impact assessment for utilizing oil palm empty fruit bunches for fuel, fiber and fertilizer-A case study of Malaysia. *Biomass and Bioenergy*, 51, 109–124.

Chiriacò, M. V., Bellotta, M., Jusić, J., & Perugini, L. (2022). Palm oil's contribution to the United Nations sustainable development goals: Outcomes of a review of socio-economic aspects. *Environmental Research Letters*, 17(6), 063007.

Chiu, R. L. H. (2003). Social sustainability, sustainable development and housing development: The experience of Hong Kong, in Housing and Social Change: East-West perspectives. In R. Forrest and J. Lee (Eds.) (New York, NY: Routledge), 1–19. doi: 10.4324/9780203402634.

Chong, K. L., Kanniah, K. D., Pohl, C., & Tan, K. P. (2017). A review of remote sensing applications for oil palm studies. *Geo-spatial Information Science*, 20(2), 184–200.

Chrisendo, D., Siregar, H., & Qaim, M. (2022). Oil palm cultivation improves living standards and human capital formation in smallholder farm households. *World Development*, 159, 106034.

Christ, K. L., & Helliar, C. V. (2021). Blockchain technology and modern slavery: Reducing deceptive recruitment in migrant worker populations. *Journal of Business Research*, 131, 112–120.

Chung, G. F. (2012). Effect of pests and diseases on oil palm yield. In *Palm oil* (pp. 163–210). AOCS Press. https://doi.org/10.1016/B978-0-9818936-9-3.50009-5

Clark, D. B., Clark, D. A., Brown, S., Oberbauer, S. F., & Veldkamp, E. (2002). Stocks and flows of coarse woody debris across a tropical rain forest nutrient and topography gradient. *Forest Ecology and Management*, 164(1–3), 237–248.

Clough, Y., Krishna, V. V., Corre, M. D., Darras, K., Denmead, L. H., Meijide, A., ... Scheu, S. (2016). Land-use choices follow profitability at the expense of ecological functions in Indonesian smallholder landscapes. *Nature Communications*, 7(1), 13137.

Coburn, C. A., & Roberts, A. C. (2004). A multiscale texture analysis procedure for improved forest stand classification. *International Journal of Remote Sensing*, 25(20), 4287–4308.

Colchester, M. (2011). Palm oil and indigenous peoples of south east Asia: Land acquisition. In *Human Rights Violations and Indigenous Peoples on the Palm Oil Frontier*. Moreton-in-Marsh, Rome: For. Peoples Prog./Int. Land Coalit.

Combres, J. C., Pallas, B., Rouan, L., Mialet-Serra, I., Caliman, J. P., Braconnier, S., ... Dingkuhn, M. (2012). Simulation of inflorescence dynamics in oil palm and estimation of environment-sensitive phenological phases: A model based analysis. *Functional Plant Biology*, 40(3), 263–279.

Comte, I., Colin, F., Grünberger, O., Whalen, J. K., Harto Widodo, R., & Caliman, J. P. (2015). Watershed-scale assessment of oil palm cultivation impact on water quality and nutrient fluxes: A case study in Sumatra (Indonesia). *Environmental Science and Pollution Research*, 22, 7676–7695.

Cooper, H. V., Evers, S., Aplin, P., Crout, N., Dahalan, M. P. B., & Sjogersten, S. (2020). Greenhouse gas emissions resulting from conversion of peat swamp forest to oil palm plantation. *Nature Communications*, 11(1), 407.

Córdoba, D., Juen, L., Selfa, T., Peredo, A. M., de Assis Montag, L. F., Sombra, D., & Santos, M. P. D. (2019). Understanding local perceptions of the impacts of large-scale oil palm plantations on ecosystem services in the Brazilian Amazon. *Forest Policy and Economics*, 109, 102007.

Corley, R. H. V., & Gray, B. S. (1976). Yield and yield components. In: Corley, R. H. V., Hardon, J. J., & Wood, B. J. (Eds.), *Oil Palm Research* (pp. 77–86). Amsterdam: Elsevier.

Corley, R. H. V., & Tinker, P. B. (2003). *The Oil Palm*. Oxford: Blackwell Science. doi:10.1007/9780470750971.

Corley, R. H. V., & Tinker, P. B. (2008). *The Oil Palm*. (4th ed.). Wiley-Blackwell, Chichester.

Corley, R. H. V., & Tinker, P. B. (2016). *The Oil Palm* (5th ed.). Hoboken, NJ: John Wiley & Sons. doi: 10.1002/9781118953297

Corley, R. H. V., Gray, B. S., & Kee, N. S. (1971). Productivity of the oil palm (Elaeis guineensis Jacq.) in Malaysia. *Experimental Agriculture*, 7(2), 129–136.

Courtenay, P. P. (1984). The diversification of Malaysian agriculture, 1950–80: Objectives and achievements. *Journal of Southeast Asian Studies*, 15(1), 166–181.

Cracknell, A. P., Kanniah, K. D., Tan, K. P., & Wang, L. (2015). Towards the development of a regional version of MOD17 for the determination of gross and net primary productivity of oil palm trees. *International Journal of Remote Sensing*, 36(1), 262–289.

Crippa, M., Janssens-Maenhout, G., Dentener, F., Guizzardi, D., Sindelarova, K., Muntean, M., ... Granier, C. (2016). Forty years of improvements in European air quality: Regional policy-industry interactions with global impacts. *Atmospheric Chemistry and Physics*, 16(6), 3825–3841.

Crippa, P., Castruccio, S., Archer-Nicholls, S., Lebron, G. B., Kuwata, M., Thota, A., ... Spracklen, D. V. (2016). Population exposure to hazardous air quality due to the 2015 fires in Equatorial Asia. *Scientific Reports*, 6(1), 37074.

Csillik, O., Cherbini, J., Johnson, R., Lyons, A., & Kelly, M. (2018). Identification of citrus trees from unmanned aerial vehicle imagery using convolutional neural networks. *Drones*, 2(4), 39.

Cui, Y., Jia, M., Lin, T. Y., Song, Y., & Belongie, S. (2019). Class-balanced loss based on effective number of samples. In *Proceedings of the IEEE/CVF Conference on Computer Vision and Pattern Recognition* (pp. 9268–9277). https://doi.org/10.48550/arXiv.1901.05555

Culman, M. A., Gomez, J. A., Talavera, J., Quiroz, L. A., Tobon, L. E., Aranda, J. M., ... Bayona, C. J. (2017, April). A novel application for identification of nutrient deficiencies in oil palm using the internet of things. In *2017 5th IEEE International Conference on Mobile Cloud Computing, Services, and Engineering (MobileCloud)*, San Francisco, CA (pp. 169–172). IEEE. doi: 10.1109/MobileCloud.2017.32

Dadap, N. C., Hoyt, A. M., Cobb, A. R., Oner, D., Kozinski, M., Fua, P. V., ... Konings, A. G. (2021). Drainage canals in Southeast Asian peatlands increase carbon emissions. *Agu Advances*, 2(1), e2020AV000321.

Dalponte, M., Ørka, H. O., Ene, L. T., Gobakken, T., & Næsset, E. (2014). Tree crown delineation and tree species classification in boreal forests using hyperspectral and ALS data. *Remote Sensing of Environment*, 140, 306–317.

Danielsen, F., Beukema, H., Burgess, N. D., Parish, F., Brühl, C. A., Donald, P. F., ... Fitzherbert, E. B. (2009). Biofuel plantations on forested lands: Double jeopardy for biodiversity and climate. *Conservation Biology*, 23(2), 348–358.

Dankers, C., (2003). *Environmental and Social Standards, Certification and Labellingfor Cash Crops*. Rome: Food and Agricultural Organisation (FAO).

Danylo, O., Pirker, J., Lemoine, G., Ceccherini, G., See, L., McCallum, I., ... Fritz, S. (2021). A map of the extent and year of detection of oil palm plantations in Indonesia, Malaysia and Thailand. *Scientific Data*, 8(1), 96.

Darmawan, S., W. Takeuchi, A. Haryati, R. N. Am, and M. Na'aim. (2016). An investigation of age and yield of fresh fruit bunches of oil palm based on ALOS PALSAR 2. *In IOP Conference Series: Earth and Environmental Science*, 8th IGRSM International Conference and Exhibition on Remote Sensing & GIS (IGRSM*2016). Kuala Lumpur, Malaysia* (Vol. 37, pp. 1–8). doi:10.1088/1755-1315/37/1/012037.

De Petris, S., Boccardo, P., & Borgogno-Mondino, E. (2019). Detection and characterization of oil palm plantations through MODIS EVI time series. *International Journal of Remote Sensing*, 40(19), 7297–7311.

De Souza, S. P., Pacca, S., De Ávila, M. T., & Borges, J. L. B. (2010). Greenhouse gas emissions and energy balance of palm oil biofuel. *Renewable Energy*, 35(11), 2552–2561.

Delgado, I. B., Colorado, L. A. M., Argüello Prada, E. J., & Martínez-Santos, J. C. (2017). Irrigation system for oil palm in Colombian internet of things approach. In Applied *Computer Sciences in Engineering: 4th Workshop on Engineering Applications, WEA 2017, Cartagena, Colombia, September 27–29, 2017, Proceedings 4* (pp. 300–311). Springer International Publishing.

Dempsey, N., Bramley, G., Power, S., & Brown, C. (2011). The social dimension of sustainable development Defining urban social sustainability. *Sustainable Development*, 19(5), 289–300.

Descals, A., Szantoi, Z., Meijaard, E., Sutikno, H., Rindanata, G., & Wich, S. (2019). Oil palm (Elaeis guineensis) mapping with details: Smallholder versus industrial plantations and their extent in Riau, Sumatra. *Remote Sensing*, 11(21), 2590.

Descals, A., Wich, S., Meijaard, E., Gaveau, D. L., Peedell, S., & Szantoi, Z. (2021). High-resolution global map of smallholder and industrial closed-canopy oil palm plantations. *Earth System Science Data*, 13(3), 1211–1231.

DeVries, B., Decuyper, M., Verbesselt, J., Zeileis, A., Herold, M., & Joseph, S. (2015). Tracking disturbance-regrowth dynamics in tropical forests using structural change detection and Landsat time series. *Remote Sensing of Environment*, 169, 320–334.

Dewi, S., Ekadinata, A., & Budidarsono, S. (2017). The role of smallholders in sustainable palm oil production. In M. Meijaard, D. Sheil, & R. Nasi (Eds.), *The Oil Palm Complex: Smallholders, Agribusiness and the State in Indonesia and Malaysia* (pp. 119–136). NUS Press. https://doi.org/10.2307/j.ctv1xz0kmE

Dewi, S., Khasanah, N. M., Rahayu, S., Ekadinata, A., & van Noordwijk, M. (2009). *Carbon Footprint of Indonesian Palm Oil Production: A Pilot Study*. Bogor: World Agroforestry Centre-ICRAF, SEA Regional Office.

Dharmawan, A. H., Mardiyaningsih, D. I., Rahmadian, F., Yulian, B. E., Komarudin, H., Pacheco, P., … Amalia, R. (2021). The agrarian, structural and cultural constraints of smallholders' readiness for sustainability standards implementation: The case of Indonesian sustainable palm oil in east Kalimantan. *Sustainability*, 13(5), 2611.

Diek, S., Fornallaz, F., Schaepman, M. E., & De Jong, R. (2017). Barest pixel composite for agricultural areas using landsat time series. *Remote Sensing*, 9(12), 1245.

Dislich, C., Keyel, A. C., Salecker, J., Kisel, Y., Meyer, K. M., Auliya, M., … Wiegand, K. (2017). A review of the ecosystem functions in oil palm plantations, using forests as a reference system. *Biological Reviews*, 92(3), 1539–1569.

Donfack, L. S., Röll, A., Ellsäßer, F., Ehbrecht, M., Irawan, B., Hölscher, D., … Zemp, D. C. (2021). Microclimate and land surface temperature in a biodiversity enriched oil palm plantation. *Forest Ecology and Management*, 497, 119480.

Dong, R., Li, W., Fu, H., Gan, L., Yu, L., Zheng, J., & Xia, M. (2020). Oil palm plantation mapping from high-resolution remote sensing images using deep learning. *International Journal of Remote Sensing*, 41(5), 2022–2046.

Dong, T., Shen, Y., Zhang, J., Ye, Y., & Fan, J. (2019). Progressive cascaded convolutional neural networks for single tree detection with google earth imagery. *Remote Sensing*, 11(15), 1786.

Dong, X., Quegan, S., Yumiko, U., Hu, C., & Zeng, T. (2015). Feasibility study of C-and L-band SAR time series data in tracking Indonesian plantation and natural forest cover changes. *IEEE Journal of Selected Topics in Applied Earth Observations and Remote Sensing*, 8(7), 3692–3699.

Donofrio, S., Rothrock, P., & Leonard, J. J. F. T. (2017). *Supply Change: Tracking Corporate Commitments to Deforestation-Free Supply Chains*. Forest Trends.

Donough, C. R., Witt, C., & Fairhurst, T. H. (2010, June). Yield intensification in oil palm using BMP as a management tool. In International Conference on Oil Palm and the Environment (pp. 23–27).

Drewer, J. et al. (2021). *Biogeosciences*, 18, 1559–1575. doi: 10.5194/bg-18-1559-2021.

Du, Z., Yu, L., Yang, J., Xu, Y., Chen, B., Peng, S., … Gong, P. (2022). A global map of planting years of plantations. *Scientific Data*, 9(1), 141.

Dufrene, E., & Saugier, B. (1993). Gas exchange of oil palm in relation to light, vapour pressure deficit, temperature and leaf age. *Functional Ecology*, 97–104.

Dufrene, E., Ochs, R., & Saugier, B. (1990). Oil palm photosynthesis and productivity linked to climatic factors. *Oleagineux*, 45(8–9), 345–355.

Duque, J. C., Patino, J. E., & Betancourt, A. (2017). Exploring the potential of machine learning for automatic slum identification from VHR imagery. *Remote Sensing*, 9(9), 895.

Earthworm (2020). https://www.earthworm.org/news-stories/ferrero-commits-to-100-percent-satellite-monitoring-of-its-palm-oil-supply-chain (accessed 3 Sept 2023).

Earthworm (2020). Resources to help oil palm plantations improve workers' and children's welfare in Indonesia. https://www.earthworm.org/news-stories/guidelines-indonesia-child-labour-casual-workers (accessed 15 Oct 2023).

Edwards, R. B. (2019). *Spillovers from Agricultural Processing Working Paper*. Hanover, NH: Dartmouth Coll. Retrieved from: https://farmlabor.ucdavis.edu/sites/g/files/dgvnsk5936/files/inline-files/Ryan%20Edwards%3B%20Ag%20Spillovers.pdf (accessed 6 Sep 2023).

Elijah, O., Rahman, T. A., Orikumhi, I., Leow, C. Y., & Hindia, M. N. (2018). An overview of Internet of Things (IoT) and data analytics in agriculture: Benefits and challenges. *IEEE Internet of Things Journal*, 5(5), 3758–3773.

Elkhrachy, I. (2018). Vertical accuracy assessment for SRTM and ASTER digital elevation models: A case study of Najran city, Saudi Arabia. *Ain Shams Engineering Journal*, 9(4), 1807–1817.

ElRaey, M., AbdelKader, A. F., Nasr, S. M., & ElGamily, H. I. (1996). Remote sensing and GIS for an oil spill contingency plan, Ras-Mohammed, Egypt. *International Journal of Remote Sensing*, 17, 2013–2026.

Enclona, E. A., Thenkabail, P. S., Celis, D., & Diekmann, J. (2004). Within-field wheat yield prediction from IKONOS data: A new matrix approach. *International Journal of Remote Sensing*, 25(2), 377–388. doi: 10.1080/0143116031000102485

ESA. (2017). *Land Cover CCI Product User Guide Version 2.* Tech. Rep. http://maps.elie.ucl.ac.be/CCI/viewer/download/ESACCI-LC-Ph2-PUGv2_2.0.pdf.

Esposito, M., Crimaldi, M., Cirillo, V., Sarghini, F., & Maggio, A. (2021). Drone and sensor technology for sustainable weed management: A review. *Chemical and Biological Technologies in Agriculture*, 8(1), 1–11.

ESRI. (2023). Understanding reclassification. *ESRI*, Nov 01. https://pro.arcgis.com/en/pro-app/latest/tool-reference/spatial-analyst/understanding-reclassification.htm#.

Estoque, R. C. (2020). A review of the sustainability concept and the state of SDG monitoring using remote sensing. *Remote Sensing*, 12(11), 1770.

ETA and TBI. (2021). *Intercropping in Oil Palm Plantations: A Technical Guide* (52 pp). Kampala: Ecological Trends Alliance and Ede: Tropenbos International.

Euler, M., Schwarze, S., Siregar, H., & Qaim, M. (2016). Oil palm expansion among smallholder farmers in Sumatra, Indonesia. *Journal of Agricultural Economics*, 67(3), 658–676.

Evaluation of MODIS NPP and GPP products across multiple biomes. Available from: https://www.researchgate.net/publication/222819866_Evaluation_of_MODIS_NPP_and_GPP_products_across_multiple_biomes (accessed 16 Oct 2023).

Fair Labor Association. (2018). *Assessing Forced Labor Risks in the Palm Oil Sector in Indonesia and Malaysia. A research Report for the Consumer Goods Forum*, November 2018. https://www.theconsumergoodsforum.com/wp-content/uploads/2018/11/201811-CGF-FLA-Palm-Oil-Report-Malaysiaand-Indonesia_web.pdf.

Fan, Y., Meijide, A., Lawrence, D. M., Roupsard, O., Carlson, K. M., Chen, H. Y., … Knohl, A. (2019). Reconciling canopy interception parameterization and rainfall forcing frequency in the Community Land Model for simulating evapotranspiration of rainforests and oil palm plantations in Indonesia. *Journal of Advances in Modeling Earth Systems*, 11(3), 732–751.

Fan, Y., Roupsard, O., Bernoux, M., Le Maire, G., Panferov, O., Kotowska, M. M., & Knohl, A. (2015). A sub-canopy structure for simulating oil palm in the Community Land Model (CLM-Palm): phenology, allocation and yield. *Geoscientific Model Development*, 8, 3785–3800.

FAO (2019). *Markets and Trade.* https://www.fao.org/markets-and-trade/commodity-policy-archive/detail/en/c/759823/.

FAO (2022). *Agricultural Production Statistics 2000–2021.* www.fao.org (accessed 15 Oct 2023).

FAO. (2022). FAOSTAT: Crops. Retrieved from https://www.fao.org/faostat/en/#data/QC.

FAOSTAT (2019). *Food and Agriculture Data.* Food and Agriculture Organization of the United Nations. Retrieved from: https://www.fao.org/faostat/en/#home.

Fargione, J., Hill, J., Tilman, D., Polasky, S., & Hawthorne, P. (2008). Land clearing and the biofuel carbon debt. *Science*, 319(5867), 1235–1238.

Farrah, M. M., Siti, A. R., Siti, L. Z., Norida, M., Nur, A. A., & Nor, A. H. (2017). Remote sensing derivation of Land Surface Temperature for insect pest monitoring. *Asian Journal of Plant Sciences*, 16(4), 160–171.

Fawcetta, D., Benjaminb, A., Hillc, T. C., Khod, L. K., Benniea, J., & Andersona, K. Unmanned aerial vehicle derived structure-from-motion photogrammetry point clouds for oil palm (Elaeis guineensis) canopy segmentation and height estimation. *International Journal of Remote Sensing*, 40(19), 7538–7560. doi: 10.1080/01431161.2019.1591651

Fernandez-Hernandez, I., Chamorro-Moreno, A., Cancela-Diaz, S., Calle-Calle, J. D., Zoccarato, P., Blonski, D., … Mozo, A. (2022). Galileo high accuracy service: Initial definition and performance. *Gps Solutions*, 26(3), 65.

Ferro, C. J. S., and T. A. Warner. 2002. Scale and Texture in Digital Image Classification. *Photogrammetric Engineering & Remote Sensing*, 68, 51–63. https://www.asprs.org/wp-content/uploads/pers/2002journal/january/2002_jan_51-63.pdf.

Feyisa, G. L., Meilby, H., Fensholt, R., & Proud, S. R. (2014). Automated water extraction index: A new technique for surface water mapping using Landsat imagery. *Remote Sensing of Environment*, 140, 23–35.

Fitrianto, A. C., Darmawan, A., Tokimatsu, K., & Sufwandika, M. (2018, April). Estimating the age of oil palm trees using remote sensing technique. In *IOP Conference Series: Earth and Environmental Science* (Vol. 148, No. 1, p. 012020). IOP Publishing. https://doi.org/10.1088/1755-1315/148/1/012020

Fitzherbert, E. B., Struebig, M. J., Morel, A., Danielsen, F., Brühl, C. A., Donald, P. F., & Phalan, B. (2008). How will oil palm expansion affect biodiversity?. *Trends in Ecology & Evolution*. 23(10), 538–545.

Fleiss, S., Hill, J. K., McClean, C., Lucey, J. M., & Reynolds, G. (2017). *Potential Impacts of Climate Change on Oil Palm Cultivation. A Science-for-Policy Paper by the SEnSOR Programme* (pp. 1–17). http://www.sensorproject.net/wp-content/uploads/2018/01/Climate-change-reportFINAL.pdf

Florinsky, I. V., Skrypitsyna, T. N., & Luschikova, O. S. (2018). Comparative accuracy of the AW3D30 DSM, ASTER GDEM, and SRTM1 DEM: A case study on the Zaoksky testing ground, Central European Russia. *Remote Sensing Letters*, 9(7), 706–714.

Forest People Programme. (2017). A comparison of leading palm oil certification standards. https://www.forestpeoples.org/en/responsible-finance-palm-oil-rspo/report/2017/comparison-leading-palm-oil-certification-standards (accessed 15 Oct 2023).

Freudenberg, M., Nölke, N., Agostini, A., Urban, K., Wörgötter, F., & Kleinn, C. (2019). Large scale palm tree detection in high resolution satellite images using U-Net. *Remote Sensing*, 11(3), 312.

Fuhrer, J. (2003). Agroecosystem responses to combinations of elevated CO2, ozone, and global climate change. *Agriculture, Ecosystems & Environment*, 97(1–3), 1–20.

Furumo, P. R., & Aide, T. M. (2017). Characterizing commercial oil palm expansion in Latin America: Land use change and trade. *Environmental Research Letters*, 12(2), 024008.

Furumo, P. R., Rueda, X., Rodríguez, J. S. & Ramos, I. K. P. (2020). Field evidence for positive certification outcomes on oil palm smallholder management practices in Colombia. *Journal of Cleaner Production*, 245, 118891.

Galileo GNSS (2014). Alternative complementary GNSS systems shown to be essential. *Galileo GNSS*, Apr. 16. https://galileognss.eu/alternative-complementary-gnss-systems-shown-to-be-essential/.

Gandaseca, S., Rosli, N., Ngayop, J., & Arianto, C. I. (2011). Status of water quality based on the physico-chemical assessment on river water at Wildlife Sanctuary Sibuti Mangrove Forest, Miri Sarawak. *American Journal of Environmental Sciences*, 7(3), 269.

García-León, D., López-Lozano, R., Toreti, A., & Zampieri, M. (2020). Local-scale cereal yield forecasting in Italy: Lessons from different statistical models and spatial aggregations. *Agronomy*, 10(6), 809.

Garcia-Torres, S., Albareda, L., Rey-Garcia, M., & Seuring, S. (2019). Traceability for sustainability-literature review and conceptual framework. *Supply Chain Management: An International Journal*, 24(1), 85–106.

Gatti, R. C., & Velichevskaya, A. (2020). Certified sustainable palm oil took the place of endangered Bornean and Sumatran large mammals habitat and tropical forests in the last 30 years. *Science of the Total Environment*, 742, 140712.

Gatti, R. C., Liang, J., Velichevskaya, A., & Zhou, M. (2019). Sustainable palm oil may not be so sustainable. *Science of the Total Environment*, 652, 48–51.

Gatto, M., Wollni, M., Asnawi, R., & Qaim, M. (2017). Oil palm boom, contract farming, and rural economic development: Village-level evidence from Indonesia. *World Development*, 95, 127–140.

Gaveau, D. L. A., & Sheil, D. (2016). Impacts of agricultural expansion on tropical forests in southeast Asia. *Current Biology*, 26(5), R195–R205.

Gaveau, D. L., Locatelli, B., Salim, M. A., Husnayaen, Manurung, T., Descals, A., … Sheil, D. (2022). Slowing deforestation in Indonesia follows declining oil palm expansion and lower oil prices. *PLoS One*, 17(3), e0266178.

Gaveau, D. L., Sheil, D., Salim, M. A., Arjasakusuma, S., Ancrenaz, M., Pacheco, P., & Meijaard, E. (2016). Rapid conversions and avoided deforestation: Examining four decades of industrial plantation expansion in Borneo. *Scientific Reports*, 6(1), 1–13.

Germer, J., & Sauerborn, J. (2008). Estimation of the impact of oil palm plantation establishment on greenhouse gas balance. *Environment, Development and Sustainability*, 10, 697–716.

Gerritsma, W. (1988). *Light Interception, Leaf Photosynthesis and Sink-Source Relations in Oil Palm*. Interim Report, Theoretical Production Ecology Department, Wageningen.

Giannakis, I., Tosti, F., Lantini, L., & Alani, A. M. (2019). Health monitoring of tree trunks using ground penetrating radar. *IEEE Transactions on Geoscience and Remote Sensing*, 57(10), 8317–8326.

Goh, K. J. (2000). Climatic requirements of the oil palm for high yields. In Managing Oil Palm for High Yields: Agronomic Principles (pp. 1–17).

Gomes, M. F., Maillard, P., & Deng, H. (2018). Individual tree crown detection in sub-meter satellite imagery using Marked Point Processes and a geometrical-optical model. *Remote Sensing of Environment*, 211, 184–195.

Gomes, V. C., Queiroz, G. R., & Ferreira, K. R. (2020). An overview of platforms for big earth observation data management and analysis. *Remote Sensing*, 12(8), 1253.

González, N. C., & Kröger, M. (2023). The adoption of earth-observation technologies for deforestation monitoring by Indigenous people: Evidence from the Amazon. *Globalizations*, 20(3), 415–431.

GPS: The Global Positioning System (2022). GPS accuracy. In *GPS: The Global Positioning System*, Mar. 3. https://www.gps.gov/systems/gps/performance/accuracy/.

Griffiths, T. & Jiwan, N. (2021). Demanding accountability: Strengthening corporate accountability and supply chain due diligence to protect human rights and safeguard the environment: Lessons from ten case studies of the Indonesian palm oil sector. *Forest Peoples Program*. https://www.forestpeoples.org/en/palm-oil-rspo/report/2021/demanding-accountability-lessons-ten-case-studies-indonesian-palm-oil.

Guerschman, J. P., Warren, G., Byrne, G., Lymburner, L., Mueller, N., & Van Dijk, A. I. J. M. (2011). *MODIS-Based Standing Water Detection for Flood and Large Reservoir Mapping: Algorithm Development and Applications for the Australian Continent*. Canberra: Water for a Healthy Country National Research Flagship Report.

Guillaume, T., Kotowska, M. M., Hertel, D., Knohl, A., Krashevska, V., Murtilaksono, K., … Kuzyakov, Y. (2018). Carbon costs and benefits of Indonesian rainforest conversion to plantations. *Nature Communications*, 9(1), 2388.

Guirado, E., Tabik, S., Alcaraz-Segura, D., Cabello, J., & Herrera, F. (2017). Deep-learning versus OBIA for scattered shrub detection with Google earth imagery: Ziziphus Lotus as case study. *Remote Sensing*, 9(12), 1220.

Gunarso, P., Hartoyo, M. E., Agus, F., & Killeen, T. J. (2013). Oil palm and land use change in Indonesia, Malaysia, and Papua New Guinea. In Killeen, T. J. & Goon, J. (Eds). *Reports from the Technical Panels of the Second Greenhouse Gas Working Group of the Roundtable on Sustainable Palm Oil*. Kuala Lumpur: Roundtable on Sustainable Palm Oil, pp. 29–63.

Gunawan, S., Budiastuti, M. T. S., Sutrisno, J., & Wirianata, H. (2021, March). The performance of oil palm productivity and management of organic materials at various rain intensity in sandy soil. *In IOP Conference Series: Earth and Environmental Science* (Vol. 709, No. 1, p. 012088). IOP Publishing.

Gutiérrez-Vélez, V. H., DeFries, R., Pinedo-Vásquez, M., Uriarte, M., Padoch, C., Baethgen, W., … Lim, Y. (2011). High-yield oil palm expansion spares land at the expense of forests in the Peruvian Amazon. *Environmental Research Letters*, 6(4), 044029.

Habib-ur-Rahman, M., Ahmad, A., Raza, A., Hasnain, M. U., Alharby, H. F., Alzahrani, Y. M., … El Sabagh, A. (2022). Impact of climate change on agricultural production; Issues, challenges, and opportunities in Asia. *Frontiers in Plant Science*, 13, 925548.

Hafizuddin-Syah, B. A. M., Shahida, S., & Fuad, S. H. (2018). Sustainability certifications and financial profitability: An analysis on palm oil companies in Malaysia. *Jurnal Pengurusan*, 54(2018), 143–154.

Hamdan, O., Aziz, H. K., & Hasmadi, I. M. (2014). L-band ALOS PALSAR for biomass estimation of Matang Mangroves, Malaysia. *Remote Sensing of Environment*, 155, 69–78.

Hamilton, R. L., Trimmer, M., Bradley, C., & Pinay, G. (2016). Deforestation for oil palm alters the fundamental balance of the soil N cycle. *Soil Biology and Biochemistry*, 95, 223–232.

Hamsa, C. S., Kanniah, K. D., Muharam, F. M., Idris, N. H., Abdullah, Z., & Mohamed, L. (2019). Textural measures for estimating oil palm age. *International Journal of Remote Sensing*, 40(19), 7516–7537.

Haniff, M. H., Henson, I. E., & Mohd Roslan, M. N. (2004, September). Continuous measurement of canopy CO2 and H2O fluxes of an oil palm plantation using eddy correlation technique. In Proceedings of the 15th Malaysian Society of Plant Phys*iology Conference, Advances in Plant Science, Port Dickson, Malaysia* (pp. 14–16).

Haniff, M. H., Ibrahim, A., Jantan, N. M., Shahabudin, N., Mos, H., & Yusup, Y. (2016). Carbon dioxide and energy fluxes above an oil palm canopy in peninsular Malaysia. *International Journal of Agronomy and Agricultural Research (IJAAR)*, 9(2), 137–146.

Hansen, M. C., Potapov, P. V., Moore, R., Hancher, M., Turubanova, S. A., Tyukavina, A., … Townshend, J. (2013). High-resolution global maps of 21st-century forest cover change. *Science*, 342(6160), 850–853.

Harjotedjo, S., Sahari, B., & Palgoenadi, B. (2013, November). Management of oil palm pests and diseases in Indonesia: Learning from the past, adapting for the future. In Proceedings of the 5th MPOB-IOPRI International Seminar, Kuala Lumpur, Malaysia (pp. 2–23).

Harrell, P. A., Kasischke, E. S., Bourgeau-Chavez, L. L., Haney, E. M., & Christensen Jr, N. L. (1997). Evaluation of approaches to estimating aboveground biomass in southern pine forests using SIR-C data. *Remote Sensing of Environment*, 59(2), 223–233.

Harsono, S. S., Grundmann, P., & Soebronto, S. (2014). Anaerobic treatment of palm oil mill effluents: Potential contribution to net energy yield and reduction of greenhouse gas emissions from biodiesel production. *Journal of Cleaner Production*, 64, 619–627.

Hartley, C. W. S. (1967). *The Oil Palm: Tropical Agriculture Series* (706 pp). London: Longmans Green.

Hartley, C. W. S. (1988). *The Oil Palm* (3rd ed., 761 pp.), Longman Group UK Limited.

Hashim, I. C., Shariff, A. R. M., Bejo, S. K., Muharam, F. M., & Ahmad, K. (2021). Machine-learning approach using SAR data for the classification of oil palm trees that are non-infected and infected with the basal stem rot disease. *Agronomy*, 11(3), 532.

Hassler, E., Corre, M. D., Tjoa, A., Damris, M., Utami, S. R., & Veldkamp, E. (2015). Soil fertility controls soil-atmosphere carbon dioxide and methane fluxes in a tropical landscape converted from lowland forest to rubber and oil palm plantations. *Biogeosciences*, 12(19), 5831–5852.

Haw, Y. H., Lai, K. W., Chuah, J. H., Bejo, S. K., Husin, N. A., Hum, Y. C., … Wu, X. (2023). Classification of basal stem rot using deep learning: A review of digital data collection and palm disease classification methods. *PeerJ Computer Science*, 9, e1325.

He, K., Zhang, X., Ren, S., & Sun, J. (2016). Deep residual learning for image recognition. In *Proceedings of the IEEE Conference on Computer Vision and Pattern Recognition*, Las Vegas, NV (pp. 770–778). doi: 10.1109/CVPR.2016.90

He, Y., Chow, Y. T., Chi, H., & Tsang, K. F. (2020, December). IoT tree healthcare with improved sensor deployment scheme. In *2020 IEEE 2nd International Workshop on System Biology and Biomedical Systems (SBBS)* (pp. 1–4). IEEE, Taichung, Taiwan, 3-4 December.

Heil, A., Langmann, B., & Aldrian, E. (2007). Indonesian peat and vegetation fire emissions: Study on factors influencing large-scale smoke haze pollution using a regional atmospheric chemistry model. *Mitigation and Adaptation Strategies for Global Change*, 12, 113–133.

Henson, I. (1993). Assessing frond dry matter production and leaf area development in young oil palms. In *Proceedings of the PORIM International Palm Oil Conference*. Progress, Prospects Challenges Towards the 21st Century. (Agriculture), Kuala Lumpur, 9–14 September.

Henson, I. E. (1991). Age-related changes in stomatal and photosynthetic characteristics of leaves of oil palm (Elaeis guineensis Jacq). *Elaeis*, 3(2), 336–348.

Henson, I. E. (2000). Modelling the effects of 'haze' on oil palm productivity and yield. *Journal of Oil Palm Research*, 12(1), 123–134.

Henson, I. E. (2003). The Malaysian National Average Oil Palm: Concept and evaluation. *Oil Palm Bulletin*, (46), 15–27.

Henson, I. E. (2009). *Modelling Dry Matter Production, Partitioning and Yield of Oil Palm: OPRODSIM: A Mechanistic Simulation Model for Teaching and Research: Technical Manual and Users' Guide*. Kuala Lumpur: Malaysian Palm Oil Board.

Henson, I. E., & Chang, K. C. (1990). Evidence for water as a factor limiting performance of field palms in West Malaysia. In Proceedings of 1989 PORIM International Palm Oil Development Conference-Agriculture Palm Oil Research Institute of Malaysia, Kuala Lumpur (pp. 487–498).

Henson, I. E., & Harun, M. H. (2005). The influence of climatic conditions on gas and energy exchanges above a young oil palm stand in north Kedah, Malaysia. *Journal of Oil Palm Research*, 17(C), 73.

Henson, I. E., & Roscoe, R. (2020). Palm oil and smallholders: The influence of policy and market access. In *Smallholders and the Non-Farm Transition in Latin America and the Caribbean* (pp. 115–143). Cham: Springer.

Henson, I. E., Harun, M. H., & Chang, K. C. (2008). Some observations on the effects of high water tables and flooding on oil palm, and a preliminary model of oil palm water balance and use in the presence of a high water table. *Oil Palm Bulletin*, 56, 14–22.

HEXNGON (2023). An Introduction to GNSS. In *HEXNGON*, Nov 01, https://novatel.com/an-introduction-to-gnss/gnss-error-sources.

Hijmans, R. J., Cameron, S. E., Parra, J. L., Jones, P. G., & Jarvis, A. (2005). Very high resolution interpolated climate surfaces for global land areas. *International Journal of Climatology: A Journal of the Royal Meteorological Society*, 25(15), 1965–1978.

Hilal, Y. Y., Yahya, A., Ismail, W. I. W., & Asha'Ari, Z. H. (2021). Neural networks method in predicting oil palm FFB yields for the Peninsular States of Malaysia. *Journal of Oil Palm Research*, 33(3), 400–412.

Hirano, T., Segah, H., Kusin, K., Limin, S., Takahashi, H., & Osaki, M. (2012). Effects of disturbances on the carbon balance of tropical peat swamp forests. *Global Change Biology*, 18(11), 3410–3422.

Hoffmann, C., Weise, C., Koch, T., & Pauly, K. (2016). From UAS data acquisition to actionable information-how an end-to-end solution helps oil palm plantation operators to perform a more sustainable plantation management. *The International Archives of the Photogrammetry, Remote Sensing and Spatial Information Sciences*, 41, 1113–1120.

Hoffmann, M. P., Vera, A. C., Van Wijk, M. T., Giller, K. E., Oberthuer, T., Donough, C., & Whitbread, A. M. (2014). Simulating potential growth and yield of oil palm (Elaeis guineensis) with PALMSIM: Model description, evaluation and application. *Agricultural Systems*, 131, 1–10.

Hooijer, A., Page, S., Canadell, J. G., Silvius, M., Kwadijk, J., Wosten, H., & Jauhiainen, wwJ. (2010). Current and future CO2 emissions from drained peatlands in Southeast Asia. *Biogeosciences*, 7(5), 1505–1514. https://doi.org/10.5194/bg-7-1505-2010

Hooijer, A., Page, S., Jauhiainen, J., Lee, W. A., Lu, X. X., Idris, A., & Anshari, G. (2012). Subsidence and carbon loss in drained tropical peatlands. *Biogeosciences*, 9(3), 1053–1071.

Hooijer, A., Vernimmen, R., Visser, M., & Mawdsley, N. (2015). Flooding projections from elevation and subsidence models for oil palm plantations in the Rajang Delta peatlands, Sarawak, Malaysia. *Deltares Report*, 1207384, 76.

Houghton, R. A., Hall, F., & Goetz, S. J. (2009), Importance of biomass in the global carbon cycle. *Journal of Geophysical Research*, 114, 1–13. doi:10.1029/2009JG000935.

Hoyt, A. M., Chaussard, E., Seppalainen, S. S., & Harvey, C. F. (2020). Widespread subsidence and carbon emissions across Southeast Asian peatlands. *Nature Geoscience*, 13(6), 435–440.

Hua, H., Li, Y., Wang, T., Dong, N., Li, W., & Cao, J. (2023). Edge computing with artificial intelligence: a machine learning perspective. *ACM Computing Surveys*, 55(9), 1–35.

Huang, B., Cova, T. J., Tsou, M. H., Bareth, G., & Silva, E. A. (2017). *Comprehensive Geographic Information Systems*. Elsevier.

Huang, H., Deng, J., Lan, Y., Yang, A., Deng, X., & Zhang, L. (2018). A fully convolutional network for weed mapping of unmanned aerial vehicle (UAV) imagery. *PLoS One*, 13(4), e0196302.

Huang, Q., Long, D., Du, M., Zeng, C., Qiao, G., Li, X., … Hong, Y. (2018). Discharge estimation in high-mountain regions with improved methods using multisource remote sensing: A case study of the Upper Brahmaputra River. *Remote Sensing of Environment*, 219, 115–134.

Huang, B., Zhang, H., & Yu, L. (2012). Improving Landsat ETM plus urban area mapping via spatial and angular fusion With MISR multi-angle observations. *IEEE Journal of Selected Topics in Applied Earth Observations and Remote Sensing*, 5, 101–109.

Hung, C., Bryson, M., & Sukkarieh, S. (2012). Multi-class predictive template for tree crown detection. *ISPRS Journal of Photogrammetry and Remote Sensing*, 68, 170–183.

Husin, H. S., Amar, N., Sajak, A. A. B., & Kassim, M. S. M. (2021, May). Distribution map of oil palm fresh fruit bunch using LiDAR. In 2021 12th International Conference on Information and Communication Systems (ICICS), Valencia (pp. 4–9). IEEE. doi: 10.1109/ICICS52457.2021.9464575

Husin, N. A., Khairunniza-Bejo, S., Abdullah, A. F., Kassim, M. S., Ahmad, D., & Azmi, A. N. (2020). Application of ground-based LiDAR for analysing oil palm canopy properties on the occurrence of basal stem rot (BSR) disease. *Scientific Reports*, 10(1), 6464.

Hutabarat, S., Slingerland, M., Rietberg, P., & Dries, L. (2018). Costs and benefits of certification of independent oil palm smallholders in Indonesia. *International Food and Agribusiness Management Review*, 21(6), 681–700.

Huth, N. I., Banabas, M., Nelson, P. N., & Webb, M. (2014). Development of an oil palm cropping systems model: Lessons learned and future directions. *Environmental Modelling & Software*, 62, 411–419.

Ibrahim, M. H., Jaafar, H. Z. E., Harun, M. H., & Yusop, M. R. (2010). Changes in growth and photosynthetic patterns of oil palm (Elaeis guineensis Jacq.) seedlings exposed to short-term CO2 enrichment in a closed top chamber. *Acta Physiologiae Plantarum*, 32, 305–313.

Ibrahim, M. S., Seman, I. A., Rusli, M. H., Izzuddin, M. A., Kamarudin, N., Hashim, K., & Abd Manaf, Z. (2020). Surveillance of Ganoderma disease in oil palm planted by participants of the smallholders replanting incentive scheme in Malaysia. *Journal of Oil Palm Research*, 32(2), 237–244.

Ibrahim, S., Zainal, A. H., & Mariamni, K. (2001). Application of optical remote sensing technology for oil palm management. In *Cutting-Edge Technologies for Sustained Competitiveness: Proceedings of the 2001 PIPOC International Palm Oil Congress, Agriculture Conference, Kuala Lumpur, Malaysia, 20–22 August 2001* (pp. 499–504). Malaysian Palm Oil Board (MPOB). https://www.cabdirect.org/cabdirect/abstract/20023008524

IDH (2023). https://www.idhsustainabletrade.com/initiative/european-sustainable-palm-oil-espo/ (date of access 11 May 2023).

Idris, A. S., Ariffin, D., & Ismail, S. (2004). Innovative technique of sanitation for controlling Ganoderma at replanting. MPOB Information Series, 213, p. 4.

Idris, A. S., Mazliham, M. S., Loonis, P., & Wahid, M. B. (2010). GanoSken for early detection of Ganoderma infection in oil palm. *MPOB Information Series*, 442, 1–4.

ISCC (2022). ISCC guidance waste and residues from palm oil mills version 3.0. Available online https://www.iscc-system.org/certification/iscc-documents/iscc-system-documents/ (accessed 19 May 2023)

Izzuddin, M. A., (2010). *Early Detection of Ganoderma Disease in Oil Palm using Field Spectroscopy*. MSc Thesis, Universiti Putra Malaysia, 130.

Izzuddin, M.A., Arof, H., bin Hashim, Z., bin Abu Seman, I., Masri, M.M., Ibrahim, S.M., Tat, E.H. & Toh, C.M. (2021). Remote sensing for detection of ganoderma disease and bagworm infestation in oil palm. *Advances in Agricultural and Food Research Journal*, 2(1).

Izzuddin, M. A., Hamzah, A., Nisfariza, M. N., Idris, A. S., Aizam, A. N., Aikal, B. S., … Shukri, I. M. (2023). Denoising of hyperspectral signal from drone for ganoderma disease detection in oil palm. *International Journal of Geoinformatics*, 19(5), 55–60.

Izzuddin, M. A., Nisfariza, M. N., Ezzati, B., Idris, A. S., Steven, M. D., & Boyd, D. (2018). Analysis of airborne hyperspectral image using vegetation indices, red edge position and continuum removal for detection of Ganoderma disease in oil palm. *Journal of Oil Palm Research*, 30, 416–428.

Izzuddin, M. A., Seman Idris, A., Nisfariza, M. N., Nordiana, A. A., Shafri, H. Z. M., & Ezzati, B. (2017). The development of spectral indices for early detection of Ganoderma disease in oil palm seedlings. *International Journal of Remote Sensing*, 38(23), 6505–6527.

Jachowski, N. R., Quak, M. S., Friess, D. A., Duangnamon, D., Webb, E. L., & Ziegler, A. D. (2013). Mangrove biomass estimation in Southwest Thailand using machine learning. *Applied Geography*, 45, 311–321.

Jacquemard, J. C., & Baudouin, L. (1998). *Oil Palm*. MacMillan, London.

Jamaludin, N. F., Ab Muis, Z., & Hashim, H. (2019). An integrated carbon footprint accounting and sustainability index for palm oil mills. *Journal of Cleaner Production*, 225, 496–509.

Jansen, L. J., & Di Gregorio, A. (2002). Parametric land cover and land-use classifications as tools for environmental change detection. *Agriculture, Ecosystems & Environment*, 91(1–3), 89–100.

Jean, N., Burke, M., Xie, M., Davis, W. M., Lobell, D. B., & Ermon, S. (2016). Combining satellite imagery and machine learning to predict poverty. *Science*, 353(6301), 790–794.

Jetz, W., McGeoch, M.A., Guralnick, R., Ferrier, S., Beck, J., Costello, M. J., … Meyer, C. (2019). Essential biodiversity variables for mapping and monitoring species populations. *Nature Ecology & Evolution*, 3(4), 539–551.

Jha, K., Doshi, A., Patel, P., & Shah, M. (2019). A comprehensive review on automation in agriculture using artificial intelligence. *Artificial Intelligence in Agriculture*, 2, 1–12.

Jiang, L., & Yu, L. (2019). Analyzing land use intensity changes within and outside protected areas using ESA CCI-LC datasets. *Global Ecology and Conservation*, 20, e00789.

Johnston, G., Riddell, A., & Hausler, G. (2017). The international GNSS service. In *Springer Handbook of Global Navigation Satellite Systems* (pp. 967–982).

Kalidas, P. (2013). Pest problems of oil palm and management strategies for sustainability. *Agrotechnology*, 11(001), 2012. doi:10.4172/2168-9881.S11-001

Kamarudin, N., & Wahid, M. B. (2010). Interactions of the bagworm, Pteroma pendula (Lepidoptera: Psychidae), and its natural enemies in an oil palm plantation in Perak. *Journal of Oil Palm Research*, 22(1), 758–764.

Kamilaris, A., Kartakoullis, A., & Prenafeta-Boldú, F. X. (2017). A review on the practice of big data analysis in agriculture. *Computers and Electronics in Agriculture*, 143, 23–37.

Kamiran, N., & Sarker, M. L. R. (2014, February). Exploring the potential of high resolution remote sensing data for mapping vegetation and the age groups of oil palm plantation. *In IOP Conference Series: Earth and Environmental Science* (Vol. 18, No. 1, p. 012181). IOP Publishing.

Kang, C. S. (2016). Microwave remote sensing for soil moisture estimation in tropical regions-a review and SMOS L2 products validation. *International Journal of Geoinformatics*, 12 (2). https://journals.sfu.ca/ijg/index.php/journal/article/view/946

Kang, C. S., & Kanniah, K. D. (2022). Land use and land cover change and its impact on river morphology in Johor River Basin, Malaysia. *Journal of Hydrology: Regional Studies*, 41, 101072.

Kang, C. S., Kanniah, K. D., & Kerr, Y. H. (2019). Calibration of SMOS soil moisture retrieval algorithm: A case of tropical site in Malaysia. *IEEE Transactions on Geoscience and Remote Sensing*, 57(6), 3827–3839.

Kanniah, K. D., Sheikhi, A., Cracknell, A. P., Goh, H. C., Tan, K. P., Ho, C. S., & Rasli, F. N. (2015). Satellite images for monitoring mangrove cover changes in a fast growing economic region in southern Peninsular Malaysia. *Remote Sensing*, 7(11), 14360–14385.

Kanniah, K. D., Tan, K. P., & Cracknel, A. P. (2012, July). UK-DMC 2 satellite data for deriving biophysical parameters of oil palm trees in Malaysia. In 2012 IEEE International Geoscience and Remote Sensing Symposium (pp. 6569–6572). IEEE.

Katayama, A., Kume, T., Komatsu, H., Saitoh, T. M., Ohashi, M., Nakagawa, M., … Kumagai, T. O. (2013). Carbon allocation in a Bornean tropical rainforest without dry seasons. *Journal of Plant Research*, 126, 505–515.

Kerr, Y. H., Waldteufel, P., Richaume, P., Davenport, I., Ferrazzoli, P., & Wigneron, J. P. (2015). SMOS level 2 processor soil moisture algorithm theoretical basis document (ATBD), SM-ESL (CBSA), Toulouse, Technical Report. SO-TN-ESL-SM-GS-0001

Kerdraon, D., Drewer, J., Chung, A. Y., Majalap, N., Slade, E. M., Bréchet, L., … Sayer, E. J. (2020). Litter inputs, but not litter diversity, maintain soil processes in degraded tropical forests-a cross-continental comparison. *Frontiers in Forests and Global Change*, 2, 90.

Khairunniza-Bejo, S., & Vong, C. N. (2014). Detection of basal stem rot (BSR) infected oil palm tree using laser scanning data. *Agriculture and Agricultural Science Procedia*, 2, 156–164.

Khalid, H., Zin, Z. Z., & Anderson, J. M. (1999a). Quantification of oil palm biomass and nutrient value in a mature plantation. I. Above-ground biomass. *Journal of Oil Palm Research*, 11(1), 23–32.

Khalid, H., Zin, Z. Z., & Anderson, J. M. (1999b). Quantification of oil palm biomass and nutrient value in a mature plantation. II. Below-ground biomass. *Journal of Oil Palm Research*, 11(2), 63–71.

Khamiz, A., Ismail, Z., & Muhammad, A. T. (2005). Nonlinear growth models for modeling oil palm yield growth. *Journal of Mathematics and Statistics*, 1(3), 225–233.

Khan, N., Kamaruddin, M. A., Ullah Sheikh, U., Zawawi, M. H., Yusup, Y., Bakht, M. P., & Mohamed Noor, N. (2022). Prediction of oil palm yield using machine learning in the perspective of fluctuating weather and soil moisture conditions: Evaluation of a generic workflow. *Plants*, 11(13), 1697.

Khatun, R., Reza, M. I. H., Moniruzzaman, M., & Yaakob, Z. (2017). Sustainable oil palm industry: The possibilities. *Renewable and Sustainable Energy Reviews*, 76, 608–619.

Kho, L. K., & Jepsen, M. R. (2015). Carbon stock of oil palm plantations and tropical forests in Malaysia: A review. *Singapore Journal of Tropical Geography*, 36(2), 249–266.

Khoon, K., Rumpang, E., Kamarudin, N., & Harun, M. H. (2019). Quantifying total carbon stock of mature oil palm. *Journal of Oil Palm Research*, 31(3), 521–527.

Khor, J. F., Ling, L., Yusop, Z., Tan, W. L., Ling, J. L., & Soo, E. Z. X. (2021). Impact of El Niño on oil palm yield in Malaysia. *Agronomy*, 11(11), 2189.

Kiew, F., Hirata, R., Hirano, T., Xhuan, W. G., Aries, E. B., Kemudang, K., … Melling, L. (2020). Carbon dioxide balance of an oil palm plantation established on tropical peat. *Agricultural and Forest Meteorology*, 295, 108189.

Kiliszek, D., & Kroszczyński, K. (2020). Performance of the precise point positioning method along with the development of GPS, GLONASS and Galileo systems. *Measurement*, 164, 108009.

Kiyono, Y., Monda, Y., Toriyama, J., Chaddy, A., Goh, K. J., & Melling, L. (2015). Destructive sampling method for estimating the biomasses of African oil palm (Elaeis guineensis) plantations on tropical peatland. *Bulletin of the Forestry and Forest Products Research Institute*, 14(3), 147–158.

Knipling, E. B. (1970). Physical and physiological basis for the reflectance of visible and near-infrared radiation from vegetation. *Remote Sensing of Environment*, 1(3), 155–159.

Koch, B., Khosla, R., Frasier, W. M., Westfall, D. G., & Inman, D. (2004). Economic feasibility of variable-rate nitrogen application utilizing site-specific management zones. *Agronomy Journal*, 96(6), 1572–1580.

Koh, L. P., & Wilcove, D. S. (2008). Is oil palm agriculture really destroying tropical biodiversity?. *Conservation Letters*, 1(2), 60–64.

Koh, L. P., Miettinen, J., Liew, S. C., & Ghazoul, J. (2011). Remotely sensed evidence of tropical peatland conversion to oil palm. *Proceedings of the National Academy of Sciences*, 108(12), 5127–5132.

Koon, L. W., & Kun, O. B. (2006). The unseen flood: Waterlogging in large oil palm plantations. *Jurutera*, 1, 28–31.

Kotowska, M. M., Leuschner, C., Triadiati, T., Meriem, S., & Hertel, D. (2015). Quantifying above-and belowground biomass carbon loss with forest conversion in tropical lowlands of Sumatra (Indonesia). *Global Change Biology*, 21(10), 3620–3634.

Krishna, V. V., Kubitza, C., Pascual, U., & Qaim, M. (2017). Land markets, property rights, and deforestation: insights from Indonesia. *World Development*, 99, 335–349.

Kuffer, M., Pfeffer, K., & Sliuzas, R. (2016). Slums from space-15 years of slum mapping using remote sensing. *Remote Sensing*, 8(6), 455.

Kurihara, J., Koo, V. C., Guey, C. W., Lee, Y. P., & Abidin, H. (2022). Early detection of basal stem rot disease in oil palm tree using unmanned aerial vehicle-based hyperspectral imaging. *Remote Sensing*, 14(3), 799.

Lachowski, H., Maus, P., Golden, M., Johnson, J., Landrum, V., Powell, J., Varner, V., Wirth, T., Gonzels, J. & Bain, S. (1995). *Guidelines for the use of digital Imagery for vegetation mapping*. US Department of Agriculture, Forest Service, Washington, D.C.

Lamade, E., & Bouillet, J. P. (2005). Lamade, E., & Bouillet, J. P. (2005). Carbon storage and global change: the role of oil palm. *Oilseeds and Fats, Crops and Lipids*, 12 (2), 154–160. https://doi.org/10.1051/ocl.2005.0154.

Langley, R. B., Teunissen, P. J., & Montenbruck, O. (2017). Introduction to GNSS. In *Springer Handbook of Global Navigation Satellite Systems* (pp. 3–23).

Larjavaara, M., & Muller-Landau, H. C. (2013). Measuring tree height: A quantitative comparison of two common field methods in a moist tropical forest. *Methods in Ecology and Evolution*, 4(9), 793–801.

Laurance, W. F., Koh, L. P., Butler, R., Sodhi, N. S., Bradshaw, C. J., Neidel, J. D., ... Vega, J. M. (2010). Improving the performance of the roundtable on sustainable palm oil for nature conservation. *Conservation Biology*, 24(2), 377–381.

Lee, J. S. H., Abood, S., Ghazoul, J., Barus, B., Obidzinski, K., & Koh, L. P. (2014). Environmental impacts of large-scale oil palm enterprises exceed that of smallholdings in Indonesia. *Conservation Letters*, 7(1), 25–33.

Lee, J. S. H., Wich, S., Widayati, A., & Koh, L. P. (2016). Detecting industrial oil palm plantations on Landsat images with Google Earth Engine. *Remote Sensing Applications: Society and Environment*, 4, 219–224.

Lees, A. C., Moura, N. G., de Almeida, A. S., & Vieira, I. C. (2015). Poor prospects for avian biodiversity in Amazonian oil palm. *PLoS One*, 10(5), e0122432.

Lelong, C. C., Roger, J. M., Brégand, S., Dubertret, F., Lanore, M., Sitorus, N. A., ... Caliman, J. P. (2010). Evaluation of oil-palm fungal disease infestation with canopy hyperspectral reflectance data. *Sensors*, 10(1), 734–747.

Lesage, C., Cifuentes-Espinosa, J. A., & Feintrenie, L. (2021). Oil palm cultivation in the Americas: Review of the social, economic and environmental conditions of its expansion. *Cahiers Agricultures*, 30, 27, https://doi.org/10.1051/cagri/2021015

Lewis, K., Rumpang, E., Kho, L. K., McCalmont, J., Teh, Y. A., Gallego-Sala, A., & Hill, T. C. (2020). An assessment of oil palm plantation aboveground biomass stocks on tropical peat using destructive and non-destructive methods. *Scientific Reports*, 10(1), 2230.

Li, D., Tong, Q., Li, R., Gong, J., & Zhang, L. (2012). Current issues in high-resolution earth observation technology. *Science China Earth Sciences*, 55, 1043–1051.

Li, K., Wan, G., Cheng, G., Meng, L., & Han, J. (2020). Object detection in optical remote sensing images: A survey and a new benchmark. *ISPRS Journal of Photogrammetry and Remote Sensing*, 159, 296–307.

Li, R., Zheng, S., Wang, E., Chen, J., Feng, S., Wang, D., & Dai, L. (2020). Advances in BeiDou Navigation Satellite System (BDS) and satellite navigation augmentation technologies. *Satellite Navigation*, 1, 1–23.

Li, T. M. (2018a). After the land grab: Infrastructural violence and the Mafia System in Indonesia's oil palm plantation zones. *Geoforum*, 96, 328–337.

Li, T. M. (2018b). Evidence-based options for advancing social equity in Indonesian palm oil: Implications for research, policy and advocacy. CIFOR Infobrief No. 208. Bogor, Indonesia: Center for International Forestry Research (CIFOR). https://doi.org/10.17528/cifor/006842.

Li, W., Dong, R., Fu, H., & Yu, L. (2018). Large-scale oil palm tree detection from high-resolution satellite images using two-stage convolutional neural networks. *Remote Sensing*, 11(1), 11.

Li, W., Fu, H., Yu, L., & Cracknell, A. (2017). Deep learning based oil palm tree detection and counting for high-resolution remote sensing images. *Remote Sensing*, 9(1), 22.

Li, W., He, C., Fang, J., Zheng, J., Fu, H., & Yu, L. (2019). Semantic segmentation-based building footprint extraction using very high-resolution satellite images and multi-source GIS data. *Remote Sensing*, 11(4), 403.

Liaghat, S., Mansor, S., Ehsani, R., Shafri, H. Z. M., Meon, S., & Sankaran, S. (2014). Mid-infrared spectroscopy for early detection of basal stem rot disease in oil palm. *Computers and Electronics in Agriculture*, 101, 48–54.

Lin, T. Y., Dollár, P., Girshick, R., He, K., Hariharan, B., & Belongie, S. (2017). Feature pyramid networks for object detection. In Proceedings of the IEEE Conference on Computer Vision and Pattern Recognition (pp. 2117–2125).

Liu, J., Xiang, J., Jin, Y., Liu, R., Yan, J., & Wang, L. (2021). Boost precision agriculture with unmanned aerial vehicle remote sensing and edge intelligence: A survey. *Remote Sensing*, 13(21), 4387.

Liu, W., Anguelov, D., Erhan, D., Szegedy, C., Reed, S., Fu, C. Y., & Berg, A. C. (2016). SSD: Single shot multibox detector. In B. Leibe, J. Matas, N. Sebe, M. Welling (Eds.), *Computer Vision – ECCV 2016. ECCV 2016*. Lecture Notes in Computer Science (Vol. 9905). Cham: Springer. https://doi.org/10.1007/978-3-319-46448-0_2.

Lodhi, V., Chakravarty, D., & Mitra, P. (2018). Hyperspectral imaging for earth observation: Platforms and instruments. *Journal of the Indian Institute of Science*, 98, 429–443.

Lu, D., and M. Batistella. (2005). Exploring TM image texture and its relationships with biomass estimation in Rondônia, Brazilian Amazon. *Acta Amazonica*, 35(2): 249–57. doi: 10.1590/S0044-59672005000200015.

Lu, X., Li, B., Yue, Y., Li, Q. and Yan, J. (2019). Grid r-cnn. In Proceedings of the IEEE/CVF Conference on Computer Vision and Pattern Recognition, Long Beach, CA (pp. 7363–7372).

Lupascu, M., Varkkey, H., & Tortajada, C. (2020). Is flooding considered a threat in the degraded tropical peatlands?. *Science of the Total Environment*, 723, 137988.

Mahabir, R., Croitoru, A., Crooks, A. T., Agouris, P., & Stefanidis, A. (2018). A critical review of high and very high-resolution remote sensing approaches for detecting and mapping slums: Trends, challenges and emerging opportunities. *Urban Science*, 2(1), 8.

Mahlein, A. K., Rumpf, T., Welke, P., Dehne, H. W., Plümer, L., Steiner, U., & Oerke, E. C. (2013). Development of spectral indices for detecting and identifying plant diseases. *Remote Sensing of Environment*, 128, 21–30.

Malaysia Central Bank (2021). *Economic and Monetary Review 2021*. https://www.bnm.gov.my/documents/20124/6458996/emr2021_en_book.pdf (accessed 15 Oct 2023).

Malaysia Palm Oil Certification Council (2019). *Accreditation Guidelines for MSPO Certification Bodies*. Retrieved from: https://www.mpocc.org.my/links/publication.

Malaysian Palm Oil Certification Council (2019). *Malaysian Sustainable Palm Oil Certification Scheme*. Retrieved from: https://www.mpocc.org.my/scheme-documents.

Malaysian Palm Oil Certification Council (2023). *MSPO Trace*. Retrieved from: https://mspo-trace.org.my/ (accessed on 25 May 2023).

Manjeri, G., Muhamad, R., & Tan, S. G. (2014). Oryctes rhinoceros beetles, an oil palm pest in Malaysia. *Annual Research & Review in Biology*, 4(22), 3429–3439. https://doi.org/10.9734/ARRB/2014/11023

Marini-Pereira, L., Pullen, S., Moraes, A. D. O., & Sousasantos, J. (2021). Ground-based augmentation systems operation in low latitudes-Part 1: Challenges, mitigations, and future prospects. *Journal of Aerospace Technology and Management*, 13, e4621.

Mansor, S. A., & Sarker, M. L. R. (2015). Remote sensing technique for estimating the age of oil palm using high resolution image. In *36th Asian Conference on Remote Sensing 2015. Philippines (pp. 24–28) October 2015*.

Marcelino, J. P., & Diaz, E. V. (2016). Frond pruning enhanced the growth and yield of eight-year-old oil palm (Jacq.) Elaeis guineensis. *Annals of Tropical Research*, 38(2), 96–105.

Martinez, A., Camberato, J. J., Owens, P., & Ashtekar, J. (2021). Using terrain algorithms on a digital elevation model to evaluate yield variability in oil palm. *Journal of Oil Palm Research*, 33(1), 84–92.

Mazmira, M., Ramlah, S., Najib, M. A., Norman, K., Kushairi, A., & Wahid, M. B. (2011). Pest management of bagworm in southern perak by aerial spraying with Bacillus thuringiensis. *Oil Palm Bulletin*, 63, 24–33.

McAlpine, C. A. et al. (2018). Forest loss and Borneo's climate. *Environmental Research Letters*, 13, 044009.

McBratney, A. Á., & Pringle, M. J., 1999. Estimating average and proportional variograms of soil properties and their potential use in precision agriculture. *Precision Agriculture*, 1, 125–152.

McCalmont, J., Kho, L. K., Teh, Y. A., Lewis, K., Chocholek, M., Rumpang, E., & Hill, T. (2021). Short-and long-term carbon emissions from oil palm plantations converted from logged tropical peat swamp forest. *Global Change Biology*, 27(11), 2361–2376.

McConnell, M. D. (2019). Bridging the gap between conservation delivery and economics with precision agriculture. *Wildlife Society Bulletin*, 43(3), 391–397.

McKenzie, S. (2004). Social sustainability: Towards some definitions. Hawke Research Institute Postgraduate Working Paper Series No 27. http://naturalcapital.us/images/Social%20 Sustainability%20-%20Towards%20Some%20Definitions_20100120_024059.pdf

McMorrow, J. (2001). Linear regression modelling for the estimation of oil palm age from Landsat TM. *International Journal of Remote Sensing*, 22(12), 2243–2264.

McMorrow, J. M. (1995). Relation of oil palm spectral response to stand age. *International Journal of Remote Sensing*, 16(16), 3203–3209.

Means, J. E., Acker, S. A., Harding, D. J., Blair, J. B., Lefsky, M. A., Cohen, W. B., … McKee, W. A. (1999). Use of large-footprint scanning airborne lidar to estimate forest stand characteristics in the Western Cascades of Oregon. *Remote Sensing of Environment*, 67(3), 298–308.

Medasani, S., & Reddy, G. U. (2018, March). Analysis and evaluation of speckle filters by using polarimetric synthetic aperture radar data through local statistics. In *Proceedings of the* 2018 Second International Conference on Electronics, Communication and Aerospace Technology (ICECA), Coimbatore (pp. 169–174). IEEE.

Meijaard, E., Brooks, T. M., Carlson, K. M., Slade, E. M., Garcia-Ulloa, J., Gaveau, D. L., … Sheil, D. (2020). The environmental impacts of palm oil in context. *Nature Plants*, 6(12), 1418–1426.

Meijaard, E., Garcia-Ulloa, J., Sheil, D., Wich, S. A., Carlson, K. M., Juffe-Bignoli, D., & Brooks, T. M. (2018). *Oil Palm and Biodiversity. A Situation Analysis by the IUCN Oil Palm Task Force*. Switzerland: IUCN.

Meijide, A., de la Rua, C., Guillaume, T., Röll, A., Hassler, E., Stiegler, C., … Knohl, A. (2020). Measured greenhouse gas budgets challenge emission savings from palm-oil biodiesel. *Nature Communications*, 11(1), 1089.

Meng, Y., Bian, L., Han, L., Lei, W., Yan, T., He, M., & Li, X. (2018, May). A global navigation augmentation system based on LEO communication constellation. In 2018 European Navigation Conference (ENC), Gothenburg, (pp. 65–71). IEEE.

Meroni, M., Waldner, F., Seguini, L., Kerdiles, H., & Rembold, F. (2021). Yield forecasting with machine learning and small data: What gains for grains?. *Agricultural and Forest Meteorology*, 308, 108555.

Merten, J., Röll, A., Guillaume, T., Meijide, A., Tarigan, S., Agusta, H., … Hölscher, D. (2016). Water scarcity and oil palm expansion: Social views and environmental processes. *Ecology and Society*, 21, 2.

Merten, J., Stiegler, C., Hennings, N., Purnama, E., Röll, A., Agusta, H., … Faust, H. (2020). Flooding and land use change in Jambi Province, Sumatra: Integrating local knowledge and scientific inquiry. *Ecology and Society*, 25(3), 14. https://doi.org/10.5751/ES-11678-250314.

Middelberg, J., Azhar, B., Khoon, K. L., & Van Der Meer, P. J. (2019). An ecosystem services analysis of oil palm and alternative land use systems on peat in Malaysia. *Journal of Oil Palm Research*, 31(93), 468–479.

Miettinen, J., Shi, C., & Liew, S. C. (2016). Land cover distribution in the peatlands of Peninsular Malaysia, Sumatra and Borneo in 2015 with changes since 1990. *Global Ecology and Conservation*, 6, 67–78.

Miettinen, J., Shi, C., & Liew, S. C. (2019). Towards automated 10–30 m resolution land cover mapping in insular South-East Asia. *Geocarto International*, 34(4), 443–457.

Migolet, P., & Goïta, K. (2020). Evaluation of FORMOSAT-2 and planetscope imagery for aboveground oil palm biomass estimation in a mature plantation in the Congo Basin. *Remote Sensing*, 12(18), 2926.

Migolet, P., Goïta, K., Ngomanda, A., & Mekui Biyogo, A. P. (2020). Estimation of aboveground oil palm biomass in a mature plantation in the Congo Basin. *Forests*, 11(5), 544.

Mills, M. B., Malhi, Y., Ewers, R. M., Kho, L. K., Teh, Y. A., Both, S., … Riutta, T. (2023). Tropical forests post-logging are a persistent net carbon source to the atmosphere. *Proceedings of the National Academy of Sciences*, 120(3), e2214462120.

Mohd Najib, N. E., Kanniah, K. D., Cracknell, A. P., & Yu, L. (2020). Synergy of active and passive remote sensing data for effective mapping of oil palm plantation in Malaysia. *Forests*, 11(8), 858.

Mohd Najib, N. E., Kanniah, K. D., Cracknell, A. P., & Yu, L. (2020). Synergy of active and passive remote sensing data for effective mapping of oil palm plantation in Malaysia. *Forests*, 11(8), 858.

Mohd Zaki, M. Z., S. H. Md Nor, A. J. Nasir, Z. Suleiman, Z. A. Hassan, and N. M. Yusoff. 2009. Remote sensing for oil palm yield forecasting. In Proceedings of Agriculture, Biotechnology & Sustainability Conference PIPOC 2009. Malaysia: Kuala Lumpur Convention Centre.

Mohd, R., & Mohd, H. H. (2004). The role of leaf area index (LAI) in oil palm. *Oil Palm Bulletin*, (48), 11–16.

Monteith, J. L. (1972). Solar radiation and productivity in tropical ecosystems. *Journal of applied ecology*, 9(3), 747–766.

Moore, T. (2022). Including indigenous peoples in geospatial services. *Frontiers in Climate*, 4, 810428.

Morales, G., Kemper, G., Sevillano, G., Arteaga, D., Ortega, I., & Telles, J. (2018). Automatic segmentation of Mauritia flexuosa in unmanned aerial vehicle (UAV) imagery using deep learning. *Forests*, 9(12), 736.

Morel, A. C., Fisher, J. B., & Malhi, Y. (2012). Evaluating the potential to monitor aboveground biomass in forest and oil palm in Sabah, Malaysia, for 2000–2008 with Landsat ETM+ and ALOS-PALSAR. *International Journal of Remote Sensing*, 33(11), 3614–3639.

Morel, A. C., Saatchi, S. S., Malhi, Y., Berry, N. J., Banin, L., Burslem, D., … Ong, R. C. (2011). Estimating aboveground biomass in forest and oil palm plantation in Sabah, Malaysian Borneo using ALOS PALSAR data. *Forest Ecology and Management*, 262(9), 1786–1798.

Morgans, C. L., Meijaard, E., Santika, T., Law, E., Budiharta, S., Ancrenaz, M., & Wilson, K. A. (2018). Evaluating the effectiveness of palm oil certification in delivering multiple sustainability objectives. *Environmental Research Letters*, 13(6), 064032.

Morton, Y. J., van Diggelen, F., Spilker Jr, J. J., Parkinson, B. W., Lo, S., & Gao, G. (Eds.). (2021). *Position, Navigation, and Timing Technologies in the 21st Century: Integrated Satellite Navigation, Sensor Systems, and Civil Applications* (Vol. 1). New York: John Wiley & Sons.

MPOC. (2007). CO2 removal through oil palm plantations contributes significantly to Malaysia's carbon sink status. https://mpoc.org.my/co2-removal-through-oil-palm-plantations-contributes-significantly-to-malaysias-carbon-sink-status/ (accessed 17 Jan 2024).

MPOC (2018). Oil palm peat ecosystem emits lower GHG. https://mpoc.org.my/oil-palm-peat-ecosystem-emits-lower-ghg-2/ (accessed 17 Jan 2024).

MPOCC (2022). Overview of the Revised MSPO Standards (MS2530:2022). https://www.mpocc.org.my/mspo-blogs/overview-of-revised-mspo-standards-ms25302022 (accessed 17 Jan 2024)

Mubin, N. A., Nadarajoo, E., Shafri, H. Z. M., & Hamedianfar, A. (2019). Young and mature oil palm tree detection and counting using convolutional neural network deep learning method. *International Journal of Remote Sensing*, 40(19), 7500–7515.

Muhadi, N. A., Abdullah, A. F., & Kassim, M. S. M. (2016, June). Quantification of terrestrial laser scanner (TLS) elevation accuracy in oil palm plantation for IFSAR improvement. *In IOP Conference Series: Earth and Environmental Science* (Vol. 37, No. 1, p. 012042). IOP Publishing.

Muhadi, N. A., Mohd Kassim, M. S., & Abdullah, A. F. (2019). Improvement of Digital Elevation Model (DEM) using data fusion technique for oil palm replanting phase. *International Journal of Image and Data Fusion*, 10(3), 232–243.

Murdiyarso, D., Van Noordwijk, M., Wasrin, U. R., Tomich, T. P., & Gillison, A. N. (2002). Environmental benefits and sustainable land-use options in the Jambi transect, Sumatra. *Journal of Vegetation Science*, 13(3), 429–438.

Murphy, D. J. (2014). The future of oil palm as a major global crop: Opportunities and challenges. *Journal of Oil Palm Research*, 26(1), 1–24.

Murphy, D. J., Goggin, K., & Paterson, R. R. M. (2021). Oil palm in the 2020s and beyond: Challenges and solutions. *CABI Agriculture and Bioscience*, 2(1), 1–22.

Mutsaers, H. J. W. (2019). The challenge of the oil palm: Using degraded land for its cultivation. *Outlook on Agriculture*, 48(3), 190–197.

Nachtergaele, F., van Velthuizen, H., Verelst, L. & Wiberg, D. (2012). *Harmonised World Soil Database Version 1.2*. Rome: FAO and Laxenburg: IIASA.

Nachtergaele, F., van Velthuizen, H., Verelst, L., Batjes, N. H., Dijkshoorn, K., van Engelen, V. W. P., Fischer, G., Jones, A., & Montanarela, L. (2010). The harmonized world soil database. In R. J. Gilkes, & N. Prakongkep (Eds.), Proceedings of the 19th World Congress of Soil Science, Soil Solutions for a Changing World, Brisbane, Australia, 1–6 August 2010 (pp. 34–37). International Union of Soil Sciences. https://edepot.wur.nl/154132.

Namanji, S., Ssekyewa, C., & Slingerland, M. A. (2021). Intercropping food crops into oil palm plantations - Experiences in Uganda and why it makes sense. Ecological Trends Alliance, Kampala, Uganda Tropenbos International, Ede, the Netherlands. Retrieved from https://www.tropenbos.org/file.php/2398/2021_brief_intercropping_oil_palm_uganda.pdf.

NASA (2004). ASTER global digital elevation map announcement. *NASA*, Sept. 07. https://asterweb.jpl.nasa.gov/gdem.asp.

NASA (2023). U.S. releases enhanced shuttle land elevation data. *NASA*, Nov. 01. https://www2.jpl.nasa.gov/srtm/.

Neuenschwander, A., Pitts, K. L., Jelley, B. P., Robbins, J., Klotz, B., Popescu, S. C., et al. (2021). Boulder, CO: NASA Natl. Snow Ice Data Cent. Distrib. Act. Arch. Cent. doi:10.5067/ATLAS/ATL08.005ATLAS/ICESat-2L3A Land and Vegetation Height, Version 5.

Ng, H. P., Chew, P. S., Goh, K. J., & Kee, K. K. (1999). Nutrient requirements and sustainability in mature oil palms-an assessment. *Planter*, 75(880), 331–345.

Ng, S. K., Cheong, T.K., Haw, K.C., Ooi, H.S.H., Yee, L.K., Kayaroganam, P., Uexküll, H.R.v., Härdter, R., 2003. Clonal oil palm production, yield performance, and nutritional requirements. In: Fairhurst, T., Härdter, R. (Eds.), *The Oil Palm-Management for Large and Sustainable Yields*. Potash & Phosphate Institute of Canada, Potash & Phosphate Institute, International Potash Institute, Singapore, pp. 99–114.

Ng, S. K., Thamboo, S., de Souza, P. (1968) Nutrient contents of oil palms in Malaya. II. Nutrients in vegetative tissues. *The Malaysian Agricultural Journal*, 46, 332–391

Ngan, S. L., Er, A. C., Yatim, P., How, B. S., Lim, C. H., Ng, W. P. Q., … Lam, H. L. (2022). Social Sustainability of Palm Oil Industry: A Review. *Frontiers in Sustainability*, 3, 855551.

Ni-Meister, W., Lee, S., Strahler, A. H., Woodcock, C. E., Schaaf, C., Yao, T., … Blair, J. B. (2010). Assessing general relationships between aboveground biomass and vegetation structure parameters for improved carbon estimate from lidar remote sensing. *Journal of Geophysical Research: Biogeosciences*, 115(G2).

Nikkels, M. J., Kumar, S., & Meinke, H. (2019). Adaptive irrigation infrastructure — linking insights from human-water interactions and adaptive pathways. *Current Opinion in Environmental Sustainability*, 40(2019), 37–42. Doi: 10.1016/j.cosust.2019.09.001

Nizalapur, V., Jha, C. S., & Madugundu, R. (2010). Estimation of above ground biomass in Indian tropical forested area using multi-frequency DLR-ESAR data. *International Journal of Geomatics and Geosciences*, 1(2), 167.

Nkongho, R. N., Feintrenie, L., & Levang, P. (2014). *The Non-Industrial Palm Oil Sector in Cameroon* (Vol. 139). CIFOR.

Nooni, I. K., Duker, A. A., Van Duren, I., Addae-Wireko, L., & Osei Jnr, E. M. (2014). Support vector machine to map oil palm in a heterogeneous environment. *International Journal of Remote Sensing*, 35(13), 4778–4794.

Noor Azmi, A. N., Bejo, S. K., Jahari, M., Muharam, F. M., Yule, I., & Husin, N. A. (2020). Early detection of Ganoderma boninense in oil palm seedlings using support vector machines. *Remote Sensing*, 12(23), 3920.

Noor, M. R. M., & Harun, M. H., 2004, The role of leaf area index (LAI) in oil palm. *Oil Palm Bulletin*, 48, 11–16.

Nordin, L., Shahruddin, A., & Mariamni, H. (2002, August). Application of AIRSAR data to oil palm tree characterization. In Proceedings of the Asian Conference on Remote Sensing (pp. 5–9).

Nourqolipour, R., Shariff, A. R. B. M., Ahmad, N. B., Balasundram, S. K., Sood, A. M., Buyong, T., & Amiri, F. (2015). Multi-objective-based modeling for land use change analysis in the South West of Selangor, Malaysia. *Environmental Earth Sciences*, 74, 4133–4143.

NRE (2015). Malaysia Biennial update report to the UNFCCC. https://unfccc.int/sites/default/files/resource/MALBUR1.pdf (accessed 17 Jan 2024).

Nunes, M. H., Ewers, R. M., Turner, E. C., & Coomes, D. A. (2017). Mapping aboveground carbon in oil palm plantations using LiDAR: A comparison of tree-centric versus area-based approaches. *Remote Sensing*, 9(8), 816.

Nunes, M. H., Ewers, R. M., Turner, E. C., & Coomes, D. A. (2017). Mapping aboveground carbon in oil palm plantations using LiDAR: A comparison of tree-centric versus area-based approaches. *Remote Sensing*, 9(8), 816.

Nuthammachot, N., & Stratoulias D. (2023). Exploring Sentinel-2 satellite imagery-based vegetation indices for classifying healthy and diseased oil palm trees. *Journal of Oil Palm Research*, 35(3), 517–527. https://doi.org/10.21894/jopr.2022.0068

Ogunkunle, A. O. (1993). Soil in land suitability evaluation: An example with oil palm in Nigeria. *Soil use and management*, 9(1), 35–39.

Okarda, B., Carolita, I., Kartika, T., & Komarudin, H. (2018, July). Mapping of smallholder oil palm plantation and development of a growth model. In *IOP Conference Series: Earth and Environmental Science* (Vol. 169, p. 012074). IOP Publishing.

Okoro, S. U., Schickhoff, U., Boehner, J., Schneider, U. A., & Huth, N. I. (2017). Climate impacts on palm oil yields in the Nigerian Niger Delta. *European Journal of Agronomy*, 85, 38–50.

Olaniyi, A. O., Ajiboye, A. J., Abdullah, A. M., Ramli, M. F., & Sood, A. M. (2015). Agricultural land use suitability assessment in Malaysia. *Bulgarian Journal of Agricultural Science*, 21(3), 560–572.

Oosterveer, P. (2015). Promoting sustainable palm oil: Viewed from a global networks and flows perspective. *Journal of Cleaner Production*, 107, 146–153.

Ordway, E. M., Naylor, R. L., Nkongho, R. N., & Lambin, E. F. (2019). Oil palm expansion and deforestation in Southwest Cameroon associated with proliferation of informal mills. *Nature Communications*, 10(1), 114.

Ortega-Reig, M., Sanchis-Ibor, C., Palau-Salvador, G., García-Mollá, M., & Avellá-Reus, L. (2017). Institutional and management implications of drip irrigation introduction in collective irrigation systems in Spain. *Agricultural Water Management*, 187, 164–172.

Osco, L. P., De Arruda, M. D. S., Junior, J. M., Da Silva, N. B., Ramos, A. P. M., Moryia, É. A. S., … Gonçalves, W. N. (2020). A convolutional neural network approach for counting and geolocating citrus-trees in UAV multispectral imagery. *ISPRS Journal of Photogrammetry and Remote Sensing*, 160, 97–106.

Ouma, Y. O., Tetuko, J., & Tateishi, R. (2008). Analysis of co-occurrence and discrete wavelet transform textures for differentiation of forest and non-forest vegetation in very-high-resolution optical-sensor imagery. *International Journal of Remote Sensing*, 29(12), 3417–3456.

Pacheco, P., Gnych, S., Dermawan, A., Komarudin, H., & Okarda, B. (2017). The palm oil global value chain: Implications for economic growth and social and environmental sustainability. In *CIFOR Working Paper No. 220. Center for International Forestry Research*, Bogor, Indonesia.

Palat, T., Chayawat, N., & Corley, R. H. V. (2012). Maximising oil palm yield by high density planting and thinning. *Planter*, 88, 241–256.

Pang, J., Chen, K., Shi, J., Feng, H., Ouyang, W., & Lin, D. (2019). Libra r-cnn: Towards balanced learning for object detection. In Proceedings of the IEEE/CVF Conference on Computer Vision and Pattern Recognition (pp. 821–830).

Paramananthan, S. (2013). Managing marginal soils for sustainable growth of oil palms in the tropics. *Journal of Oil Palm, Environment and Health (JOPEH)*, 4.

Paramananthan, S. 2003. Oil palm - Management for large and sustainable yields. In Fairhurst, T. H. and R. Härdter, (eds). Singapore: Potash & Phosphate Institute/Potash& Phosphate Institute of Canada (PPI/PPIC) and International Potash Institute (IPI). pp. 27–57.

Pardon, L., Bessou, C., Nelson, P. N., Dubos, B., Ollivier, J., Marichal, R., ... Gabrielle, B. (2016). Key unknowns in nitrogen budget for oil palm plantations. A review. *Agronomy for Sustainable Development*, 36, 1–21.

Parris, D. L., Dapko, J. L., Arnold, R. W., & Arnold, D. (2016). Exploring transparency: A new framework for responsible business management. *Management Decision*, 54(1), 222–247.

Pasaribu, S. I., & Vanclay, F. (2021). Children's rights in the Indonesian oil palm industry: Improving company respect for the rights of the child. *Land*, 10(5), 500.

Pasmans, T. 2017. Local stories in a global commodity new insights on palm oil smallholders in Indonesia. In *Report of a Learning Project with Musim Mas*, Indonesia. 37 pp.

Paterson, R. R. M. (2007). Ganoderma disease of oil palm-A white rot perspective necessary for integrated control. *Crop Protection*, 26(9), 1369–1376.

Paterson, R. R. M., Kumar, L., Taylor, S., & Lima, N. (2015). Future climate effects on suitability for growth of oil palms in Malaysia and Indonesia. *Scientific Reports*, 5(1), 14457.

Pathak, H. S., Brown, P., & Best, T. (2019). A systematic literature review of the factors affecting the precision agriculture adoption process. *Precision Agriculture*, 20, 1292–1316.

Patthanaissaranukool, W., Polprasert, C., & Englande Jr, A. J. (2013). Potential reduction of carbon emissions from Crude Palm Oil production based on energy and carbon balances. *Applied Energy*, 102, 710–717.

Pavón-Pulido, N., López-Riquelme, J. A., Torres, R., Morais, R., & Pastor, J. A. (2017). New trends in precision agriculture: A novel cloud-based system for enabling data storage and agricultural task planning and automation. *Precision Agriculture*, 18, 1038–1068.

Pavuluri, C. M., & Kawamura, K. (2018). Organic aerosols in South and East Asia: Composition and sources. *Land-Atmospheric Research Applications in South and Southeast Asia*, 379–408.

Payne, M. (2021). Satellite remote sensing of deforestation for oil palm. *Nature Reviews Earth & Environment*, 2(4), 230–230.

Pekel, J. F., Cottam, A., Gorelick, N., & Belward, A. S. (2016). High-resolution mapping of global surface water and its long-term changes. *Nature*, 540(7633), 418–422.

Peladarinos, N., Piromalis, D., Cheimaras, V., Tserepas, E., Munteanu, R. A., & Papageorgas, P. (2023). Enhancing smart agriculture by implementing digital twins: A comprehensive review. *Sensors*, 23(16), 7128.

Pendrill, F., Persson, U. M., Godar, J., & Kastner, T. (2019). Deforestation displaced: Trade in forest-risk commodities and the prospects for a global forest transition. *Environmental Research Letters*, 14(5), 055003.

Pervan, B. (2021). Ground-Based Augmentation Systems for Aviation Applications. In Y. Jade Morton, Frank van Diggelen, James J. Spilker Jr., Bradford W. Parkinson, Sherman Lo, Grace Gao (Eds.), *Position, Navigation, and Timing Technologies in the 21st Century: Integrated Satellite Navigation, Sensor Systems, and Civil Applications* (Vol. 1, pp. 259–276), Wiley-IEEE Press.

Phua, M. H., Chong, C. W., Ahmad, A. H., & Hafidzi, M. N. (2018). Understanding rat occurrences in oil palm plantation using high-resolution satellite image and GIS data. *Precision Agriculture*, 19, 42–54.

Piekutowska, M., Niedbała, G., Piskier, T., Lenartowicz, T., Pilarski, K., Wojciechowski, T., … Czechowska-Kosacka, A. (2021). The application of multiple linear regression and artificial neural network models for yield prediction of very early potato cultivars before harvest. *Agronomy*, 11(5), 885.

Pillay, R., Venter, M., Aragon-Osejo, J., González-del-Pliego, P., Hansen, A. J., Watson, J. E., & Venter, O. (2022). Tropical forests are home to over half of the world's vertebrate species. *Frontiers in Ecology and the Environment*, 20(1), 10–15.

Pirker, J., Mosnier, A., Kraxner, F., Havlík, P., & Obersteiner, M. (2016). What are the limits to oil palm expansion?. *Global Environmental Change*, 40, 73–81.

Pitman, A. J. (2003). The evolution of, and revolution in, land surface schemes designed for climate models. *International Journal of Climatology: A Journal of the Royal Meteorological Society*, 23(5), 479–510.

Pittman, A. M., Carlson, K. M., Curran, L. M., & Ponette-Gonzalez, A. (2013). NASA satellite data used to study the impact of oil palm expansion across Indonesian Borneo. *The Earth Observer*, 25(5), 12–15.

Pohl, C., & Loong, C. K. (2016, April). In-situ data collection for oil palm tree height determination using synthetic aperture radar. In *IOP Conference Series: Earth and Environmental Science* (Vol. 34, No. 1, p. 012027). IOP Publishing.

Poortinga, A., Tenneson, K., Shapiro, A., Nquyen, Q., San Aung, K., Chishtie, F., & Saah, D. (2019). Mapping plantations in Myanmar by fusing Landsat-8, Sentinel-2 and Sentinel-1 data along with systematic error quantification. *Remote Sensing*, 11(7), 831.

Potapov, P., Hansen, M. C., Laestadius, L., Turubanova, S., Yaroshenko, A., Thies, C., … Esipova, E. (2017). The last frontiers of wilderness: Tracking loss of intact forest landscapes from 2000 to 2013. *Science Advances*, 3(1), e1600821.

Potapov, P., Li, X., Hernandez-Serna, A., Tyukavina, A., Hansen, M. C., Kommareddy, A., … Hofton, M. (2021). Mapping global forest canopy height through integration of GEDI and Landsat data. *Remote Sensing of Environment*, 253, 112165.

Potapov, P., Yaroshenko, A., Turubanova, S., Dubinin, M., Laestadius, L., Thies, C., … Zhuravleva, I. (2008). Mapping the world's intact forest landscapes by remote sensing. *Ecology and Society*, 13(2), 16. doi:10.5751/ES-02670-130251

Pouliot, D. A., King, D. J., Bell, F. W., & Pitt, D. G. (2002). Automated tree crown detection and delineation in high-resolution digital camera imagery of coniferous forest regeneration. *Remote Sensing of Environment*, 82(2–3), 322–334.

Prabowo, D., Maryudi, A., & Imron, M. A. (2017). Conversion of forests into oil palm plantations in West Kalimantan, Indonesia: Insights from actors' power and its dynamics. *Forest Policy and Economics*, 78, 32–39.

Prabowo, W. E., Darras, K., Clough, Y., Toledo-Hernandez, M., Arlettaz, R., Mulyani, Y. A., & Tscharntke, T. (2016). Bird responses to lowland rainforest conversion in Sumatran smallholder landscapes, Indonesia. *PLoS One*, 11(5), e0154876.

Prakash, R., Singh, D., & Pathak, N. P. (2011). A fusion approach to retrieve soil moisture with SAR and optical data. *IEEE Journal of Selected Topics in Applied Earth Observations and Remote Sensing*, 5(1), 196–206.

PrecisionHawk (2017). *Automated Weed Detection With Drones*.

Pribadi, D. O., Rustiadi, E., Nurdin, M., Saad, A., Pravitasari, A. E., Mulya, S. P., & Ermyanyla, M. (2023). Mapping smallholder plantation as a key to sustainable oil palm: A deep learning approach to high-resolution satellite imagery. *Applied Geography*, 153, 102921.

Pu, R., & Landry, S. (2012). A comparative analysis of high spatial resolution IKONOS and WorldView-2 imagery for mapping urban tree species. *Remote Sensing of Environment*, 124, 516–533.

Purwanto, E. (2020). *New ISPO: A New Hope to Strengthen Oil-Palm Governance? Info brief - October, 2020.* Bogor: Tropenbos Indonesia. Available on https://www.tropenbos-indonesia.org/file.php/2198/202010_infobrief_ispo-english.pdf (accessed 2 Sep 2023).

Putra, Y. C., & Wijayanto, A. W. (2023). Automatic detection and counting of oil palm trees using remote sensing and object-based deep learning. *Remote Sensing Applications: Society and Environment*, 29, 100914.

Putra, Y. C., Wijayanto, A. W., & Chulafak, G. A. (2022). Oil palm trees detection and counting on Microsoft Bing Maps Very High Resolution (VHR) satellite imagery and Unmanned Aerial Vehicles (UAV) data using image processing thresholding approach. *Ecological Informatics*, 72, 101878.

Putro, L. H. S. (2022). Emissions of CH4 and CO2 from wastewater of palm oil mills: A real contribution to increase the greenhouse gas and its potential as renewable energy sources. *Environment and Natural Resources Journal*, 20(1), 61–72. Doi:10.32526/ennrj/20/202100149.

Puttanapong, N., Martinez Jr, A., Bulan, J. A. N., Addawe, M., Durante, R. L., & Martillan, M. (2022). Predicting poverty using geospatial data in Thailand. *ISPRS International Journal of Geo-Information*, 11(5), 293.

Puttemans, S., Van Beeck, K., Goedemé, T., Imai, F., Tremeau, A., & Braz, J. (2018, January). Comparing boosted cascades to deep learning architectures for fast and robust coconut tree detection in aerial images. In Proceedings of the 13th International Joint Conference on Computer Vision, Imaging and Computer Graphics Theory and Applications (Vol. 5, pp. 230–241). ScitePress.

Pyle, J. A., Warwick, N. J., Harris, N. R. P., Abas, M. R., Archibald, A. T., Ashfold, M. J., … Young, P. J. (2011). The impact of local surface changes in Borneo on atmospheric composition at wider spatial scales: Coastal processes, land-use change and air quality. *Philosophical Transactions of the Royal Society B: Biological Sciences*, 366(1582), 3210–3224.

Qaim, M., Sibhatu, K. T., Siregar, H., & Grass, I. (2020). Environmental, economic, and social consequences of the oil palm boom. *Annual Review of Resource Economics*, 12(1), 321–344. https://doi.org/10.1146/annurev-resource-110119-024922

Qin, Y., Xiao, X., Dong, J., Zhou, Y., Wang, J., Doughty, R. B., … Moore III, B. (2017). Annual dynamics of forest areas in South America during 2007–2010 at 50-m spatial resolution. *Remote Sensing of Environment*, 201, 73–87.

Qiu, F., Berglund, J., Jensen, J., Thakkar, P. & Ren, D. (2004). Speckle noise reduction in SAR imagery using a local adaptive median filter. *GIScience & Remote Sensing*, 41, 244–266.

Rahmani, M. K. I. (2022). Blockchain technology: Principles and algorithms. In *Blockchain Technology and Computational Excellence for Society 5.0* (pp. 16–27). IGI Global.

Rahmani, M. K. I., Shuaib, M., Alam, S., Siddiqui, S. T., Ahmad, S., Bhatia, S., & Mashat, A. (2022). Blockchain-based trust management framework for cloud computing-based internet of medical things (IoMT): A systematic review. *Computational Intelligence and Neuroscience*, 2022.

Rakib, M. R. M., Bong, C. F. J., Khairulmazmi, A., Idris, A. S., Jalloh, M. B., & Ahmed, O. H. (2017). Association of copper and zinc levels in oil palm (Elaeis guineensis) to the spatial distribution of Ganoderma species in the plantations on peat. *Journal of Phytopathology*, 165(4), 276–282.

Rao, A. K. (1990). Basal stem rot (Ganoderma) in oil palm smallholdings. In Proceedings of the Ganoderma Workshop, Palm Oil Research Institute of Malaysia, Bangi, Selangor, Malaysia (pp. 113–131).

Rao, V., Tittinutchanon, P., Nakharin, C., & Corley, R. H. V. (2008). The Univanich oil palm breeding programme and progeny trial results from Thailand. *Planter*, 84(989), 519–531.

Rashid, T., Taqvi, S. A. A., Sher, F., Rubab, S., Thanabalan, M., Bilal, M., & ul Islam, B. (2021). Enhanced lignin extraction and optimisation from oil palm biomass using neural network modelling. *Fuel*, 293, 120485.

Reddy, C. S. (2021). Remote sensing of biodiversity: What to measure and monitor from space to species?. *Biodiversity and Conservation*, 30(10), 2617–2631.

Rees, A.R. (1964). The apical organization and phyllotaxis of the oil palm. *Annals of Botany*, 28(1), pp. 57–69.

Rees, R. W., Flood, J., Hasan, Y., Potter, U. & Cooper, R. M. (2009). Basal stem rot of oil palm (Elaeis guineensis); mode of root infection and lower stem invasion by Ganoderma boninense. *Plant Pathology*, 58(5), 982–989

Reiche, J., Mullissa, A., Slagter, B., Gou, Y., Tsendbazar, N. E., Odongo-Braun, C., ... Herold, M. (2021). Forest disturbance alerts for the Congo Basin using Sentinel-1. *Environmental Research Letters*, 16(2), 024005.

Reiche, J., Verhoeven, R., Verbesselt, J., Hamunyela, E., Wielaard, N., & Herold, M. (2018). Characterizing tropical forest cover loss using dense Sentinel-1 data and active fire alerts. *Remote Sensing*, 10(5), 777.

Reichstein, M., Falge, E., Baldocchi, D., Papale, D., Aubinet, M., Berbigier, P., ... Valentini, R. (2005). On the separation of net ecosystem exchange into assimilation and ecosystem respiration: Review and improved algorithm. *Global Change Biology*, 11(9), 1424–1439.

Reijnders, L., & Huijbregts, M. A. (2008). Palm oil and the emission of carbon-based greenhouse gases. *Journal of Cleaner Production*, 16(4), 477–482.

Ren, S., He, K., Girshick, R., & Sun, J. (2015). Faster r-cnn: Towards real-time object detection with region proposal networks. *Advances in Neural Information Processing Systems*, 28.

ReportLinker (2023). https://www.reportlinker.com/clp/global/15670#block-indicator (date of access 11 May 2023).

Ritchie, H., & Roser, M. (2021a). *Forests and Deforestation. Our World in Data.*

Ritchie, H., & Roser, M. (2021b). *Palm oil. Our World in Data.*

Rizeei, H. M., Shafri, H. Z., Mohamoud, M. A., Pradhan, B., & Kalantar, B. (2018). Oil palm counting and age estimation from WorldView-3 imagery and LiDAR data using an integrated OBIA height model and regression analysis. *Journal of Sensors*, 2018, 1–13.

Romero, H. M., Guataquira, S., & Forero, D. C. (2022). Light interception, photosynthetic performance, and yield of oil palm interspecific OxG Hybrid (Elaeis oleifera (Kunth) Cortés x Elaeis guineensis Jacq.) under Three Planting Densities. *Plants*, 11(9), 1166.

Rosenqvist, A., Shimada, M., Ito N., & Watanabe, M. (2007). ALOS PALSAR: A pathfinder mission for global-scale monitoring of the environment. *IEEE Transactions on Geoscience and Remote Sensing*, 45, 3307–3316.

RSPO (2020). Principles and Criteria for the production of sustainable palm oil 2018, Available online at https://rspo.org/updated-principles-and-criteria-2018-include-the-revised-supply-chain-requirements/ (Accessed 14 May 2023).

RSPO(2022).Impactreport2022.https://rspo.org/wp-content/uploads/RSPO-Impact-Report-2022.pdf (accessed 1 Sep 2023).

RSPO Smallholders Task Force (2023). *RSPO Smallholders*. RSPO. https://rspo.org/smallholders#definition (accessed 23 Aug 2023).

Ruslan, A. A., Salleh, S. M., Hatta, S. F. W. M., & Sajak, A. A. B. (2021). IoT soil monitoring based on LoRa module for oil palm plantation. *International Journal of Advanced Computer Science and Applications*, 12(5). https://doi.org/10.14569/ijacsa.2021.0120527

Rutten, R., Bakker, L., Alano, M. L., Salerno, T., Savitri, L. A., & Shohibuddin, M. (2019). Smallholder bargaining power in large-scale land deals: a relational perspective. In *De-centring Land Grabbing* (pp. 194–220). Routledge.

Sahr, K., White, D., & Kimerling, A. J. (2003). Geodesic discrete global grid systems. *Cartography and Geographic Information Science*, 30(2), 121–134.

Santika, T., Wilson, K. A., Budiharta, S., Law, E. A., Poh, T. M., Ancrenaz, M., ... Meijaard, E. (2019). Does oil palm agriculture help alleviate poverty? A multidimensional counterfactual assessment of oil palm development in Indonesia. *World Development*, 120, 105–117.

Santika, T., Wilson, K. A., Law, E. A., St John, F. A., Carlson, K. M., Gibbs, H., ... Struebig, M. J. (2021). Impact of palm oil sustainability certification on village well-being and poverty in Indonesia. *Nature Sustainability*, 4(2), 109–119.

Santika, T., Wilson, K. A., Meijaard, E., Budiharta, S., Law, E. E., Sabri, M., ... Poh, T. M. (2019). Changing landscapes, livelihoods and village welfare in the context of oil palm development. *Land Use Policy*, 87, 104073.

Santoro, M., Cartus, O., Carvalhais, N., Rozendaal, D., Avitabilie, V., Araza, A., ... Willcock, S. (2020). The global forest above-ground biomass pool for 2010 estimated from high-resolution satellite observations. *Earth System Science Data Discussions*, 2020, 1–38.

Santos, A. A. D., Marcato Junior, J., Araújo, M. S., Di Martini, D. R., Tetila, E. C., Siqueira, H. L., ... Gonçalves, W. N. (2019). Assessment of CNN-based methods for individual tree detection on images captured by RGB cameras attached to UAVs. *Sensors*, 19(16), 3595.

Santos, C., & Messina, J. P. (2008). Multi-sensor data fusion for modeling African palm in the Ecuadorian Amazon. *Photogrammetric Engineering & Remote Sensing*, 74(6), 711–723.

Santoso, H., & Winarna, W. (2021). Eksplorasi Pendugaan Hara Daun Tanaman Kelapa Sawit Menggunakan Pesawat Tanpa Awak dan Kamera Multispektral. *Jurnal Penelitian Kelapa Sawit*, 29(1), 49–62.

Santoso, H., Gunawan, T., Jatmiko, R. H., Darmosarkoro, W., & Minasny, B. (2011). Mapping and identifying basal stem rot disease in oil palms in North Sumatra with QuickBird imagery. *Precision Agriculture*, 12, 233–248.

Santoso, H., Tani, H & Wang, X. (2017). Random Forest classification model of basal stem rot disease caused by Ganoderma boninense in oil palm plantations. *International Journal of Remote Sensing*, 38(16), 4683–4699

Santoso, H., Tani, H., Wang, X., Prasetyo, A. E., & Sonobe, R. (2019). Classifying the severity of basal stem rot disease in oil palm plantations using WorldView-3 imagery and machine learning algorithms. *International Journal of Remote Sensing*, 40(19), 7624–7646.

Sapalina, F., Farrasati, R., Wiratmoko, D., Rahutomo, S., Santoso, H., Ginting, E. N., ... Hidayat, F. (2022). Oil palm intercropping system: A potential nature-based solution to improve soil biology activities in North Sumatra plantation, Indonesia. *Malaysian Journal of Microbiology*, 18(2). https://doi.org/10.21161/mjm.211275

Sarkar, M. S. K., Begum, R. A., & Pereira, J. J. (2020). Impacts of climate change on oil palm production in Malaysia. *Environmental Science and Pollution Research*, 27, 9760–9770.

Sarker, L. R., & Nichol, J. E. (2011). Improved forest biomass estimates using ALOS AVNIR-2 texture indices. *Remote Sensing of Environment*, 115(4), 968–977.

Sarzynski, T., Giam, X., Carrasco, L., & Lee, J. S. H. (2020). Combining radar and optical imagery to map oil palm plantations in Sumatra, Indonesia, using the Google Earth Engine. *Remote Sensing*, 12(7), 1220.

Saswattecha, K., Kroeze, C., Jawjit, W., & Hein, L. (2015). Assessing the environmental impact of palm oil produced in Thailand. *Journal of Cleaner Production*, 100, 150–169.

Satriawan, H., Harahap, E. M., & Karim, A. (2015). Effectiveness of soil conservation to erosion control on several land use types. *Agriculture (Pol'nohospodárstvo)*, 61(2), 61–68.

Sayer, E. J., Joseph Wright, S., Tanner, E. V. J., Yavitt, J. B., Harms, K. E., Powers, J. S., ... Turner, B. L. (2012). Variable responses of lowland tropical forest nutrient status to fertilization and litter manipulation. *Ecosystems*, 15, 387–400. https://doi.org/10.1007/s10021-011-9516-9

Schoneveld, G. C., Ekowati, D., Andrianto, A., & Van Der Haar, S. (2019). Modeling peat-and forestland conversion by oil palm smallholders in Indonesian Borneo. *Environmental Research Letters*, 14(1), 014006.

Selvaraj, S., & Sundaravaradhan, S. (2020). Challenges and opportunities in IoT healthcare systems: A systematic review. *SN Applied Sciences*, 2(1), 139.

Septiarini, A., Sunyoto, A., Hamdani, H., Kasim, A. A., Utaminingrum, F., & Hatta, H. R. (2021). Machine vision for the maturity classification of oil palm fresh fruit bunches based on color and texture features. *Scientia Horticulturae*, 286, 110245.

Septiwibowo, B., Hadiwijaya, B., & Caliman, J. P. (2019, October). CO2 balance on oil palm agrosystem in Sumatra, Indonesia. In *IOP Conference Series: Earth and Environmental Science* (Vol. 336, No. 1, p. 012022). IOP Publishing.

Setyamidjaja, D. (2006). Kelapa Sawit, Teknik Budidaya, Panen dan Pengolahan. *Kanisius. Yogyakarta*, 127.

Sha, Z., Bai, Y., Li, R., Lan, H., Zhang, X., Li, J., … Xie, Y. (2022). The global carbon sink potential of terrestrial vegetation can be increased substantially by optimal land management. *Communications Earth & Environment*, 3(1), 8.

Shafri, H. Z., Anuar, M. I., Seman, I. A., & Noor, N. M. (2011). Spectral discrimination of healthy and Ganoderma-infected oil palms from hyperspectral data. *International journal of remote sensing*, 32(22), 7111–7129.

Shafri, H. Z., Hamdan, N., & Saripan, M. I. (2011). Semi-automatic detection and counting of oil palm trees from high spatial resolution airborne imagery. *International Journal of Remote Sensing*, 32(8), 2095–2115.

Shaharum, N. S. N., Shafri, H. Z. M., Ghani, W. A. W. A. K., Samsatli, S., & Yusuf, B. (2019). Mapping the distribution of oil palm using landsat 8 data by comparing machine learning and non-machine learning algorithms. *Pertanika J Sci Technol*, 27, 123–135.

Shaheen, M., Soma, M. K., Zeba, F., & Aruna, M. (2020). Precision agriculture in India - challenges and opportunities. *International Journal of Agricultural Resources Governance and Ecology*, 16, 223–246.

Shamshiri, R. R., Weltzien, C., Hameed, I. A., Yule, J. I., E Grift, T., Balasundram, S. K., … Chowdhary, G. (2018). Research and development in agricultural robotics: A perspective of digital farming.

Shashikant, V., Shariff, A. R. M., Nordin, L., & Pradhan, B. (2012, December). Estimation of above ground biomass of oil palm trees by PALSAR. In *2012 IEEE Colloquium on Humanities, Science and Engineering (CHUSER)* (pp. 838–841). IEEE.

Shashikant, V., Shariff, A. R. M., Wayayok, A., Kamal, M. R., Lee, Y. P., & Takeuchi, W. (2023). Strategic short note: Comparing soil moisture retrieval from water cloud model and neural network using PALSAR-2 for oil palm estates. In *IoT and AI in Agriculture: Self-sufficiency in Food Production to Achieve Society 5.0 and SDG's Globally* (pp. 367–371). Singapore: Springer Nature Singapore.

Shea, K. (2022). Measuring the impact of monitoring: How we know transparent near-real-time data can help save the forests. In *Transformational Change for People and the Planet: Evaluating Environment and Development* (pp. 263–273).

Shimada, M., Isoguchi, O., Tadono, T., & Isono, K. (2009). PALSAR radiometric and geometric calibration. *IEEE Transactions on Geoscience and Remote Sensing*, 47(12), 3915–3932.

Sibhatu, K. T. (2023). Oil palm boom: Its socioeconomic use and abuse. *Frontiers in Sustainable Food Systems*, 7, 1083022.

Silalertruksa, T., Gheewala, S. H., Pongpat, P., Kaenchan, P., Permpool, N., Lecksiwilai, N., & Mungkung, R. (2017). Environmental sustainability of oil palm cultivation in different regions of Thailand: Greenhouse gases and water use impact. *Journal of Cleaner Production*, 167, 1009–1019.

Singh, M., Malhi, Y., & Bhagwat, S. (2014). Biomass estimation of mixed forest landscape using a Fourier transform texture-based approach on very-high-resolution optical satellite imagery. *International Journal of Remote Sensing*, 35(9), 3331–3349.

Singh, S. L., Sahoo, U. K., Kenye, A., & Gogoi, A. (2018). Assessment of growth, carbon stock and sequestration potential of oil palm plantations in Mizoram, Northeast India. *Journal of Environmental Protection*, 9(09), 912.

Sinha, S., Sharma, L. K., & Nathawat, M. S. (2015). Improved land-use/land-cover classification of semi-arid deciduous forest landscape using thermal remote sensing. *The Egyptian Journal of Remote Sensing and Space Science*, 18(2), 217–233.

Sishodia, R. P., Ray, R. L., & Singh, S. K. (2020). Applications of remote sensing in precision agriculture: A review. *Remote Sensing*, 12(19), 3136.

Skidmore, A. K., Coops, N. C., Neinavaz, E., Ali, A., Schaepman, M. E., Paganini, M., … Wingate, V. (2021). Priority list of biodiversity metrics to observe from space. *Nature Ecology & Evolution*, 5(7), 896–906.

Slingerland, M. A., Khasanah, N. M., van Noordwijk, M., Susanti, A., & Meilantina, M. (2019). Improving smallholder inclusivity through integrating oil palm with crops. In *Exploring Inclusive Palm Oil Production* (No. 59, pp. 147–154). ETFRN and Tropenbos International, Wageningen.

Sodhi, N. S., Posa, M. R. C., Lee, T. M., Bickford, D., Koh, L. P., & Brook, B. W. (2010). The state and conservation of Southeast Asian biodiversity. *Biodiversity and Conservation*, 19, 317–328.

Sodri, A., & Septriana, F. E. (2022). Biogas Power Generation from Palm oil mill effluent (POME): Techno-economic and environmental impact evaluation. *Energies*, 15(19), 7265.

Song, Y., & Wu, P. (2021). Earth observation for sustainable infrastructure: A review. *Remote Sensing*, 13, 1528.

Song, Y., Wu, P., Hampson, K., & Anumba, C. (2021). Assessing block-level sustainable transport infrastructure development using a spatial trade-off relation model. *International Journal of Applied Earth Observation and Geoinformation*, 105, 102585.

Spalding, M. (2010). *World Atlas of Mangroves*. Routledge.

Srestasathiern, P., & Rakwatin, P. (2014). Oil palm tree detection with high resolution multi-spectral satellite imagery. *Remote Sensing*, 6(10), 9749–9774.

Statistica (2023). Contribution of the palm oil industry to the gross domestic product (GDP) of Malaysia from 2015 to 2021. https://www.statista.com/statistics/952996/malaysia-palm-oil-share-of-gdp/ (accessed 26 Aug 2023).

Statistica (2023). https://www.statista.com/topics/6079/global-palm-oil-industry/ (accessed 14 Oct 2023).

Suharyanti, N. A., Mizuno, K., & Sodri, A. (2020). The effect of water deficit on inflorescence period at palm oil productivity on peatland. In E3S Web of Conferences (Vol. 211, p. 05005). EDP Sciences.

Suksong, W., Tukanghan, W., Promnuan, K., Kongjan, P., Reungsang, A., Insam, H., & Sompong, O. (2020). Biogas production from palm oil mill effluent and empty fruit bunches by coupled liquid and solid-state anaerobic digestion. *Bioresource Technology*, 296, 122304.

Sumarga, E., Hein, L., Hooijer, A., & Vernimmen, R. (2016). Hydrological and economic effects of oil palm cultivation in Indonesian peatlands. *Ecology and Society*, 21(2).

Sumiani, Y., & Sune, B. H. (2007). Feasibility study of performing an life cycle assessment on crude palm oil production in Malaysia. *International Journal of Life Cycle Assessment*, 12(1), 50–58.

Sunaryathy, P. I., Suhasman, S., Kanniah, K. D., & Tan, K. P. (2015). Estimating aboveground biomass of oil palm trees by using the destructive method. *World Journal of Agricultural Research*, 3(1), 17–19.

Susila, W. R. (2004). Contribution of oil palm industry to economic growth and poverty alleviation in Indonesia. *Jurnal Penelitian dan Pengembangan Pertanian*, 23(3), 107–113.

Sustainable Certification in the Malaysian Palm Oil Industry. Available from: https://www.researchgate.net/publication/353731397_Sustainable_Certification_in_the_Malaysian_Palm_Oil_Industry (accessed Oct 14 2023).

Syahrinudin. (2005). The potential of oil palm and forest plantations for carbon sequestration on degraded land in Indonesia. In: P.L.G. Vlek, M. Denich, C. Martius, C. Rodgers, & N. van de Giesen (Eds.), *Ecology and Development Series No. 28*. Gottingen: Cuvillier Verlag.

Taheripour, F., Hertel, T. W., & Ramankutty, N. (2019). Market-mediated responses confound policies to limit deforestation from oil palm expansion in Malaysia and Indonesia. *Proceedings of the National Academy of Sciences*, 116(38), 19193–19199.

Tan, K. P., Kanniah, K. D., & Cracknell, A. P. (2012). A review of remote sensing based productivity models and their suitability for studying oil palm productivity in tropical regions. *Progress in Physical Geography*, 36(5), 655–679.

Tan, K. P., Kanniah, K. D., & Cracknell, A. P. (2013). Use of UK-DMC 2 and ALOS PALSAR for studying the age of oil palm trees in southern peninsular Malaysia. *International Journal of Remote Sensing*, 34(20), 7424–7446.

Tan, K. P., Kanniah, K. D., & Cracknell, A. P. (2014). On the upstream inputs into the MODIS primary productivity products using biometric data from oil palm plantations. *International Journal of Remote Sensing*, 35(6), 2215–2246.

Tan, X. J., Cheor, W. L., Yeo, K. S. & Leow, W. Z. (2022). Expert systems in oil palm precision agriculture: A decade systematic review. *Journal of King Saud University – Computer and Information Sciences*, 34, 1569–1594

Tang, K. H. D. (2019). Climate change in Malaysia: Trends, contributors, impacts, mitigation and adaptations. *Science of the Total Environment*, 650, 1858–1871.

Tao, H. H., Snaddon, J. L., Slade, E. M., Henneron, L., Caliman, J. P., & Willis, K. J. (2018). Application of oil palm empty fruit bunch effects on soil biota and functions: A case study in Sumatra, Indonesia. *Agriculture, Ecosystems & Environment*, 256, 105–113.

Teh, C. B. S., & Cheah, S. S. (2018a). Modelling crop growth and yield in palm oil cultivation. In: Rival, A. (Ed.) *Achieving Sustainable Cultivation of Oil Palm* (Vol. 1, pp. 1–45).

Teh, C. B. S., & Cheah, S. S. (2018b). Modelling crop growth and yield in palm oil cultivation. In: Rival, A. (Ed.) *Achieving Sustainable Cultivation of Oil Palm* (Vol. 1, pp. 183–227). Cambridge: Burleigh Dodds Science Publishing.

Teng, K. C., Koay, J. Y., Tey, S. H., Lim, K. S., Ewe, H. T., & Chuah, H. T. (2015). A dense medium microwave backscattering model for the remote sensing of oil palm. *IEEE Transactions on Geoscience and Remote Sensing*, 53(6), 3250–3259.

Teoh, C. H. (2002). The palm oil industry in Malaysia: From seed to frying pan. *Plantation Agriculture - WWF Malaysia.* https://pdfs.semanticscholar.org/aebb/021cdcdf799e098f2f1c6abc0917cc6eb18f.pdf?_ga=2.262695243.636229517.1578446253-1275365382.1569230019. (accessed 4 Sep 2023).

The Edge Malaysia (2023). https://theedgemalaysia.com/article/sime-darby-plantation-plans-fully-use-genomeselect-seeds-oil-palm-replanting-2023 (accessed 11 July 2023).

The Edge Malaysia (2023). Malaysian Genomics partners DLF Engineering for distribution of screening tests, lab services in Singapore, https://theedgemalaysia.com/node/674423 (accessed 16 Oct 2023).

The European Space Agency, 2021, Halting deforestation from space https://www.esa.int/Enabling_Support/Preparing_for_the_Future/Space_for_Earth/Halting_deforestation_from_space.

Thenkabail, P., GangadharaRao, P., Biggs, T., Krishna, M., & Turral, H. (2007). Spectral matching techniques to determine historical land-use/land-cover (LULC) and irrigated areas using time-series 0.1-degree AVHRR pathfinder datasets. *Photogrammetric Engineering & Remote Sensing*, 73(10), 1029–1040.

Thenkabail, P. S., Biradar, C. M., Noojipady, P., Dheeravath, V., Li, Y., Velpuri, M., … Dutta, R. (2009). Global irrigated area map (GIAM), derived from remote sensing, for the end of the last millennium. *International Journal of Remote Sensing*, 30(14), 3679–3733.

Thenkabail, P. S., Stucky, N., Griscom, B. W., Ashton, M. S., Diels, J., Van der Meer, B., & Enclona, E. (2004). Biomass estimations and carbon stock calculations in the oil palm plantations of African derived savannas using IKONOS data. *International Journal of Remote Sensing*, 25(23), 5447–5472.

Toh, C. M., Ewe, H. T., Tey, S. H., & Tay, Y. H. (2019). A study on oil palm remote sensing at L-Band with dense medium microwave backscattering model. IEEE Transactions on Geoscience and Remote Sensing, 57(10), 8037–8047.

Toth, C., & Jóźków, G. (2016). Remote sensing platforms and sensors: A survey. *ISPRS Journal of Photogrammetry and Remote Sensing*, 115, 22–36.

Tsoumis, G. (1991). *Science and Technology of Wood: Structure, Properties, Utilization*. New York: Van Nostrand Reinhold. 494 pp.

Tsouurs, G. (1991). *Scimu and Tbchnolog of Wood: Stntcture, hopntie\tltikzatiun*. New York: Chapman and Hall.

Tucker, C. J., Vanpraet, C., Boerwinkel, E., & Gaston, E. A. (1983). Satellite remote sensing of total dry matter production in the Senegalese Sahel. *Remote Sensing of Environment*, 13(6), 461–474.

Turner, D. P., Ritts, W. D., Cohen, W. B., Gower, S. T., Zhao, M., Running, S. W., … Munger, J. W. (2003). Scaling gross primary production (GPP) over boreal and deciduous forest landscapes in support of MODIS GPP product validation. *Remote Sensing of Environment*, 88(3), 256–270.

Uexküll, H.v., Henson, I.E., & Fairhurst, T. (2003). Canopy management to optimize yield. In: Fairhurst, T., Härdter, R. (Eds.), *Oil Palm Management for Large and Sustainable Yields*. Potash & Phosphate Institute of Canada, Potash & Phosphate Institute, International Potash Institute, Singapore, pp. 163–180.

UNDP (2019). Global Food and Agricultural Commodity Systems. https://www.undp.org/facs/publications/indonesia-glance-country-guide (accessed 26 Aug 2023).

USDA (2023). https://apps.fas.usda.gov/psdonline/app/index.html#/app/advQuery (accessed 15 June 2023).

Vadivelu, S., Ahmad, A., & Choo, Y. H. (2014). Remote sensing techniques for oil palm age classification using landsat-5 tm satellite. *Science International*, 26(4), 1547–1551.

Vaidyanathan, G. (2011a). Blue carbon plan takes shape. *Nature*, 20, 60–62.

Vaidyanathan, G. (2011b). Counting the carbon cost of peatland conversion. *Nature*. https://doi.org/10.1038/news.2011.139

Vali, A., Comai, S., & Matteucci, M. (2020). Deep learning for land use and land cover classification based on hyperspectral and multispectral earth observation data: A review. *Remote Sensing*, 12(15), 2495.

Van Der Weide, R., Blankespoor, B., Elbers, C., & Lanjouw, P. (2022). How accurate is a poverty map based on remote sensing data? *An Application to Malawi. An Application to Malawi*. https://ssrn.com/abstract=4462694

Van Gelder, J. W. (2004). *Greasy Palms: European buyers of Indonesian Palm Oil*. London: Friends of the Earth.

Van Kraalingen, D. W. G., Breure, C. J., & Spitters, C. J. T. (1989). Simulation of oil palm growth and yield. *Agricultural and Forest Meteorology*, 46(3), 227–244.

Van Noordwijk, M., & Lusiana, B. (1999). WaNuLCAS, a model of water, nutrient and light capture in agroforestry systems. In *Agroforestry for Sustainable Land-Use Fundamental Research and Modelling with Emphasis on Temperate and Mediterranean Applications: Selected Papers from a Workshop Held in Montpellier, France, 23–29 June 1997* (pp. 217–242). Springer Netherlands.

Vassilaki, D. I., & Stamos, A. A. (2020). TanDEM-X DEM: Comparative performance review employing LIDAR data and DSMs. *ISPRS Journal of Photogrammetry and Remote Sensing*, 160, 33–50.

Verbesselt, J., Hyndman, R., Newnham, G., & Culvenor, D. (2010). Detecting trend and seasonal changes in satellite image time series. *Remote sensing of Environment*, 114(1), 106–115.

Verheye, W. (2010). Growth and production of oil palm. In *Land Use, Land Cover and Soil Sciences*. UNESCO-EOLSS Publishers.

Vijay, V., Reid, C. D., Finer, M., Jenkins, C. N., & Pimm, S. L. (2018). Deforestation risks posed by oil palm expansion in the Peruvian Amazon. *Environmental Research Letters*, 13(11), 114010.

Villela, A. A., D'Alembert, B. J., Rosa, L. P., & Freitas, M. V. (2014). Status and prospects of oil palm in the Brazilian Amazon. *Biomass and Bioenergy*, 67, 270–278.

von Uexkull, H., Fairhurst, T., & Henson, I. E., (2003). Canopy management to optimize yield. In T. Fairhurst, & R. Härdter (Eds.), *Oil Palm Management for Large and Sustainable Yields*. Potash & Phosphate Institute of Canada, Potash & Phosphate Institute, International Potash Institute, Singapore, pp. 163–180.

Wang, X., Girshick, R., Gupta, A., & He, K. (2018). Non-local neural networks. In Proceedings of the IEEE Conference on Computer Vision and Pattern Recognition (pp. 7794–7803).

Wang, X., Shrivastava, A., & Gupta, A. (2017). A-fast-rcnn: Hard positive generation via adversary for object detection. In Proceedings of the IEEE Conference on Computer Vision and Pattern Recognition (pp. 2606–2615).

Wang, Y., Ma, J., Xiao, X., Wang, X., Dai, S., & Zhao, B. (2019). Long-term dynamic of Poyang Lake surface water: A mapping work based on the Google Earth Engine cloud platform. *Remote Sensing*, 11(3), 313.

Wang, Y., Wang, H., Dang, Y., Ma, H., Xu, C., Yang, Q., ... Fang, S. (2022). BDS and Galileo: Global ionosphere modeling and the comparison to GPS and GLONASS. *Remote Sensing*, 14(21), 5479.

Wang, Y., Zhu, X., & Wu, B. (2019). Automatic detection of individual oil palm trees from UAV images using HOG features and an SVM classifier. *International Journal of Remote Sensing*, 40(19), 7356–7370.

Watmough, G. R., Marcinko, C. L., Sullivan, C., Tschirhart, K., Mutuo, P. K., Palm, C. A., & Svenning, J. C. (2019). Socioecologically informed use of remote sensing data to predict rural household poverty. *Proceedings of the National Academy of Sciences*, 116(4), 1213–1218.

Watson-Hernández, F., Gómez-Calderón, N., & da Silva, R. P. (2022). Oil palm yield estimation based on vegetation and humidity indices generated from satellite images and machine learning techniques. *AgriEngineering*, 4(1), 279–291.

Watts, J. D., Pasaribu, K., Irawan, S., Tacconi, L., Martanila, H., Wiratama, C. G. W., ... Manvi, U. P. (2021). Challenges faced by smallholders in achieving sustainable palm oil certification in Indonesia. *World Development*, 146, 105565.

Wen, P. P., & Sidik, M. J. (2011). Impacts of rainfall, temperature and recent El Ninos on fisheries and agricultural products in The West Coast of Sabah. *Bor Science*, 28(2011), 73–85.

Wessel, B., Huber, M., Wohlfart, C., Marschalk, U., Kosmann, D., & Roth, A. (2018). Accuracy assessment of the global TanDEM-X Digital Elevation Model with GPS data. *ISPRS Journal of Photogrammetry and Remote Sensing*, 139, 171–182.

Winarni, R. R., Sutrisno, E., & Jiwan, N. (2014, October). Beyond EU, RSPO and ISPO sustainability requirements. In *Ecosystem Alliance Agro-commodity Workshop, Amsterdam and the Hague* (pp. 27–31).

Witt, C., Fairhurst, T. H., & Griffiths, W. (2005). Key principles of crop and nutrient management in oil palm. *Better Crops*, 89(3), 27–31.

Woittiez, L. S., Van Wijk, M. T., Slingerland, M., Van Noordwijk, M., & Giller, K. E. (2017). Yield gaps in oil palm: A quantitative review of contributing factors. *European Journal of Agronomy*, 83, 57–77.

Wood, B. J. & Chung, G.F. (2003). A critical review of the development of rat control in Malaysian agriculture since the 1960s. *Crop Protection*, 22 (3), 445–461.

Wood, B. J., & Fee, C. G. (2003). A critical review of the development of rat control in Malaysian agriculture since the 1960s. *Crop Protection*, 22(3), 445–461.

Wood, B. J., & Kamarudin, N. (2019). Bagworm (Lepidoptera: Psychidae) Infestation in the centennial of the Malaysian oil palm industry-A review of causes and control. *Journal of Oil Palm Research*, 31(3), 364–380.

Wood, B. J., Corley, R. H. V., & Goh, K. H. (1973). Studies on the effect of pest damage on oil palm yield. 58.

World Rainforest Movement (2022a). Oil palm plantations and water grabbing: Ivory Coast and Gabon, https://www.wrm.org.uy/bulletin-articles/oil-palm-plantations-and-water-grabbing-ivory-coast-and-gabon (accessed 14 Oct 2023).

World Rainforest Movement (2022b). From Asia to Africa: Tentacles of oil palm plantations are squeezing communities dry. Available online at https://www.wrm.org.uy/bulletin-articles/from-asia-to-africa-tentacles-of-oil-palm-plantations-are-squeezing-communities-dry (accessed 28 Aug 2023).

Wösten, J. H. M., Clymans, E., Page, S. E., Rieley, J. O., & Limin, S. H. (2008). Peat-water interrelationships in a tropical peatland ecosystem in Southeast Asia. *Catena*, 73(2), 212–224.

Wulder, M., Niemann, K. O., & Goodenough, D. G. (2000). Local maximum filtering for the extraction of tree locations and basal area from high spatial resolution imagery. *Remote Sensing of Environment*, 73(1), 103–114.

Xia, M., Li, W., Fu, H., Yu, L., Dong, R., & Zheng, J. (2019, May). Fast and robust detection of oil palm trees using high-resolution remote sensing images. In *Automatic Target Recognition XXIX* (Vol. 10988, pp. 65–73). SPIE.

Xiao, C., Qin, R., & Huang, X. (2020). Treetop detection using convolutional neural networks trained through automatically generated pseudo labels. *International Journal of Remote Sensing*, 41(8), 3010–3030.

Xin, Y., Sun, L., & Hansen, M. C. (2021). Biophysical and socioeconomic drivers of oil palm expansion in Indonesia. *Environmental Research Letters*, 16(3), 034048.

Xu, H. (2006). Modification of normalised difference water index (NDWI) to enhance open water features in remotely sensed imagery. *International Journal of Remote Sensing*, 27(14), 3025–3033.

Xu, Y., Ciais, P., Yu, L., Li, W., Chen, X., Zhang, H., … Gong, P. (2021). Oil palm modelling in the global land surface model ORCHIDEE-MICT. *Geoscientific Model Development*, 14(7), 4573–4592.

Xu, Y., Yu, L., Ciais, P., Li, W., Santoro, M., Yang, H., & Gong, P. (2022). Recent expansion of oil palm plantations into carbon-rich forests. *Nature Sustainability*, 5(7), 574–577.

Xu, Y., Yu, L., Li, W., Ciais, P., Cheng, Y., & Gong, P. (2020). Annual oil palm plantation maps in Malaysia and Indonesia from 2001 to 2016. *Earth System Science Data*, 12(2), 847–867.

Xu, Y., Yu, L., Peng, D., Zhao, J., Cheng, Y., Liu, X., … Gong, P. (2020). Annual 30-m land use/land cover maps of China for 1980-2015 from the integration of AVHRR, MODIS and Landsat data using the BFAST algorithm. *Science China Earth Sciences*, 63, 1390–1407.

Xu, Y., Yu, L., Zhao, F. R., Cai, X., Zhao, J., Lu, H., & Gong, P. (2018). Tracking annual cropland changes from 1984 to 2016 using time-series Landsat images with a change-detection and post-classification approach: Experiments from three sites in Africa. *Remote Sensing of Environment*, 218, 13–31.

Yadav, S., Padalia, H., Sinha, S. K., Srinet, R., & Chauhan, P. (2021). Above-ground biomass estimation of Indian tropical forests using X band Pol-InSAR and Random Forest. *Remote Sensing Applications: Society and Environment*, 21, 100462.

Yang, L., Meng, X., & Zhang, X. (2011). SRTM DEM and its application advances. *International Journal of Remote Sensing*, 32(14), 3875–3896.

Yang, M., Yarra, R., Zhang, R., Zhou, L., Jin, L., Martin, J. J. J., & Cao, H. (2022). Transcriptome analysis of oil palm pistil during pollination and fertilization to unravel the role of phytohormone biosynthesis and signaling genes. *Functional & Integrative Genomics*, 22(2), 261–278.

Yang, Y., Anderson, M., Gao, F., Xue, J., Knipper, K., & Hain, C. (2022). Improved daily evapotranspiration estimation using remotely sensed data in a data fusion system. *Remote Sensing*, 14(8), 1772.

Yang, Y., Mao, Y., & Sun, B. (2020). Basic performance and future developments of BeiDou global navigation satellite system. *Satellite Navigation*, 1(1), 1–8.

Yara (2023). Crop nutrition oil palm. https://www.yara.my/crop-nutrition/oil-palm/ (accessed 14 Oct 2023).

Yara (2023). Oil palm. https://www.yara.my/crop-nutrition/oil-palm/how-to-increase-oil-palm-yield---understanding-nutrient-role/ (accessed 31 March 2023).

Yavasli, D. D. (2012). Recent approaches in above ground biomass estimation methods. *Aegean Geographical Journal*, 21(1), 39–51.

Yu, L., & Gong, P. (2012). Google Earth as a virtual globe tool for Earth science applications at the global scale: Progress and perspectives. *International Journal of Remote Sensing*, 33(12), 3966–3986.

Yu, L., Cao, Y., Cheng, Y., Zhao, Q., Xu, Y., Kanniah, K., ... Gong, P. (2023). A study of the serious conflicts between oil palm expansion and biodiversity conservation using high-resolution remote sensing. *Remote Sensing Letters*, 1–15.

Yu, Z., Wang, J., Li, Y., Shum, C. K., Wang, B., He, X., ... Zhou, B. (2022). Remote sensing of suspended sediment in high turbid estuary from sentinel-3A/OLCI: A case study of Hangzhou Bay. *Frontiers in Marine Science*, 9, 1008070.

Yulianti, N. (2009). *Cadangan karbon lahan gambut dari agroekosistem kelapa sawit PTPN IV Ajamu, kabupaten Labuhan Batu Sumatera Utara (Thesis)*. Sekolah Pascasarjana Institut Pertanian Bogor. Bogor.

Yulianti, N., Sabiham, S., Ardiansyah, M., Murtilaksono, K., Sutarta, E. S., & Darmosarkoro, W. (2010). Allometric equation of oil palm: an estimation approach of biomass carbon stock in tropica peatland. In *International Symposium and Workshop on tropical Peathand Management*, Palangkaraya (pp. 159–162).

Yusoff, M. Z. M. (2019). Loose fruit collector machine in Malaysia: A review. *International Journal of Engineering Technology and Sciences*, 6(2), 65–75.

Yusuf, A. A., Roos, E. L., & Horridge, J. M. (2018). Indonesia's moratorium on palm oil expansion from natural forests: Economy-wide impacts and the role of international transfers. *Asian Development Review*, 35(2), 85–112.

Zang, J., Ni, W., & Zhang, Y. (2023). Spatially-explicit mapping annual oil palm heights in peninsular Malaysia combining ICESat-2 and stand age data. *Remote Sensing of Environment*, 295, 113693.

Zeileis, A. (2005). A unified approach to structural change tests based on ML scores, F statistics, and OLS residuals. *Econometric Reviews*, 24(4), 445–466.

Zhao, K., Wulder, M. A., Hu, T., Bright, R., Wu, Q., Qin, H., ... Brown, M. (2019). Detecting change-point, trend, and seasonality in satellite time series data to track abrupt changes and nonlinear dynamics: A Bayesian ensemble algorithm. *Remote Sensing of Environment*, 232, 111181.

Zheng, J., Li, W., Xia, M., Dong, R., Fu, H., & Yuan, S. (2019, July). Large-scale oil palm tree detection from high-resolution remote sensing images using faster-rcnn. In *IGARSS 2019-2019 IEEE International Geoscience and Remote* Sensing Symposium (pp. 1422–1425). IEEE.

Zheng, D., Rademacher, J., Chen, J., Crow, T., Bresee, M., Le Moine, J., & Ryu, S. (2004). Estimating aboveground biomass using Landsat 7 ETM+ data across a managed landscape in Northern Wisconsin, USA. *Remote Sensing of Environment*, 93, 402–411.

Zhu, J. Y., Krahenbuhl, P., Shechtman, E., & Efros, A. A. (2015). Learning a discriminative model for the perception of realism in composite images. In *Proceedings of the IEEE International Conference on Computer Vision* (pp. 3943–3951).

Index